Early
Flat Earth
Writings

A Compilation of Writings
by Various Authors
of the Late 1800's and Early 1900's

Early Flat Earth Writings

Published by Theonomos (theonomos.com) (The views expressed in this book are not necessarily the views of the publisher.)

ISBN 978-0-9861305-6-4

Table of Contents

Astronomy and the Bible Reconciled by "Vox" 1
    Dialogue 2
    Astronomy or the Bible? 9
    Why Does a Ship's Hull Disappear Before the Mast? 10
    Extracts and Illustrative Diagrams from "Parallax" and other Writers 11
    Great Circle Sailing 15
    Stations and Distances 16
    Analogy in Favour of Rotundity Criticized 20

Fifty Scientific Facts by Ebenezer Breach 23
    Preface 24
    Proclamation on Fifty Facts 26

Bible Astronomy 35
    The Firmament, or Heavens 35
    Sun, Moon, and Stars 36
    The Foundations, or Pillars of the Earth 37

The Bible V Neo-Science by "Iconoclast" 39
Biblical Cosmography 45
Chart & Compass – A Letter of Remonstrance 61
Collapse of the Globular Theory by E. Clifton 69
The Coming Man – Modern Astronomy by "Zetetes" 73
Compare the Following Startling Discrepancies 89
The Dauntless Astronomy by Ebenezer Breach 93
Does the Earth Rotate? No! by William Westfield 101
The Earth a Plane by John Quinlan 109
Earth not a Globe but Positively a Plane by John T. Lawson 115
Experimental Proofs by "Parallax" 129

The Terrestrial Plane by Frederick H. Cook 155
    Introduction 156
    Nebula Philosophy and Gravity 157
    Discovery of Neptune 161
    Eclipses 165
    The Sun's Distance 168
    "Cui Bono" 171

The Enlightenment of the World by John G. Abizaid 173
    Preface 177
    The Water 178
    The Land 183
    The Sun, Moon and Stars 186
    Charts 188

The Midnight Sun by "Zetetes" 207
    The Journey 208
    Strange Nights 208
    The Sun's Motion 209
    How the Sun is Seen 210
    Nature Asleep in Sunshine 210
    Civilization North 211
    A Farewell View 211
    Proof the Earth is not a Globe 212
    The Plane Truth 215
    The Sun's Spiral Path 216
    A Faithful Witness 217

One Hundred Proofs from the Scriptures 219
The Popularity of Error & Unpopularity of Truth by John Hampden 229
The Shape of the World by A. E. Skellam 249
The Southern Midnight Sun by "Zetetes" 253

The Sun Standing Still by Albert Smith **257**
    Four Leading Theories 260
    The Latest Exposition 263

Cranks by "Zetetes" **269**
    Modern Astronomy 270
    Water Level 271
    Our Belief 273
    Scientific Blasphemy 274

Truth – The Earth is Flat by S. G. Fowler **277**
    Forward 279
    The Earth is Flat 281
    Fowler's Flat Earth Sea Plane Map 286
    Fowler's Sydney Meridian Sky-Plane Projection 287

The Vanishing Ship by "Search Truth" **291**
Is the Newtonian Theory True? By William Carpenter **293**
    Globularity 294
    Eclipses 302
    Gravity 303
    The Globe's Motions 304
    Confession of Copernicus 305
    Atmospheric Pressure 305
    Light 307
    Mystery 307
    Ship on the Horizon 308
    Common Sense 308

The Bible and Science **309**
    The Higher Criticism 311
    Moses and Geologists 313
    The Torah of Moses 316
    Geology 319
    The Hebrew and Greek 321

Nuts for Agnostics 323
Science Versus Christianity 325
    Astronomy 327
    Revival of Ancient Magic? 333
Patriarchal Longevity by "Paralax" 341
Selections from *The Earth* 369
    Explanation of Middleton's Attempted Dimensions of the Earth 370
    Celestial Phenomena 372
    Stellar Motion 377
    The Midnight Sun – N. and S. 381
    South and North: Their Respective Stars and Motions, etc. 389
    Note from Lady Blount 393
    The Book of Job in Connection with "science" 403
    The Antarctic Expedition 408
    Middleton's Attempted Dimensions of the Earth 412
    Extract from Dr. Hasting's Bible Dictionary 413
    Practical Proof that the Earth is not a Globe 415
    The Earth: Is it a Globe? 420
    The Mutual Relations of the Sun and Earth 424
    South Pole 434
    The Sun and Moon Miracle 438
    Celestial Motions 443
    Converging Meridians Down South 452
    The Earth: Is it a Globe? 454
    Stretched out upon the Waters 457
    The Bedford Level 460
    The Sun's Motions North and South 465
    Darwin's Theories Denied and Refuted by his own Followers 471
    What did "the Discovery" Men Discover? 471
    Homley Talks – The Real Reason Why 472

Astronomy and the Bible
RECONCILED.

A PLAIN DISCOURSE

UPON THE SUBJECT OF THE

EARTH,

SUN, MOON AND STARS;

SHOWING THE POSITION THE EARTH OCCUPIES IN

CREATION.

BY

VOX.

AUTHOR OF
("A CHRISTIAN HISTORICAL POEM: QUEEN VICTORIA.")

EXTRACTS AND ILLUSTRATIVE DIAGRAMS FROM

"PARALLAX,"

AND OTHER WRITERS.

NEW EDITION.

ASTRONOMY AND THE BIBLE RECONCILED.

"Firm stands the foot
That treads the solid ground of Scripture."

It is really magnificent to contemplate with what despatch God does His work. In the beginning of creation He has a multitude of glorious thoughts in His mind, and a host of things to do with His hands; He sets to work, and in six days all is finished!

DIALOGUE.

A.—Sir, I have been taught from my earliest days that the earth is a globe—a planet—rotating upon its own axis, and travelling round the sun at an enormous rate;* that the sun is the centre of the solar system, itself stationary, while all the planets and other heavenly bodies are moving in their orbits round it. But, as I am told you object to all this, I shall be glad to hear your objections and have your opinions.

B.—Before entering upon this interesting subject, then, I inquire—Do you accept the Bible as Divine authority? For that is my royal warrant and sheet-anchor.

A.—I profess to do so.

B.—Well then, formerly I believed as you do, and with no better reason than that which you render—namely, because I was in that way taught. Nevertheless, theoretic astronomy has always presented an insurmountable difficulty to my implicit reception of it, on account of its total inability to reconcile its conclusions with the revealed Word of God. This it was that ultimately determined me to lay aside all preconceptions, and to form my judgment anew from the plain statements made upon the subject of the Creation in the inspired writings. Sceptics make the daring assertion that "the Bible is not true, because (they say) it contradicts science." To this it may be replied that a reputed science presuming to contradict the Word of God is, in reality, not science at all, but is "falsely so called." For brilliant ideas I admit that modern astronomy is a splendid system. Notwithstanding, if it be not really true, then it is but a "splendid lie" palmed upon poor, credulous mankind.

A.—Allow me here to ask you a question, although you may justly consider it somewhat aside from the subject before us.

---

* 600 miles per hour is the supposed velocity in the latitude of England.

B.—Certainly. Yet it is advisable to keep to the point as much as possible. What is the question?

A.—Have you obtained from the Scriptures any reasonable idea as to when God created the angels?

B.—The Bible does not give what may be called creative dates. Yet we may learn from its contents that when the Almighty, in the eternal counsels, had decreed creation, He created all things. Let us then begin with the statement that God created "the deep," which is the opening scene for the operations of the Creator, described in Genesis i., 1, thus :—" In the beginning God created the heaven and the earth." And I would suggest to you that the heaven here spoken of is not the same as that mentioned in v. 8, for the reason that *this* was created "In the beginning," and *that* was made on the second day. The heaven in v. 1, then, I conceive to be the *third* heaven, or Paradise, with all its living hosts.* In Nehemiah ix., 5—6, we read :—" Blessed be Thy glorious name, which is exalted above all blessing and praise. Thou, even Thou, art Lord alone ; Thou hast made heaven, the heaven of heavens, with all their host, the earth and all things that are therein, the seas and all that there is therein, and Thou preservest them all ; and the host of heaven worshippeth Thee." The host here referred to is evidently a host of intelligent beings — angels, capable of worshipping Jehovah. This brings me to your question as to "when God created the angels?" The answer, inferentially, is— at, or "in the beginning" of His work of creation.

A.—I imagine there was a long duration between the creation of the third heaven and the creation of the earth, although these distinct creations are stated in one short verse.

B.—Indeed it may be so, for the Spirit of God in His narrative passes on very rapidly, and to say much in a few words is a scriptural characteristic. I would now at once call your attention to Genesis i., 2—5, and ask you to get in your own mind, a clear, distinct idea of the " deep " there spoken of. The great, vast deep. It is a historical fact that the actual position of the earth, at its first stage of creation, was *beneath* the surface of the waters ; created in the womb of the " deep," and awaiting the full time to be brought forth.

A.—And its primeval condition was—?

B.—Well, as to its shape, so to speak, it was shapeless ; a chaos, unfit, *as* it was and *where* it was, for organic life, animal or vegetable. It contained no "fulness ;" that is, it was empty— "without form and void." Darkness was upon the "deep," and the Spirit of God moved upon the face of the waters, and God said, "Let there be light, and there was light." Now, it is a circumstance worthy of notice that the darkness is not all dissipated ; but just where the Spirit moved, there light was created, and only there. The idea is, an inner circle of light and an outer circle of darkness. Of course, God is over the darkness as well, to the utmost limit of creation. He is omnipresent

---

* That there are three heavens is evident, for Paul says :—He " was caught up to the third heaven."—II. Corinthians xii., 2.

according to Psalm cxxxix.; but I mean, the Spirit, within an inner circle only, moved with creative energy. For where He did not move, darkness remained. Thus we are told, in Job xxxviii., 9, that "thick darkness is a swaddling band for the sea." And God, co-operating with the Spirit, divides the light from the darkness.

A.—Some philosophers think that light proceeds from a point, while others believe it is an undulating ether.

B.—Yes; "so many men, so many opinions." But observe, if you please, the order of creation thus far: First, the third heaven; secondly, the earth; and thirdly, light.

A.—I perceive that natural light—that is, daylight—is created before the sun, as a vessel of light, exists.

B.—It is so. But let us be brief. The work of the first day is finished, and we come to the second. "And God said, let there be a firmament."* That is, the air we breathe, and upward. And mark, where does God make the firmament, or, as we shall call it, the first heaven? Is it made above the "deep"? No, but in the middle of it—"in the midst of the waters." In thinking of this passage, the action of God seems to be, that He puts His almighty arms deep down into the waters, and raises the upper half of the great deep, and places it a long way off over our heads. And with this material He appears to shape the sky, which becomes, in the words of Job xxxvii., 18, "strong, and as a molten looking-glass." And the sky is, I suggest, the second heaven, and the framework of all the luminaries. It may be that those waters are the source of the "fruitful rain from heaven." It is, however, admitted that vapours arise from the waters below, and from the earth, and descend again in the form of rain. I may be told that this is speculative, and I allow it is. But, passing over this instance, it is my aim, in this discourse, to deal more with matters of fact.

A.—Do the Scriptures anywhere state that there are waters above?

B.—Yes; in Jeremiah x., 13, it is written, "There is a multitude of waters in the heavens." Further observe, that by means of the "expansion" God divides the waters *under* the firmament from the waters *above* the firmament. And God called the expansion "Heaven; and the evening and the morning were the second day."

A.—Am I to understand you to mean that the earth was created below the surface of the "deep?"

B.—Exactly so, for God divided the waters in the "midst," and yet the earth was not visible.

A.—Proceed, I am attentive.

B.—God's next work appears to be to separate the waters that were before mingled and mixed. He, as it were, passes through, or by the way of, the paths of the "deep," and, in

---

* Hebrew, *an expansion.*

the appropriate language of Psalm xxxiii., 7, "He layeth up the deep in storehouses"—or, as some translate, "in bags"—and, shall I say, localises the salt water and the fresh, and *we* have given them names. And, mark, sir, it is now for the first time that the earth is *seen*, when God says, "Let the dry land appear." And thus "He bringeth forth the hidden things of darkness."

A.—The earth, you say then, appears on or in the waters?

B.—I do. It is stretched out upon the deep.* That is the position the earth occupies in the Creation.

A.—It just occurs to my memory that somewhere in Scripture it says, "He hangeth the earth upon nothing."

B.—Yes, it is in Job xxvi., 7; but they who are learnèd say that it is an incorrect rendering. Dr. Adam Clarke, for instance, in his commentary on this passage says, the literal translation is: "He layeth the earth upon the waters, nothing sustaining it." And this reading, you will perceive, harmonizes with the passages already quoted.

A.—I think in the *Book of Common Prayer* there is the expression "the round world."

B.—There is. But then, as you know, the *Book of Common Prayer* is not the Bible, and it is *that* we have to face. Now, by a very simple illustration, I think you will perceive that the doctrine of the rotundity of the earth is at variance with, and curiously inconsistent with, the general language of the Prophets and Apostles. I will suppose that I have an orange in my hand, and am engaged in describing its shape. Would you not deem the language inconsistent with the object described were I to speak of its length, breadth, and ends?

A.—I should think you would choose the words "circumference" and "diameter" as more properly applying.

B.—Certainly. And tell me, is not the Holy Spirit, think you, consistent in the choice of words?

A.—That is not to be questioned.

B.—The Bible, as a matter of fact, never uses the words "circumference" and "diameter" in speaking of the earth. It does, however, speak of the "length, breadth, and ends of the earth." In short, the assertion that the earth is a sphere is a mere assumption, worse than useless in a Scriptural sense; it is a stumbling-block to many who desire to be rightly guided. Both the Bible and sound common sense point equally in the opposite direction. The popular idea that, in the "Land of Topsy-turveydom," as it is called, the Antipodes are walking with their feet opposite to ours, or, in other words, with their heads downwards, is ridiculous in the extreme; and the law of gravitation is an unsatisfactory explanation of the supposed phenomenon. It is beyond doubt scientifically demonstrable, as well as scripturally evident, that this

---

* Not as represented on our globes, but stretched out without reference to it being a spherical body.

firm earth on which we "live, move, and have our being" really rests in, or on, the waters; and beneath the gigantic weight of which, with all its rocks, mountains, hills, plains, and valleys, the great deep "couches as a lion." In proof of this assertion we cite Psalm xxiv., 1, 2: "The earth is the Lord's; He hath founded it upon the seas, and established it upon the floods." Again, in Deuteronomy v., 8:—"Thou shalt not make thee any graven image, or any likeness of anything that is in heaven above, or that is in the earth beneath, or that is in the waters beneath the earth." And again we read: O give thanks to Him "that stretcheth out the earth above the waters" (Psalm cxxxvi., 6).

A.—You say, then, that the earth does not roll in an orbit round the sun?

B.—In my opinion—an opinion arrived at after years of careful study of the Word of God, on this and other subjects—the true figure of the earth is that of a plane, without axial or orbital motion. It is fixed, or has only a slight fluctuation, a slow sinking and rising in the waters, thereby causing the tides to ebb and flow. Whereas, the sun is not fixed but actually travels; as the sweet psalmist of Israel sang: "It is as a bridegroom coming out of his chamber, and rejoiceth as a strong man to run a race. His going forth is from the end of the heaven, and his circuit unto the ends of it" (Psalm xix., 5—6). On one occasion the Lord permitted His servant Joshua to arrest its course. Joshua said, in the sight of Israel, "Sun, stand thou still upon Gibeon; and the sun stood still in the midst of heaven, and hasted not to go down about a whole day" (Joshua x., 12, 13). On another, the Lord himself, without the agency of man, turned the sun back again in its own path; which thing was indicated by the shadow returning ten degrees in the sundial of Ahaz. (Isaiah xxxviii., 8.)

A.—Is there not, think you, a strong probability that God spoke on these matters according to appearances and the then ignorance of men?

B.—Such a presumption is, I consider, without foundation. No; it has been the fashion of the time to call the earth a globe, and men have employed their genius in foolishly deranging nature's plans. In fact, our sciences have deceived us and we must turn again to the Scriptures and to first principles. Error may, for a time, obscure truth, but cannot annihilate it.

A.—I confess that this argument is working a revolution in my astronomical ideas.

B.—My dear friend, let me here ask you a simple question. If, as the Newtonians say, the sun is the centre of the solar system, and the earth revolves round the sun, what, in the name of reason, did the earth revolve round before the sun was made? For the earth was created "in the beginning," and the sun was not made till the fourth day!

A.—For my own part, I am at a loss to answer. The query, indeed, seems to me to place the Newtonians in a dilemma, though doubtless their ingenuity will find a way of escape.

B.—A way of escape? Yes, some of our leading scientific men, when they find the Holy Scriptures cross their pet theories, have an easy way of setting the Scriptures aside altogether. Others, not liking to do this, turn and twist their meaning to suit their own purpose. Thus, the mind of God and the mind of man on scientific subjects — especially astronomical — take different views and opposite directions. Is it ever hinted in the Scriptures that God, having "founded the earth upon the seas, and established it upon the floods," afterwards lifted it from its original position and threw it out into infinite space to revolve in a plane round the sun? Or that He plucked the sun from its first place and dashed it into the sea in order to repulse the earth to a distance of over ninety millions of miles away? How is it that there is no mention of so marvellous a feat, if ever performed?

A.—Let us now refer to the moon. Do you deny that she is an opaque body? That is, has no light of her own, but is, as it is said, a reflector of the light of the sun. Do you, I say, deny this?

B.—If the moon were a mere reflector it is only reasonable to conclude that she would reflect whatever she received. That is to say, if she received heat, she would throw off heat; whereas her beams are proverbially cold. Therefore, her light, which is her own, is different in its nature from that of the sun. The moon is self-luminous, and, according to Genesis i., 16, she is "a great light," that, as a poet says, "warms not but illumes.'

A.—How about the stars? Are not some of these worlds, like our own?

B.—Whoever holds the theory of the plurality of worlds is undoubtedly "wise above what is written" in Scripture. The plain declaration is, that God made the sun, moon and stars lights, and "set them in the heavens to give light on the earth."

A.—O, but the Apostle Paul, in his Epistle to the Hebrews, speaks of the plurality of worlds. For instance, "By whom also He made the worlds" (i., 2); and again, "The worlds were framed by the word of God" (xi., 3). How do you account for that?

B.—I account for it in this way. Prior to the introduction of the Copernican system of astronomy, the word in the original Greek which is now translated by the plural noun "worlds" was then translated in the singular number, "world." And it is interesting to note that the Roman Catholics continue to render the word in the singular in their Latin Testament, as follows:—

> Per quem etiam mundum conditit.—Hebrews i., 2.
> Per fidem intelligimus constructum fuisse mundum verbo Dei.—Hebrews xi., 3.

And also in the French thus:—

> Par lequel aussi il a fait le monde.—Hebrews i., 2.
> Le monde a été fait par la parole de Dieu.—Hebrews xi., 3.

And the same word is translated by some "age." It follows, then, that the new reading was probably introduced in deference to the new theory—a theory in which our opponents are totally mistaken in their unwarrantable assumptions. And the disgrace incurred by a partiality shown for their own conceits at the expense of truth, and a proper regard for the written Word of God, will be acutely felt when their delusive theories vanish like bubbles in the air, which they surely will do sooner or later. It is claimed for the Bible that it is the grandest group of writings in the world, and it is accepted by the purest minds as Divine authority. And all, whose minds are open, are here appealed to "to search the Scriptures" as to the probable shape of the earth, and its position in the universe; and to accept their simple, sober teaching in preference to the teaching of uninspired men, however learnèd and clever they may be. Glad indeed are we to know that there are Christians who are courageous enough to reject the ingenious but fallacious astronomical presentions of the age, and to accept the simple, yet beautiful account of the Creation given in the pages of inspiration. There is an incommunicable secret between the heart of God and the obedient Christian, and the latter knows that, whatever modern astronomy, or her sister-science geology, may assert, the light of all the sciences in the world put together is less reliable than the teaching of the "good old Book." And let the pious student but drink in its unpolluted truth, as it were, into the system of the new man, and the effect must be to brace up the human soul, and make it valiant for his Master.

This little paper is written with the full conviction that the alleged antagonism between the Scriptures and true science is erroneous—the discrepancies arising from a false mode of reasoning, admitting as facts things unproved, and that have no foundation in truth. The Zetetic* process, advocated in this pamphlet, is one demanding that propositions shall be *proved*. The statement that the *surface of all standing water is horizontal* is one that has been *proved*, and therefore it is admitted as stating a veritable fact. It should be remarked too, that the results of numerous practical experiments, with this mode of procedure, have confirmed all the statements contained in the Holy Scriptures as to the earth's position. With these results before us, having the weight of proof, it is absolutely safe to affirm, that this method is in reality the only known, simple, and natural way of RECONCILING ASTRONOMY WITH THE BIBLE.

Though treating this subject briefly and decisively, the author has been desirous to avoid the use of language, or manifest a spirit, that might be considered objectionable while advancing these opinions; deeming it becoming to regard the educational prejudices of our opponents with consideration and respect. Abuse and sarcasm are not the weapons we like, nor are these necessary to the argument. We quarrel with no man, but with the "methods" and "systems" of some, and "thrice is he armed that hath his quarrel just."

---

* Zetetic, from the Greek verb, *Zeteo*, to seek, to search, to examine.

NOTES AND EXTRACTS.

ASTRONOMY OR THE BIBLE?

If error is opposed to truth, if darkness is opposed to light, then it may be easily shown that, in letter and in spirit, Astronomy and the Bible are strangely opposed to each other. The Infidel knows this as a fact, and plumes himself upon it. Yet it may be logically maintained that the truths of science, if pursued upon purely zetetic principles—that is, the method of simple inquiry—and the creatorial truths of the Bible, do not controvert each other, but, side by side, like telegraph wires, they run in parallel lines. It is utterly impossible to reconcile the hypotheses of Newton with the Bible. Let the sciences be pursued without the admission of fanciful data, facts proved, and nothing *assumed* "for the sake of argument," then mountains of difficulties will be removed, and the path made easy to the understanding of each, with due honour to both. But our opponents ask—"What has astronomy to do with the Bible, or *vice versâ*?" This seemingly extraordinary question may be answered thus: Inasmuch as astronomy treats of God's creation, by so much the Bible justly claims the right to criticise astronomical teaching. The book of Genesis unequivocally states, that the earth was first created in the "deep;" that afterwards, it was "founded upon the seas, and established upon the floods." Psalm civ. declares it "should not be removed;" and the II. Peter, iii., 5, speaks of the earth "standing out of the water and in the water," as it were, like a ship at anchor. On the contrary, the Newtonians conceive the earth a globe, and hurl it out into infinite space, there to travel round the sun at a rate of over 1,000 miles a minute. * Now, since these two classes of ideas are contradictory of each other, they cannot both be true. And Satan uses these conflicting statements to damage simple souls, to agitate the minds of Biblical students, to diminish the comfort of believers, and shake, if not destroy, their accustomed confidence in the Word of God. Is it not surprising that really pious and intelligent Christians should be so mistaken as to countenance a theory, a declared ignorance of which has been made the subject of an attack upon the personal divinity of our blessed Lord Himself? In a lecture, delivered before a large audience, in the Masonic Hall, Birmingham, and reported in the *Birmingham Daily Gazette*, May 2nd, 1871, the Rev. C. Voysey said: "A being so necessarily unacquainted with the laws of nature as Christ was could not have been the Lord of Nature; and one that did not even know the earth was a globe could scarcely have been its omnipotent Creator." Such sentiments as these, uttered by a professedly religious teacher, in total disregard of the assurance of Job xxxvi., 4, that "He is perfect in knowledge," to say the least, shock the sensibilities of that divine nature which every true Christian possesses. Christians feel keenly the insults offered, in this too scientific age, to the memory of Christ, whose heart they know to have been the truest and tenderest that ever throbbed, and his mind the most comprehensive, and the purest of the pure. Jesus Christ once, speaking of heaven, calls it "God's throne, and the earth God's footstool." Would

---

* The earth travels at the rate of 68,000 miles an hour.—*Chambers' Information for the People*, vol. I., p. 12.

he have used this language of the earth if it had been a globe rolling away in space?* Certainly not. We say, with the good M. St. Pierre, that "we respect Newton for his genius and virtues, but we respect truth much more. The Bible contains ideas of a much sounder philosophy." But it is refreshing to be assured that all erroneous ideas will ultimately give place to truth, according to the word of Jesus when He said: "Every plant, which my heavenly Father hath not planted shall be rooted up." Truly the language of Scripture is inconsistent with the notion of the earth being a revolving globe, but it is quite consistent with the idea of its being a nearly motionless plane. Which then, gentle reader, will you decide for—NEWTONIAN ASTRONOMY OR THE BIBLE?

Mr. Gladstone has lately added his weighty testimony to the sad truthfulness of our statement, that some of our sciences and no little of our current literature have done much to bring the Holy Scriptures into contempt in the popular mind. That distinguished author, in an article in *Good Words*,† says that, "the conclusions of science, as to natural objects, have shaken or destroyed the assertions of the early Scriptures."

WHY DOES A SHIP'S HULL DISAPPEAR BEFORE THE MAST?

This is simply explained by the natural and everywhere visible law of perspective. In a case, such as the one referred to in Job xxxvii., 10., when "by the breath of God frost is given, and the breadth of the waters is straightened," the icy surface would undoubtedly be *horizontal*, not convex.

In a work on "The Lighthouses of the World," there is the following remark:—"The light on the Spurn Point Lighthouse, at the mouth of the Humber, has been seen thirty miles off." Now, according to the theory of curvature, the light would be, allowing for all deductions, 323 feet below the horizon! A still more startling statement appeared in the *St. James's Gazette* of July, 1886. "A nearly extinct volcano which, capped with eternal snow, towers over Teheran, may be seen at a distance of 200 miles!" It is almost superfluous to state that these accounts are utterly at variance with the globular theory.

In the *Girls' Own Paper*, February 13th, 1886, under the heading "Educational," the Editor notices a paper sent to him by an "Enquirer" unknown to us, "calling in question the spherical shape of the earth." The Editor, apparently a little puffed up by a fancied possession of a superior knowledge, says: "If this tract had not been professedly Christian, and our correspondent one who would appeal to the divinely-inspired Scriptures, we should take no notice of either. But as one simple text, in support of the facts above named, is not supposed by the writer to exist in the Bible, we must refer him to Isaiah xl., 22, 'It is he that sitteth

---

* "The earth's orbit is 514,800,000 miles. Its speed is therefore $18\frac{1}{4}$ miles per second: forty-five times that of a cannon ball. The sun is $1\frac{1}{4}$ million times the size of the earth." These are some of the inconceivable, bewildering figures of modern astronomy.

† April, 1890.

on the circle of the earth.'" Did the learned Editor really think we were ignorant of this passage? Why did he not also quote v. 28 of the same chapter? "The Lord, the Creator of the ends of the earth." Evidently it was because he knew any reference to the "ends of the earth" would spoil his own *scientific* interpretation of v. 22, for a circle has no "ends." It did not strike the Editor that a circle is not necessarily a globe! The luminaries describe circles in the heavens, the rainbow too, though we rarely see more than an arc of it. The sun's motion also, being concentric with the polar centre, describes a circle upon the earth beneath. Judging from the context, the meaning of the text quoted is more—He is the supreme sovereign ruler of the earth, possessing attributes that cannot be adequately represented. The Editor then, overflowing with wit, "recommends his friends to take a few voyages round the world, and observe the constellations above them changing, and reappearing as they return to each starting point." But his argument proves nothing, because it is well known that *similar* appearances would occur upon a plane! He next inquires for "what the Yankees call the jumping-off place." Now, the adventurous Editor may find a convenient spot for a "jumping-off" experiment in the Antarctic Ocean, called "Termination Land." There is nothing hurtful in his criticism; in fact, it is a little amusing.

The antediluvian earth was considered by the ancients to have been a smooth, uniform plane, without mountains. Whether this was the case or not, it is remarkable that no mountains are spoken of till the time of the Flood. This even plane, they say, was broken into pieces at the Deluge, and sank into the abyss, when "all the fountains of the great deep were opened." Further, that when the earth, in its altered form, again rose to the surface, it had its present deformities and incommodiousness. And these will remain, subject to volcanic eruptions, till the great conflagration (II. Peter, iii., 10) that will precede the "consummation devoutly to be wished"—the bringing in of another and happier order of things—"new heavens and a new earth." Christ, little esteemed now in the world, will then be "the joy of the whole earth."

The zetetic process—which means, inquiry before conclusion, and is the only course that can lead to simple, unalterable truth—is ably set forth in a remarkable book written by a gentleman and scholar under the *non de plume* of "Parallax," a copy of which should be in every Public Library. The work is entitled—"EARTH NOT A GLOBE."

EXTRACTS AND ILLUSTRATIVE DIAGRAMS FROM "PARALLAX" AND OTHER WRITERS.

"Parallax" says: If the earth is a globe, and is 25,000 English statute miles in circumstance, the surface of all standing water must have a certain degree of convexity—every part must be an arc of a circle. From the summit of any such arc there will exist a curvature or declination of 8 inches in the first statute mile. In the second mile the fall will be 32 inches; in the third 72 inches, or six feet. In every mile, after the first, the curvature downwards from the summit of an arc increases as the square of the distance multiplied by 8 inches. The rule, however, requires to be modified

NOTES AND EXTRACTS.

after the first thousand miles. * The following table will show, at a glance, the amount of curvature, in round numbers, in different distances up to 1,000 miles :—

Curvature in 1 statute mile, 8 inches.
,, 2 ,, 32 ,,
,, 3 ,, 6 feet.
,, 4 ,, 10 ,,
,, 5 ,, 16 ,,
,, 6 ,, 24 ,,
,, 7 ,, 32 ,,
,, 8 ,, 42 ,,
,, 9 ,, 54 ,,
,, 10 ,, 66 ,,
,, 20 ,, 266 ,,
,, 30 ,, 600 ,,
,, 40 ,, 1,066 ,,
,, 50 ,, 1,666 ,,
,, 60 ,, 2,400 ,,
,, 70 ,, 3,266 ,,
,, 80 ,, 4,266 ,,
,, 90 ,, 5,400 ,,
,, 100 ,, 6,666 ,,

Many instances could be given of lights being visible at sea for distances which would be impossible upon a globular surface of 25,000 miles in circumference. The following is one example:—The coal fire (which was once used) on the Spurn Point Lighthouse, at the mouth of the Humber, which was constructed on a good principle for burning, has been seen thirty miles off.

Allowing 16 feet for the altitude of the observer (which is more than is considered necessary,† 10 feet being the standard; but 6 feet may be added for the height of the eye above the deck), 5 miles must be taken from the 30 miles, as the distance of the horizon. The square of 5 miles, multiplied by 8 inches, gives 416 feet; deducting the altitude of the light (93 feet), we have 323 feet as the amount this light should be *below the horizon.* The above calculation is made on the supposition that statute miles are intended, but it is very probable that nautical measure is understood, and if so, the light would be depressed fully 600 feet (pp. 29, 30).

The completion of the great ship canal, which connects the Mediterranean Sea with the Gulf of Suez, on the Red Sea, furnishes another instance of entire discrepancy between the earth's rotundity and the results of practical engineering. The canal is 100 English statute miles in length, and is entirely without locks; so that the water within it is really a continuation of the Mediterranean Sea to the Red Sea. "The average level of the Mediterranean is six inches above the Red Sea; but the flood tides in the Red Sea rise four feet above the highest, and its ebbs fall nearly three feet below the lowest, in the Mediterranean." The *datum* line is twenty-six feet below the level of the Mediterranean, and is continued horizontally

---

* Any work on geometry or geodesy will furnish proofs of this declination.

† By all the figures given is meant the minimum distance from which the light can be seen in clear weather from a height of 10 feet above the sea level.—*Lighthouses of the World*, pp. 9 and 32. Laurie, London.

from one sea to the other; and throughout the whole length of the work the surface of the water runs parallel with this *datum*, as shown in the following section, published by the authorities.

FIG. 1

A A A, is the surface of the canal, passing through several lakes, from one sea to the other; D D, the bed of the canal, or horizontal *datum* line to which the various elevations of land, &c., are referred, but parallel to which stands the surface of the water throughout the entire length of the canal; thus proving that the half-tide level of the Red Sea, the 100 miles of water in the canal, and the surface of the Mediterranean Sea, are a continuation of one and the same horizontal line.

If the earth is globular, the water in the centre of the canal, being fifty miles from each end, would be the summit of an arc of a circle, and would stand at more than 1,600 feet above the Mediterranean and Red Sea ($50^2 \times 8$ inches$=1,666$ feet 8 inches), as shown in diagram.

FIG. 2.

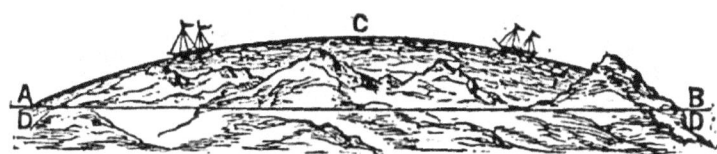

A, the Mediterranean Sea; B, the Red Sea; and A C B, the arc of water connecting them; D D, the horizontal *datum* which, if the earth is globular, would really be the chord of the arc A C B.

> Right lines, running parallel with each other, appear to approach in the distance.
> The eye-line, and the surface of the earth and sky, run parallel with each other;
> > *Ergo*, the earth and sky appear to approach in the distance.
> Lines which appear to approach in the distance are parallel lines.
> The surface of the earth appears to approach the eye-line;
> > *Ergo*, the surface of the earth is parallel with the eye-line.
> The eye-line is a right line.
> The surface of the earth is parallel, or equi-distant;
> > *Ergo*, the surface of the earth is *a right line*—a plane.
>
> That part of any receding body which is nearest to the surface upon which it moves, contracts, and becomes invisible before the parts which are further away from such surface.

NOTES AND EXTRACTS.

The hull of a ship is nearer to the water—the surface on which it moves—than the mast-head;
Ergo, the hull of an outward-bound ship, must be the first to disappear.

This will be seen mathematically in the following diagram—

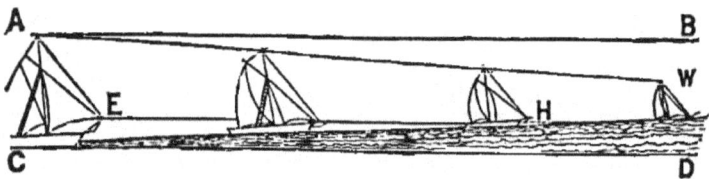

The line A B represents the altitude of the mast-head; E H, of the observer; and C D, of the horizontal surface of the sea. By the law of perspective, the surface of the water appears to ascend towards the eye-line, meeting it at the point H, which is the horizon. The ship appears to ascend the inclined plane C H, the hull gradually becoming less until, on arriving at the horizon H, it is apparently so small that its vertical depth subtends an angle, at the eye of the observer, of less than one minute of a degree, and it is therefore invisible; whilst the angle subtended by the space between the mast-head and the surface of the water is considerably more than one minute, and therefore, although the hull has disappeared in the horizon as the vanishing point, the mast-head is still visible *above* the horizon. But the vessel continuing to sail, the mast-head gradually descends in the direction of the line A W, until at length it forms the same angle of one minute at the eye of the observer, and then becomes invisible.

The following outline sketch represents a contracted section of the London and North-Western Railway from London to Liverpool, through Birmingham.

Fig. 4.

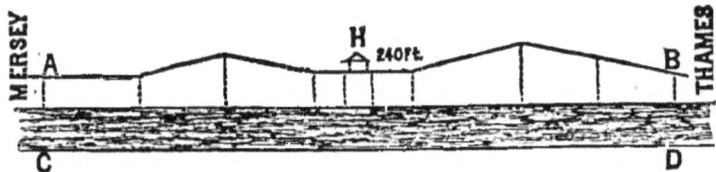

The line A B is the surface, with its various inclines and altitudes, and C D is the *datum* line, from which all the elevations are measured; H is the station at Birmingham, the elevation of which is 240 feet above the *datum* line C D, which line is a continuation of the level of the river Thames at D, to the level of the river Mersey at C. The direct length of the line is 180 miles; and it is a right or absolutely straight line, in a vertical sense, from London to Liverpool. Therefore, the station at Birmingham is 240 feet above the level of the Thames, continued in a right line throughout the whole length of the railway. But, if the earth is a globe, the *datum* line will be the *chord* of the arc D D D,

Fig. 5.

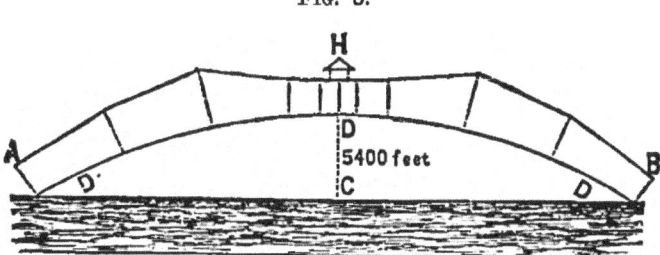

and the summit of the arc at D, will be 5,400 feet above the *chord* C; added to the altitude of the station H, (240 feet), the Birmingham station (H) would be, if the earth is a globe, 5,640 feet above the horizontal *datum* D D, or vertically above the Trinity high-water mark at London Bridge.

It is found practically, and in fact, not to be more than 240 feet; hence, the theory of rotundity must be a fallacy. Sections of all other railways will give similar proofs that the earth is in reality a plane (p. 47).

GREAT CIRCLE SAILING.

Among landsmen a great amount of misconception prevails as to what is really meant by the so-called "great circle sailing," and notwithstanding that the subject is very imperfectly understood, the "project" or hypothesis—for it is nothing more—is often very earnestly advanced as an additional proof of the earth's rotundity. But, like all the other "proofs" which have been given, there is no necessary connection between the facts adduced and the theory sought to be proved. Although professional mariners are familiar with several modes of navigation, "parallel sailing," "plane sailing," "traverse sailing," "middle latitude sailing," "Mercator sailing," and "great circle sailing," the "Mercator" and "great circle" methods are now the favourites. Many persons suppose that the words "great circle sailing" simply mean that the mariner, instead of sailing in a direct line from one place to another, on the same latitude, takes a circuitous path to the south or north of this direct line, where the degrees of longitude being smaller, the distance passed over, although apparently greater, is actually *less*. It is then falsely argued that, as "the greatest distance round is the nearest path," the degrees of longitude *must* be smaller, and therefore the earth *must* be a globe. This is another instance of the self-deception practised by many of the advocates of rotundity. The contraction or convergence of the degrees of longitude beyond the equator is unproved; and again, if they were convergent there could not be a single inch of gain in taking a so-called great circle course between any two places east and west of each other. Let the following experiment be tried in proof of this statement:—On an artificial globe mark out a great circle path, between, say, Cape Town and Sydney, or Valparaiso and Cape Town. Take a strip of sheet lead, and bend it to the form of this path; and, after making it straight, measure its length as compared with the parallel of latitude between the places. The result will fully satisfy the experimenter that *this* view of great circle sailing is contrary to known geometrical principles. The great circle sailing is not the shortest route possible, but merely shorter than several other routes which have been theoretically suggested and adopted; and to affirm that the results are confirmatory or demonstrative of the earth's rotundity is in the highest degree illogical (pp. 269-284).

STATIONS AND DISTANCES

The author of "Lessons in Elementary Astronomy" says (p. 15)—"The most complete proof that the earth is a globe consists in the fact that travellers over the surface, whether by sea or land, always find the distance between different stations exactly such as agree with the calculated distances."

The above sentence is such a compound of childish fable, and either unwarrantable assurance or ignorance, that were it not that the author is an ardent and extensive but not a careful or over-scrupulous writer, in defence of the Newtonian astronomy, it would really be unworthy of criticism. It is one of those utterances which indicate a desperate determination to support a cause at all hazards, and without regard to any evidence but such as agrees with a foregone conclusion. So great is the number of those who advocate the earth's rotundity, who do not hesitate to show the same spirit, that it is really a difficult thing to feel that respect for them which persons who merely differ in opinion ought at all times to show and feel towards each other. What can be more misleading, or illogical, or even more the reverse of fact, than to say that "travellers always find the distance between different stations exactly such as agree with the calculated distances, and therefore the earth is a globe?" A mariner at sea, coming in contact with new land, immediately ascertains the latitude by taking the sun's altitude at noon, and the longitude by the local meridian time in relation to the meridian time at Greenwich. Neither the altitude of the sun, nor the time by chronometer, has any logical connection with the shape of the earth. It is true elements connected with the supposition of the earth's rotundity may be mixed up with the mode of finding latitude and fixing longitude; and anyone may afterwards readily find the places again by sailing until the sun's altitude and the time by chronometer are the same as those first published, when, of course, they must have arrived at the same position, whether the earth is a globe or a plane. It is altogether wrong to say that places, either on land or sea, are found by calculation, except that when places have already been found, and their latitudes and longitudes given, calculation—which is merely the use of formulæ resulting from previous observation—may be used to find them again. But, primarily and essentially, places are found by observation, and not by calculation. If any one will read the reports of the leading circumnavigators and travellers of different countries, they will find many instances where calculation has failed to agree with observations, and where renewed observations have had to be made before anything like the proper position of places in the maps could be fixed. In the majority of instances, where calculations, even when mixed up with some amount of observation, have been relied on, errors have been found. The following passage is quoted from "South Sea Voyages," by Captain Ross; vol. 1., p. 285 :—"By noon (March 9th, 1840) we were in latitude 64° 20′ S., and longitude 164° 20′ E., and therefore about seventy miles north of the land laid down by Lieutenant Wilkes, and not far from the spot from which he must have supposed he saw it ; but having now searched for it at a distance varying from fifty to seventy miles from it, to the north, south, east, and west, as well as having sailed directly over its assigned position, we were compelled to infer that it has no real existence."

THE SUN'S PATH EXPANDS AND CONTRACTS DAILY FOR SIX MONTHS ALTERNATELY. This is a matter of absolute certainty, proved by what is called, in technical language, the northern and southern declination, which is simply saying that the sun's path is nearest the polar centre in summer, and farthest away from it in winter. Thus, day and night, long and short days and nights, morning and evening twilight, winter and summer, the long periods of alternate light and darkness at the northern or polar centre of the earth, arise from the expansion and contraction of the sun's path ; and are all a part of one and the same general phenomenon (pp. 108—115).

FIG. 6.
Showing the earth a plane surrounded by ice and the sun moving over it.

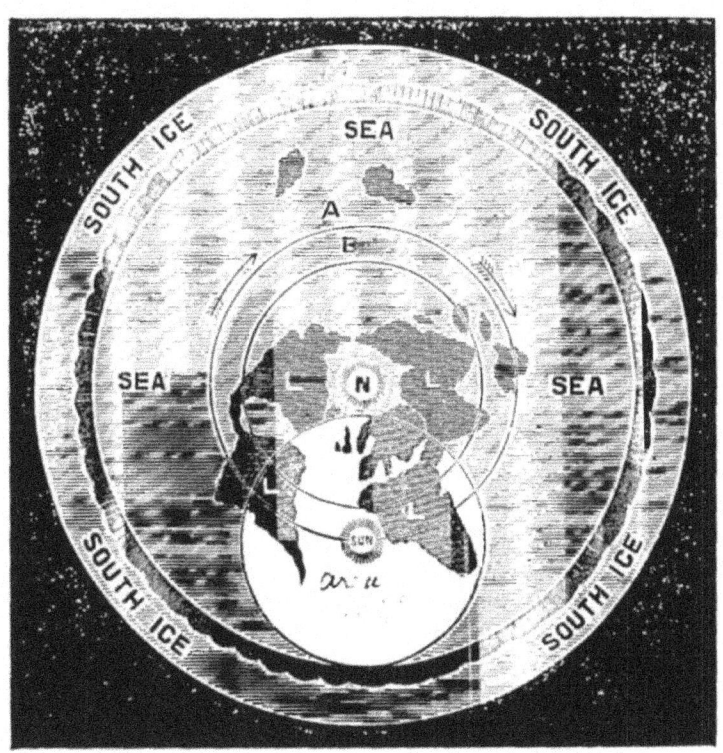

NOTES AND EXTRACTS.

The sun describes the circle A on the 21st of December in one day, or 24 hours. Hence, in that period, mid-day and mid-night, and morning and evening twilight, occur to every part of the earth, *except within the arctic circle*, N. There it is more or less in darkness for several months in succession, or until the sun, by gradually coming nearer to the inner circle, throws his light more and more over the centre. At every place underneath a line drawn across the circle of the sun's light (which radiates equally in all directions) it is noonday; and beyond the northern centre, on the same line, it is midnight. From the 21st of December the sun's path begins to contract every day for six months, until the 21st of June, when it reaches the inner circle B, and it is evident that the same extent of sunlight as that which radiates from the outer circle A, will reach over or beyond the northern centre N, when morning, noon, evening, and night occur as before; but the light continuing, during the daily motion of the sun, to reach over the northern centre, that centre will be continually illuminated for several months together, as before it was in constant darkness.

THE TRUE DISTANCE OF THE SUN may be readily and most accurately ascertained by the simplest possible process. The operation is one in plane trigonometry, which admits of no uncertainty, and requires no modification or allowance for probable influences. This method of measuring distances applies equally to the moon and stars; and it is easy to demonstrate, to place it beyond the possibility of error, so long as assumed premises are excluded, that the moon is nearer to the earth than the sun, and that all the visible luminaries in the firmament are contained within a vertical distance of 1,000 statute miles.

From which it unavoidably follows that the magnitude of the sun, moon, stars, and comets is comparatively small—much smaller than the earth from which they are measured, and to which, therefore, they must of necessity be secondary and subservient. They cannot, indeed, be anything more than "centres of action," throwing down light and chemical products upon the earth (pp. 99—104).

QUERY :—HOW IS IT THE EARTH IS NOT AT ALL TIMES ILLUMINATED ALL OVER ITS SURFACE?—First, if no atmosphere existed no doubt the light of the sun would diffuse over the whole earth at once, and alternations of light and darkness could not exist. Secondly, as the earth is covered with an atmosphere of many miles in depth, the density of which gradually increases downwards to the surface, all the rays of the light, except those which are vertical, as they enter the upper stratum of air, are arrested in their course of diffusion, and by refraction bent downwards towards the earth; and as this takes place in all directions round the sun — equally where density and other conditions are equal, and *vice versâ*—the effect is a comparatively distinct disc of sunlight (p. 123).

It has been demonstrated that the earth is a plane, the surface-centre of which is immediately underneath the star called, "Polaris;" and the extremities of the earth are bounded by a vast region of ice and water, and irregular masses of land, which

NOTES AND EXTRACTS.

bear evidence of fiery origin and action. The whole terminates in fog and darkness, where snow and driving hail-piercing sleet and boisterous winds, howling storms, madly mounting waves, and clashing icebergs are almost constant (p. 177).

Vasco de Gama says, in his "Voyages to the South"—The waves rise like mountains in height; ships are heaved up to the clouds, and apparently precipitated by circling whirlpools to the bed of the ocean. The winds are piercing cold, and so boisterous that the pilot's voice can seldom be heard, whilst a dismal and almost continual darkness adds greatly to the danger."

How far in the gloom and darkness of the south this wilderness of storm and battling elements extends there is at present no evidence. All that we can say is that man, with all his mightiest daring and power of endurance, has only succeeded in reaching the threshold of this restless, dark, and forbidding region of the material world. The earth rests upon and within the waters of the "great deep." It is a vast floating island, buoyed up by the waters, and held in its place by long "spurs" of land shooting into the icy barriers of the southern circumference. Geological researches demonstrate that it was originally a stratified structure, definite and regular in form and extent, and that all the confused and irregular formations observable in every part have resulted from internal convultions (pp. 179-180).

The southern region of the earth is not central, but circumferential; and therefore there is no southern pole, no south polar star, and no southern circumpolar constellations. All statements to the contrary are doubtful, inconsistent with known facts, and therefore not admissible as evidence (p. 290).

Mr. Elliott, an American aëronaut, in a letter giving an account of his ascension from Baltimore, thus speaks of the appearance of the earth from a balloon:—"I don't know that I ever hinted heretofore that the aëronaut may well be the most sceptical man about the rotundity of the earth. Philosophy imposes the truth upon us; but the view of the earth from the elevation of a balloon is that of an immense terrestrial basin, the deeper part of which is that directly under one's feet. As we ascend, the earth beneath us seems to recede, while the horizon gradually and gracefully lifts a diversified slope, stretching away farther and farther to a line that, at the highest elevation, seems to close with the sky. Thus, upon a clear day, the aëronaut feels as if suspended at about an equal distance between the vast blue oceanic concave above, and the equally expanded terrestrial basin below."

The zetetic process forbids that, because an assumption of the earth's rotundity and diurnal motion seem to explain certain phenomena, therefore the assumption becomes, and must be admitted to be, a fact. This is intolerable, even in an abstract sense, but in practice must be unconditionally repudiated.

19

NOTES AND EXTRACTS.

ANALOGY IN FAVOUR OF ROTUNDITY CRITICISED.

To those who are not strictly logical, a favourite "argument" in support of the earth's globular form is "that as all the heavenly bodies are worlds, and visibly round, may not the earth be so necessarily, seeing it is one of the same category?" This is only seemingly plausible. In reality it is a piece of self-deception. It must be *proved* that the stars are worlds; and to do this, or to make it even possible that they are so, it must be proved that they are millions of miles distant from the earth, and from each other, and hundreds or thousands of miles in diameter. By plane trigonometry, in special connection with carefully *measured* base lines, it has been *demonstrated*—placed beyond all power of doubt—that the sun, moon, stars, comets, and meteors of every kind, are all within a distance of a few thousand miles from the sea level of the earth; that therefore they are very small objects, therefore, not worlds, and therefore, from analogy, offer no logical reason or pretext for concluding that this world is globular (p. 300.)

The Copernican or Newtonian theory of astronomy is an "absurd composition of truth and error;" and as admitted by its founder, "not necessarily true, nor even probable;" that instead of its being a general conclusion derived from known and admitted facts, it is a heterogeneous compound of assumed premises, isolated truths, and variable appearances in nature. Its advocates are challenged to show a single instance wherein a phenomenon is explained, a calculation made, or a conclusion advanced, without the aid of an avowed or implied assumption! The very construction of a theory at all, but especially such as the Copernican, is a complete violation of that natural and legitimate mode of investigation to which the term "Zetetic" has been applied. The doctrine of the universality of gravitation is a pure assumption, made only in accordance with that "pride and ambition which has led philosophers to think it beneath them to offer anything less to the world than a complete and finished system of nature." It was said, in effect, by Newton, and has ever since been insisted upon by his disciples: "Allow us, without proof, which is impossible, the existence of two universal forces—centrifugal and centripetal, or attraction and repulsion—and we will construct a theory which shall explain all the leading phenomena and mysteries of nature." An apple falling from a tree, or a stone rolling downwards, and a pail of water tied to a string and set in motion, were assumed to be types of the relations existing among all the bodies in the universe. The moon was assumed to have a tendency to fall towards the earth, and the earth and moon together towards the sun. The same relation was assumed to exist between all the smaller and larger luminaries in the firmament; and it soon became necessary to extend these assumptions to affinity. The universe was parcelled out into systems—co-existent and illimitable. Suns, planets, satellites, and comets were assumed to exist infinite in number and boundless in extent; and to enable the theorists to explain alternating and constantly recurring phenomena, which were everywhere observable, these numberless and for-ever-extending objects were assumed to be spheres. The earth we inhabit was called *a planet*, and because it was thought to be reasonable that the luminous objects in the firmament, which were called planets, were spherical and had motion, so it was only reasonable to suppose that as the earth was a planet it must also be spherical and have motion—*ergo*, the earth is a globe, and moves upon axes and in an orbit round the sun ! And as the earth is a globe and is inhabited, so again it is only reasonable to conclude that the planets are worlds like the earth, and are inhabited by sentient beings. What reasoning ! What shameful perversion of intellectual gifts ! The very foundation of this complicated theory is false, incapable of proof, and contrary to known possibilities. The human mind cannot possibly conceive of its truth and application. To assume the existence of two opposite, equal, universal forces is to seek to make true things or ideas which are necessarily contradictory; to make black and white, hot and cold, up and down, life and death, and truth and falsehood, one and the same. Can anyone by any known possibility conceive of two opposite equal powers, acting simultaneously, producing change of position or motion in that which is thus acted upon? Do not two opposite forces, when equal in intensity and operating at the same moment, neutralize each other ? There is nothing in practical science to gainsay this conclusion ; and in the earliest days of the Newtonian astronomy this contradiction was quickly perceived, but as the assumption was an essential part of the system it was not rejected. An attempt was made to overcome the fatal objection that from two opposite equal forces, acting simultaneously on the earth, *no motion whatever* could arise, by the further assumption that when the earth was first made, the Creator threw it out into space, at right angles to the two forces which had been assumed to exist universally, and that, then the conjoint action of attraction and repulsion, with the "primitive impulse," resulted in a parabolic orbit round the sun (pp. 347—349).

It will scarcely be believed that La Place (La Place le Grand) actually entered into an elaborate calculation with a view to determine at what particular point the Creator held the earth at the time of giving the grand push, and that, after a most profound investigation he arrived at the sublime and never-to-be-forgotten conclusion, that, when the primitive impulse was imparted, the earth was held exactly twenty-five miles from the centre ; "and hence," quoth La Place, "the earth revolved upon her axis in twenty-four hours." If she had been held a little nearer to the centre, our days would have been longer, and if a little further off, she would have revolved with greater velocity, and our days would have been shorter.—*Electrical Theory of the Universe*, by T. S. Mackintosh.

For the learning, the patience, the perseverance, and devotion for which philosophers have ever been examples, honour and applause need not be withheld ; but their false reasoning, the advantages they have taken of the general ignorance of mankind in respect to astronomical subjects, and the unfounded theories they have advanced and defended, cannot be otherwise than regretted, and ought to be by every possible means uprooted (p. 351).

NOTES AND EXTRACTS.

To those who possess a copy of "Parallax"'s book it is, as the author himself says, "most important to the reader that he should thoroughly understand the bearings of the various explanations which have been given of the phenomena which the Newtonian philosophers have hitherto relied on as proofs of their hypotheses. They have assumed certain conditions to exist in order to explain certain phenomena; and because the explanations of such phenomena have appeared plausible, they have thought themselves justified in concluding that their assumptions must be looked upon as veritable facts. The contrary, or Zetetic process, has necessitated that the foundations be demonstrated; that the earth be proved by special and direct experiments to be a plane, irrespective of all consequences, regardless of whether numerous or any phenomena can be understood in connection with it or not.
** Wherever doubt shall exist as to the sufficiency of the phenomenal explanations offered, the mind must at once fall back upon the grand reserved proposition that *water is horizontal*, and, therefore, any want of satisfaction in explaining phenomena must be met by further efforts in that direction, and not by the mentally suicidal process of denouncing a proved foundation."

The precision of astronomy arises, not from theories, but from prolonged observations, and the regularity of the motions, or the ascertained uniformity of their irregularities.—*Million of Facts*, by Sir Richard Phillips (p. 358).

No particular theory is required to calculate eclipses; and the calculations may be made with great accuracy *independent of every theory.*—*Somerville's Physical Sciences* (p. 46).

The flat earth floating tremulously on the sea, the sun moving always over it, giving day when near enough, and night when too far off; the self-luminous moon, with a semi-transparent invisible moon created to give her an eclipse now and then; the new law of perspective, by which the vanishing of the hull before the mast, usually thought to prove the earth globular, *really proves it flat;* all these and other things are well fitted to form excercises in learning the elements of astronomy. "Parallax," though confident in the extreme, neither impeaches the honesty of those whose opinions he assails, nor allots them any future inconvenience.—Augustus de Morgan, Professor of Mathematics in Cambridge University, President of the Royal Astronomical Society, F.R.A.S. *Athenæum Journal*, October 12th, 1872.

In Christian circles it is acknowledged that there is no real harmony between Modern Astronomy and the Holy Scriptures, while the "Zetetic Philosophy" is steadily gaining favour with the people.

The Greatest Event of the Age.
THE
→DOWNFALL OF MODERN ASTRONOMY.←

No objections from Greenwich, nor Cambridge.

FIFTY SCIENTIFIC FACTS

For Surrender to Nature's Fixed Truths.

PROCLAMATION BY

Mr. E. BREACH, C.S.,

Author of "100 Proofs of Fixed Earth and Travelling Sun"; with Royal Patronage seven times.

"GOD SAVE THE QUEEN."

"If our premises be disputed and our facts challenged, the whole range of Astronomy does not contain the proof of its own accuracy; . . . and the whole science of Astronomy must fall to the ground."—PROFESSOR WOODHOUSE, late Professor of Astronomy and Geometry, Cambridge.

Level Seas, Level Walls, Level Railways, prove a Level Earth, as the great wall of China, 1500 miles long, proves it.

SECOND THOUSAND.

Price One Penny each; 9d. per doz.; 5s. per 100.

PUBLISHED BY—

Messrs. John Williams, 32, Bankside, London, S.E.; Rigler, 53, Commercial Rd., Portsmouth: Claxton, 146, Kingston Rd., Buckland.

REEVES & BRIGDEN, PRINTERS, RUSSELL ST., PORTSMOUTH.

PREFACE.

"HE IS THE FREE MAN WHOM THE TRUTH MAKES FREE, AND ALL ARE SLAVES BESIDE."

This year, 1896, being the jubilee year of my acquaintance with the science of Astronomy—for it was when the writer was a lad ten years of age that a gentleman who kept a boarding school at Islington, and visited our house annually, gave me as a present, one year, a monthly volume of the Tract Society on the Solar System—that my first interest was taken in that subject, being very pleased with the picture representing a ship going to sea, tumbling over an orange, it led me to close examination of that side of the subject for 25 years, when a strong conviction entered my mind that the sun had been mismeasured, as *no light* is ever made *larger* than the place to be enlightened. It made me examine the matter thoroughly; and it proved that the accepted theory which I had obstinately adhered to for a quarter-of-a-century, was *all wrong*, and had to be thrown overboard to lighten the ship. "I will purely purge away thy dross and take away all thy tin." Now it is my sole desire after another quarter-of-a century of certifying nature's fixed truths, to release as many captives as possible in this year of jubilee.

Nearly all the educated classes of civilized nations are still slaves to the abominable system of Copernicus, Galileo, and Newton. Copernicus said "it was not necessary that hypothesis be true so long as calculations agree with calculations," he did it for speculation; Galileo for accommodation to regulate the planets; and Newton for gravitation, which only accounts for one half the earth's supposed motion, and he died leaving the problem unsolved. Clairant, who followed, gave it up in despair; but Buffon, the naturalist, thought it would not do to let such a sly fox go, so hunted it out of its hole and it has been kept on the course ever since by scientific huntsmen and misled packs. Professor Rawson in a vote of thanks at The Scientific Institute, said: "The centre of gravitation as worked out by that thinker, Sir Isaac, only explained one half of the earth's motion." We reject the other half, so the thing is gone, and we announce its downfall, so we free all captives.

We were led to compile these 50 scientific facts upon the suggestion of W. Green, Esq., R.N. (Retired Paymaster), Kingston Crescent, who thought such facts would help the cause. Men mistake the electric light of science for the glorious sunshine of nature's fixed truths.

Colonel T. and several on his side think fit to deem Moses and the Prophets fools, and such as we, who believe them, "donkeys." Let them remember that Balaam's ass saw more than his master, and displayed more common sense in her short speech than Balaam did in all his actions. "A fool is wiser in his own eyes than seven men that can render a reason."—*Prov. xxvi.* 16.

> The Earth in shape a wedding cake
>   With ice cliffs all around,
> From forty to four hundred feet,
>   Each soaring from the ground.
>
> The North Pole is the centre true,
>   With North Star up above,
> The South all round the earth we view
>   Nor does it ever move.

THE FITTEST MUST SURVIVE.

I remain, dear Fellow Countrymen,

Yours sincerely,

EBENEZER BREACH.

20, Northam Street, Portsmouth.

N.B.—Who will help spread the light? Send to the Author's address, who will write, and circulate literature on the subject.

"*Declare His works with rejoicing.*"

Proclamation on Fifty Facts.

1—The word Astronomy is taken from the Greek words signifying the Laws of the Stars. Astron, a star, Nomos, a law. Nemo, to regulate, a science that explains the regulation of the stars. It is quite distinct from the science of Physical Geography, which belongs entirely to the Earth. For Astronomy we must *look up*; for Geography, *look down*.

2—The Earth belongs not at all to the science of Astronomy as it is not a planet, a star, nor a heavenly body. The mixing of things by speech, which by nature are distinct, causes contradiction, circumlocution and all absurdity.

3—The Earth was created three days before the planets, therefore it is not in the order of planets; it is not a globe, and has never revolved a mile. It must be as fixed as a sun-dial to accomplish the remarkable phenomena, the precision of the equinoxes.

4—A planet is a wandering star; the Earth is not a wandering star. A star is a burning lamp; the Earth is not a burning lamp. The heavenly bodies are in the firmament of the sky; the Earth is not in the sky, nor in the plane of the sun, else when on the Equator you would see the sun in a straight line before you, but it is seen right over head.

> The Planets well might blush to own for company
> The Earth so vile and stained.

5—The planets have a very eccentric motion. They occasionally slacken their pace, then stop, move backwards on their track, stop again, and finally resume their onward motion. If this occurred to the earth as a planet, everything must go to confusion and destruction, revolving as is supposed several ways at one time.

6—No revolving body whatever was ever made, constructed, or intended for habitation. Show us a revolving body, we show you an uninhabited body. Show us an inhabited body, we show you an unrevolving body.

Ministers of the Church of England, see Homily xxix, paragraph 15, which you vow to abide by at your ordination.

7—The non-revolution of the earth is every moment self-evident; but people have been so hoodwinked that false theories are believed before the testimony of the eyes. If true, no railway engine could keep the metals, as the hand cannot be kept on the drum of a revolving shaft.

DOWNFALL OF MODERN ASTRONOMY.

8—Every solid building must rest on solid foundations. That the earth is on its foundations is declared by the Master-Builder 30 times over; 12 times it is said to be established or fixed: 4 times on its pillars, once on its sure base that it should not move at any time.

9—At least three-fourths of the world's surface is water; water always finds its level, and moves till it does. Every calm sea is a true level, therefore three-fourths of the world's surface being level, of what use is it to suppose the other quarter globular.

10—The late R. A. Proctor said, "If the sea is proved to be a level there is something wrong in modern Astronomy." Ryde Pier will prove it is, as it should be 16-ft. under sea level, 5 miles across, 5 by 5- 25 by 8-in. for rotundity; but it can be seen above the surface of the ocean from Southsea Beach or Point. With the testimony of the aurenot, R.A.P. remarked. "Ah! there they have got us, as *the earth always appears concave.*"

11—The River Nile drops but a foot in 1000 miles. The Island of St. Helena can be seen 100 miles at sea. The national flag of a ship can be discerned 15 miles at sea. "Most certainly, Sir, the sea is a level," said a life-long mariner at Point one day. We have heard enough of the baseless vanishing ship theory to make us feel sea-sick all our life, if we live to be 100 years old. There is nothing in it but an optical illusion.

12—To suppose a globular ocean, a globular and revolving earth, and an earth full set in the ecliptic or sun's pathway, is a trio of folly not to be tolerated in the mind of any creature under heaven. Lord, slay England's darling sins of ignorance. *Leviticus iv. 13.*

13—Modern Astronomy is a suicidal science, for if its hypothesis were put in force, no creature could live ten minutes. There is no Nadir, nor point of the heavens immediately under our feet. There are no Antipodes. So Botany Bay is not under London Bridge, as the old lady said she had heard.

14—Nature is never absurd, monstrous, nor ridiculous, nor does it ever arrange for, or supply the unnecessary; but the modern Astronomers do, and have compelled the Press to issue tons of the unnecessary, the absurd, and the ridiculous; have thoroughly humbugged ministers, colleges, schools, platforms, and families in all countries.

15—If ships go out to sea sailing over an orange-shaped ocean, they should as often be seen mountains high above the range of vision as they are constantly seen hull down, which phenomena is owing entirely to angular vision. Mr. Keith says we can only see distinct for 3 miles. By telescope the ship will appear again as on a sheet of glass. I have proved this by the Nab light-ship, &c.

16 - The earth is the largest body with a solid crust in the universe. The stability of the earth is due to the immobility of matter, unless acted upon by some exterior force for a given period. If the earth were anything but an irregular plane the Jews must have known it,

DOWNFALL OF MODERN ASTRONOMY.

as their scientific knowledge was direct from God, through Moses. Daniel declares that "the greatness of the kingdom of the universe is under the whole heaven," not above it.—*Daniel vii. 27.* The ancient Jews always considered the earth as a plane. They *would* laugh at our astronomers.

17—Mr. Keith says "The voyages of circumnavigators as Cook, Anson and Drake, have been frequently adduced by writers in geography to prove that the earth is a sphere: but when we reflect that all of them sailed westward, and not northward or southward, it is evident that they might have performed the same voyages had the earth been in the form of a drum or cylinder, which it is."

18—Navigation is the art of directing a course, or ascertaining a position (when there are no land marks) by means of objects external to the earth, as the stars, etc.; also by taking a meridian altitude of the sun by a simple addition or subtraction. Mariners would not trust to artificial globes, but to charts as "flat as the surface of the sea."—Keith, p.p. 415.

19—The foundation of modern Astronomy rests on the fiat of Sir Isaac Newton: "The Sun is the centre of the Solar System, and immovable." This foundation was removed when Sir Wm. Herschell discovered the motion of the sun towards Hercules, and the downfall of modern Astronomy should then have been proclaimed.

20 - All Astronomers now declare the constant revolution of the sun through the circuit of the heavens and the 12 signs of the Zodiac, which is the sole cause of the regularity of the seasons, and is so proclaimed by divine authority. See *Job xxxviii. 32, R.V. marg.* If the astronomers arrangement of the seasons were correct, four North Stars would be required instead of one, to keep the centre of the earth in its four alternate positions.

21—As the late R. A. Proctor gives a correct and beautiful plate of the "sun's path" through the 12 signs for every month and day in the year, in his picture of the seasons, he had no need of giving another plate to show an "earth's orbit" through the same signs, as such a phenomena has no existence. "Sun's path" cancels "earth's orbit." Nature never allows the unnecessary.

22—The imaginary orbit of the earth was first estimated by Copernicus at 600,000,000 miles; but a successor struck off 410,000,000 thus making it 190,000,000, so mother earth's course was exceedingly curtailed. When we mentioned this to Professor Ranson on Easter Monday, he laughed most heartily, as he did before when we reminded him that one Astronomer had struck off twenty trillions from Bessel's measurement of the star 61 Cigni. The fixed stars cannot be magnified, not even with the Lick telescope in California.

23—The Astronomers have lately struck off 4,000,000 in the apparent distance of the sun from the earth, making it 91,000,000 instead of 95,000,000. Professor E. tells us that they had lately discovered an error in the parallax of the sun's distance of 100,000,000 miles, which makes our measurement of the sun's distance of 5000 miles correct. We shall never require to strike off 4,000,000.

24—Millions, Bil'ions, and Trillions are trucked about by the Astronomers as though they meant tens, scores, or hundreds; but these all fall to the ground like dead birds, with fright, at a total eclipse of the sun at Portugal, some 200 years ago, when it was darker than midnight.

25—Astronomers have varied about the distance of the sun. Some have estimated it at only three millions: but to suppose the sun at any distance to have power without motion is absurd. Or to suppose a rising and setting sun without *perpetual* motion, is also absurd.

26—The Sun is a concentrated body of light, heat, and attraction, not an expanded substance at all. It is *anima mundi*, the soul of the universe, and as no corporeal being has more than one soul, so the corporeal universe has but *one sun*. The stars are not suns as has been foolishly supposed.

27—The Sun, as we measured it on the ecliptic, and by eclipses last August, is only 5000 miles in diameter, half the diameter of the earth, and not more than 5000 miles—or its own diameter—distant. It could not be seen at all if it were at the distance alleged by modern astronomy, nor could the moon, as the greatest reflector in the universe, derive its light therefrom.

28—Though the prediction of eclipses are correct by the metonic cycle, as used by the Chaldeans, the arrangement for them by modern astronomy is out of all proportions, and rendered impossible.

29—The Moon's diameter is estimated by Joyce at 2,200 miles, and the Sun's diameter is estimated at 882,000 miles. How can a disc that is only 2,200 miles, eclipse a disc the size of the sun at 95,000,000 miles distant. Impossible and absurd. Two bodies to eclipse each other must bear the proportion of at least 3 to 5.

30—The earth's measurement in comparison with the sun is estimated as a mere speck, a dot, a point. A point has position in space, but no magnitude. How can that which has *no* magnitude eclipse *all* magnitude, which the sun is made to represent? Impossible! "We refuse to live on a speck," said an *Evening News* reporter.

31—As the Moon is but a fourth part the diameter of the earth, she is consequently but the fourth part of a speck, a dot, a point. What can be done with that? How can a grain of sand eclipse a pyramid; a speck, St. Paul's; a point of a pin, the Town Hall of Portsmouth? Again we say impossible!

32—It is supposed that an eclipse of the moon is caused by the earth intervening between the sun and moon. The earth is reckoned to travel 1,100 miles per minute; how long would it be passing the moon, travelling herself at 180 miles per minute? Not four minutes. Yet the last eclipse of the moon, on February 28th, lasted 4½ hours; so it could not be the earth intervening, as both luminaries were above the horizon when the eclipse commenced, and the spots of the moon

DOWNFALL OF MODERN ASTRONOMY.

could be seen distinctly through the shadow; the moon was also seen among the stars. Lieut. Pearce agrees with the writer in the distance of the nearest star at 5,000 miles. How can your hat be trillions of miles away when it is on your head?

33—The Greeks noticed that the stars, planets, and all the heavenly bodies more or less eclipse each other, which proves how near they must be to each other. Mr. Proctor tells us "There are many extinct stars always floating in space which are called dead suns," and says "it might be the case with our sun in 17,000,000 years." All that time for a sun to die; yet it is to be as good as dead at the time of the second advent, which is likely to take place in a few years. What will have become of their millions then? Mr. Proctor and all Rationalistic men mismeasure the days of creation, which are measured, entirely as solar days. If not, then the age of Adam, who lived on the 6th and 7th solar day, must be altered to "And all the days of Adam were five thousand, or five million, 930 years, and he died." However the extinct stars or some unknown planet may effect the eclipse of the moon, or it might be a shadow cast from the surface of the earth, as the angle of refraction always equals the angle of incidence. This is the only phenomena we are not quite satisfied about.

34—The Moon must be on the Ecliptic at the time of the eclipse. The ecliptic is the pathway of the sun, therefore both bodies are at the same distance from the earth at the time, and not more than 5000 miles. There must also be a coincidence of the sun and moon at the time, and the moon must be in one of her nodes, a point in which her orbit intersects that of the sun. She is also among the fixed stars of the Zodiac through which the ecliptic runs, which includes over 1,200 stars; therefore the nearest fixed star Centani, is not 7,600,000,000 miles off, which would make the railway fare, at 1d. per 100 miles, come to £103,000,000, as lately stated by Sir R. Ball, in Portsmouth Town Hall. We should only have to charge 4s. 2d.

35—The Ecliptic being the pathway of the Sun in the sky, it was sheer madness of Galileo to sit down and lift the earth, in imagination, into the ecliptic. It never was, never is, and never will be there. We might as well suppose a railway engine lifted into the moon. Therefore we strike it off the ecliptic straight.

36—It is stated by astronomers that the sun is 3,000,000 miles nearer to us in the winter than in summer; but it is not so. The sun is, in reality, at least 180 degrees farther away to the south of us on the 21st December than on the 21st June. The sky is like to a molten looking-glass, and the sun travels in a parallel spiral course continually, as Capt. Pary, Dr. Nansen, and others have seen the sun circulate on the northern horizon for 24 hours. So there is only one sun to do it; seeing is believing. See *Westminster Gazette*, February 14th, 1896.

37—There is but one sun required to enlighten the whole of a circular plane; it makes the arc of a circle over one-half of the plane, then another arc of a circle over the other half, setting to Europe and America, while rising to India, Asia and Australia at the same time.

DOWNFALL OF MODERN ASTRONOMY.

If the earth were a revolving globe we must see all the stars of southern constellations once in a year; but they can only be seen by those living on the Equator.

38—No one is a true Astronomer that does not thoroughly understand the sun, but can only be reckoned an amateur Astronomer, whatever books may be written. At present, the greatest amount of assumption makes the greatest Astronomer. A man, or mind, without assumption, is no Astronomer; but one that assumes that Noah's flood never occurred, though the earth is three-fourths water to prove it, and is also loaded with the greatest amount of mathematical assumption, he is the greatest Astronomer of the age, and is allowed F.R.A.S. to his name—which might stand for "Fellow of the Royal Assumption Society." *versus* Astronomical. Euclid is only assumed demonstration. They make great discoveries of things that never existed, and spend 30 years of their life to explain phenomena that never occurred, as the 2nd axis of rotation.

39—Mathematics is a science of certainty when applied to abstractions in numbers, form, and quality, but pregnant with absurdity when applied to qualities, or metaphysical subjects in which the data is uncertain, or hypothetical or assumed, as in the science of Astronomy.

40—"In the science of Astronomy we only deal with the possible," said Professor Egerton, "but in the science of Chemistry with the actual; we have it before us, therefore it is unassumed." But mathematical Astronomy is a futile source of popular error in which all common and uncommon sense is misled, ever since Galileo announced the mobility of the earth and in imagination beheld it sweeping round the heavens in the precise track followed by the sun. All that the sun lost, the earth was supposed to gain.

41—Sir Richard Phillips in his Million Facts, says: "Nothing therefore can be more impertinent than the assertion of modern writers that the accuracy of astronomical predictions arises from any modern theory." Astronomy is strictly a science of observation, and far more indebted to the false theory of Astrology, than to the equally false and fanciful theory of any modern.

42—We find that four or five thousand years ago, the mean motion of the Sun, Moon and Planets were known to a second, just as at present, and the moon's nodes, the latitudes of the planets, &c., were all adopted by Astrologers in preparing horoscopes for any time past or present. Ephemerides of the planets places, of eclipses, &c., have been published for above 600 years, and were at first nearly as precise as at present.

43—Sir Richard Phillips is utterly opposed to the fanciful theory of gravitation, and says: "It is waste of time to break a butterfly on a wheel, but as astronomy and all science is beset with fancies about attraction and repulsion, it is necessary to eradicate them. Every species and variety of attraction and repulsion are absurd." Gravitation is only another name for weight. What causes the apple to fall on the earth from the tree? Why its own weight, to be sure; nothing more required. Nature never appoints the unnecessary. If a law or object is unnecessary, it does not exist.

44—The stars, or heavenly bodies, are not kept in their orbits by hugging, tugging, and pulling at each other, but are as independent of each other as pedestrians walking the streets. Aristotle declared a grand truism, that between the laws which regulate the celestial and terrestrial systems there is no shadow of affinity whatever.

45—Professor Rawson states "That in some problems of astronomy they are obliged to consider the earth as perfectly at rest;" then why not in all; else they are deceiving people. But since our last lecture at the Albert Hall, he does not believe the earth is a globe. He is not a dogmatical astronomer, and is willing, I understand, to come over on our side at last. So we gained when we
  Measured the Sun, and measured the Moon,
  Examined the Stars, and were home before noon !

46—No wonder that Professor Woodhouse should say at Cambridge one day that "We have the senses, Scripture, and facts on our side, which they have not, and if their mathematical theories were attacked, the whole range of astronomy does not contain the proof of its own accuracy ; and if the public lost confidence in them as the proper authorities, the whole range of modern astronomy must fall to the ground." See *The Earth-not a Globe-Review*, January 1896. In mensuration there must be two sides of a triangle to attain the third by a proper instrument. What the Astronomers have is a vanishing base. We make a fixed base of 10,000 miles, and by equalateral triangle heaven itself is 10,000 miles distant.

47—As the writer (Mr. E. Breach, Poet and Author. with Royal patronage seven times) discovered, and proved by analogy, in the month of November, 1871, that the sun is not larger than the earth. but only half its diameter, 5000 miles ; also proved by measurement of the sun on the Ecliptic, in September, 1895, that the sun and all had been mismeasured and misunderstood by Astronomers, as everything is measured in the solar system on the basis of the sun's measurement. We never make a light larger than the place to be enlightened, nor is it ever placed farther away than is necessary ; nor do we carry the room round the light, but the light round the room. We therefore declare the DOWNFALL OF MODERN ASTRONOMY.

48—And as on the 1st January, 1896, we discovered and proved immense disproportions and errors in the arrangements for eclipses, though they have always been correctly predicted by the metonic cycle, but which has no more to do with the arrangement and measurement of earth. sun, and moon, than Old Moore's arrangement for the weather has to do with the metrological predictions day by day. We therefore most unhesitatingly and unflinchingly declare the Downfall of Modern Astronomy—a very darling, but most erroneous science—and advise all our fellow-countrymen to do their utmost to tread it under their feet henceforth and for ever.

49—In *The National Reformer* some time ago, the great leader, Mr. Bradlaugh, said "That modern science had surely sapped the foundation of all supernatural religion." We now beg to assert on the contrary, that supernatural religion has at last sapped the foundation of all modern science, or oppositions of science, falsely so called. The science of God will save you from Rationalism. See 100 Proofs of Fixed Earth and Travelling Sun, price 2d., by E. Breach, to be obtained of Smith and Seal, 71, King's Road, Southsea.

50—Lord Bacon rejected the Copernican theory with scorn, and compared it to a sleek, well-shaped hide, stuffed with rubbish, but containing nothing to eat. He complained that Astronomy had with great injury been separated from natural philosophy, of which it was one of the noblest provinces, and annexed to the domain of mathematics. The world stood in need, he said, of a very different astronomy, of a living astronomy; of an astronomy that should set forth the nature, the motions, and the influences of the heavenly bodies as they really are. Of what value is a theory which is true only on a supposition in the highest degree extravagant? Sweep all such leaven out of your houses for ever. Amen. In the inhabited earth to come, they shall walk no more after the stubbornness of their evil heart.

> Let Modern Astronomy go to the winds,
> And Natural Astronomy have silver wings;
> Since all the king's horses and all the king's men,
> Can ne'er set Galileo in triumph again.

50a—As for the Royal Astronomer and all his clan, we shall come Lord Nelson over them—"England expects every man this day to do his duty,"—and at once to change the present accepted theory of the shape of the earth. As the President of the Geographical Society stated at the Portland Hall when Capt. Nares went to the North Pole, "We have found tropical vegetation as far north as Disco. If we find it any farther north we shall have to change the present accepted theory of the shape of the earth." Do it at once, lads, and let's have no more bother. We thank God we are as free from error as we are of the gipsies. But we are determined to rout these stupid, extravagant, outrageous errors imposed upon the public for over 300 years, and thereby one priest has been the means of deceiving the whole educated world.

(Signed) In the name of the Chaldeans, the real founders of Astronomy; the Egyptians, Chinese, Greeks, and Romans, the true promoters; in the name of Hipparchus, the father and prince of Astronomers; in the name of Ptolemy, whose system continued 1400 years unopposed; in the name of Tycho Brahe, the greatest observational astronomer the world ever saw, who built an observatory on purpose to oppose that system; and in the name of Lord Bacon, and all that have not been misled, this day and for ever.

Bible Astronomy.

The works of the Lord are great, sought out of all them that have pleasure therein."— Ps. cxi. 2.

I.—The Firmament, or Heavens.

(These terms, like the word TENT, *are often used to denote the space enclosed, as well as the structure enclosing it).*

" And God *made* the *firmament*, and divided the waters which were under the firmament from the waters which were above the firmament." GEN. i. 7.

" The windows (or ' flood-gates', *margin*) of heaven were opened." GEN. vii. 11. So that the waters from above the firmament poured forth at the time of the flood.

" The heavens declare the glory of God, and the firmament sheweth His *handy-work*." Ps. xix. 1.

" Canst thou with Him *spread* out the *sky* which is *strong* as a *molten mirror?*" JOB xxxvii. 18, R.V.

" The likeness of a firmament, like the colour of the terrible *crystal* (or ' ice', *margin*, R.V.), stretched forth over their heads above." EZEK. i. 22.

" He that *buildeth* His *chambers* in the *heaven*, and hath *founded* His *vault* upon the earth." AMOS ix. 6, R.V.

" He *walketh* on the *vault* of heaven." JOB xxii. 14, *margin*, R.V.

" He that *created* the *heavens* and stretched them forth." Is. xlii. 5, R.V.

" He that *sitteth* above the *circle* of the earth . . . that stretcheth out the heavens as a *curtain*, and spreadeth them out as a *tent* to dwell in." Is. xl. 22, R.V.

"He hath described a *boundary* upon the face of the waters, unto the *confines* of light and darkness." Job xxvi. 10, R.V. Descriptive of the outer circumference of the world with its impassable ice-barriers, beyond which the light of the sun never reaches.

"When He *established* the heavens, I was there. when He set a circle upon the face of the deep: when He made *firm* the *skies* above." Prov. viii. 27, 28, R.V.

"My right hand hath *spread out* the heavens." Is. xlviii. 13.

Do not these verses describe the firmament, not as unlimited space, but as a *firm solid structure* resting upon foundations, (see 2 Sam. xxii. 8; and Job xxvi. 11) a lofty *dome* or vault of marvellous workmanship, stretched out over the *circular plane* of the earth, and enclosing it "as a tent to dwell in?"

II.—Sun, Moon, and Stars.

"And God *made* two great lights; the greater light to *rule* the day, and the lesser light to rule the night; the *stars* also. And God set them *in the firmament* of the heaven *to give light* upon *the earth*, and to rule over the day and over the night, and to divide the light from the darkness." Gen. i. 16-18.

"In them (the heavens) hath He set a tabernacle for the *sun*, which is as a bridegroom *coming out* of his chamber, and rejoiceth as a strong man to *run his course.*" His going forth is from the end of the heaven, and his *circuit* unto the ends of it. Ps. xix. 4-6, R.V.

"The sun also *ariseth* (*zarach*, 'bursts forth') and the sun *goeth down* (*bo*, 'goes in') and hasteth to his place where he ariseth." Eccles. i. 5. See *Young' Crit. Concord.*

"To Him that made great lights . . . . the sun to *rule* by day . . . . the *moon* and *stars* to rule by night" . . . . Ps. cxxxvi. 7-9.

"The sun and moon *stood still* in their *habitation*." HAB. iii. 11; also see JOSH. x. 12-14.

What do these verses teach but that God made the sun, moon, and stars, to serve the earth, (in comparison of which they are all probably very small) and that they *circle* in the firmament or dome of heaven, bringing light *consecutively* to every part of the world?

"Under the sun"; "under heaven." ECCLES. i. 3, and *passim*.

In the book of Ecclesiastes the former expression occurs *thirty* times; the latter three times. What possible meaning can the words convey to the mind of a Newtonian philosopher? Do they not imply a real definite "*up*" and "*down*," "*above*" and *below*?

III.—The Foundations, or Pillars of the Earth.

"Of old hast thou *laid* the *foundation* of the earth." Ps. cii. 25.

"Where wast thou when I laid the foundations of the earth? Whereupon were the foundations ('sockets,' *margin*) thereof *fastened*? Or who laid the *corner-stone* thereof?" JOB xxxviii. 4, 6. See also PROV. viii. 29.

"The *pillars* of the earth are the Lord's, and He hath set the *world upon them*." 1 SAM. ii. 8.

"Which shaketh the earth out of her *place*, and the *pillars* thereof *tremble*." JOB ix. 6.

"Ye *enduring* foundations of the earth." MICAH vi. 2.

"Who laid the foundations of the earth, (Heb. He founded the earth upon her *bases*) that it should *not* be *moved* for ever." Ps. civ. 5. See *margin*, R.V.

"Thou hast *established* the earth, and it *abideth*" (or "standeth", *margin*). Ps. cxix. 90.

"The world also is stablished that it *cannot be moved*." Ps. xciii. 1; and xcvi. 10.

"That *spreadeth abroad* the *earth*." Is. xliv. 24, R.V.

"He that spread abroad the earth, and that which cometh out of it." Is. xlii. 5, R.V.

"To Him that spread forth the earth above the waters." Ps. cxxxvi. 6. See also GEN. xlix. 25.

"Heaven *above* . . . earth *beneath* . . . waters *under* the earth." DEUT. v. 8. Also EXOD. xx. 4.

Not the sea lying upon the earth, but the earth resting in, and upon, the waters of the mighty deep.

"He hath founded it upon the *seas*, and established it upon the *floods*." Ps. xxiv. 2.

How can these oft-repeated declarations (which might be further multiplied) concerning the *fixed foundations* of the *outspread earth*, at all consist with the idea that the earth is a revolving globe, rushing through infinite space at the rate of more than a thousand miles every minute? Or what can be the meaning of "the *ends* of the earth," so frequently mentioned in Scripture?

In conclusion, if, out of regard to the theories of modern astronomers, we take the many passages quoted above to imply only the very opposite of what they appear to mean, are we not helping to bring the word of God into increasing disregard? If we admit a part of the Bible to be so evidently untrustworthy that we cannot accept its teaching, must it not tend to a weakening of general confidence in the whole?

Yet all the many and various references given above are confirmed by the incontrovertible facts of nature; while there are no inspired statements of a contrary character.

"Seek ye out of the Book of the Lord and read." Is. xxxiv. 16.

This Leaflet can be had from E. H., 19, Fairmead Road, Holloway. N. Price 2d. per dozen, 2½d. post free; or 1s. 6d. per 100.

The Bible v Neo-Science.

By ICONOCLAST.

A few extracts for reasonable and consistent Christians to meditate upon, after which they may be led to inquire, whether The Bible is not *truly scientific*, and therefore the assumptions of Modern Theoretical Astronomy and, the (so-called) Sciences ramifying from it, *are in direct antagonism* with the Book on which their Christianity is based.

" For if ye believed Moses, ye would believe Me, for he, wrote of Me, but if ye believe not his writings, how shall ye believe My words?" *John* V. 46-47.

" In the beginning God *created* the heaven and the *earth*." *Gen.* I. 1.

" And *the evening* and *the morning* were *the first day*." *Gen.* I. 5.

" And God said, Let there be *light*! and there was *light*; "

" And God saw the *light*, that it was good, and God divided the *light* from the darkness." *Gen.* I. 3-4.

We here have LIGHT WITHOUT THE SUN, that orb not being created until the FOURTH DAY; Modern Astronomy assumes ALL LIGHT TO emanate from the sun.

" And God *made* two great lights, the greater light to *rule* the day, and the lesser light to *rule* the night; the stars also. And God set them in the firmament of heaven to give light upon the earth, and to *rule* over the day and over the night, and *to divide the light* from the darkness." *Gen.* I. 16-18.

" *The Sun* to rule by day . . . *The Moon* and stars to rule by night . . . *Ps.* CXXXVI. 8-9.

" So God created man." . . . *Gen.* I. 27.

" And God saw *everything that He had made*, and, behold, it was very good." . . . *Gen.* I. 31.

" The *pillars* of the earth are the Lord's, and *He hath set the world upon them*." 1 *Sam.* II. 8.

" Of old hast Thou laid the *foundation* of the earth." . *Ps.* CII. 25.

" Ye enduring *foundations* of the earth." *Micah* VI. 2.

" Thou hast *established the earth*, and it abideth." (or standeth) *Ps* CXIX 90.

" The world also is established that *it cannot be moved*." *Ps.* XCVI. 10. *Ps.* XCIII. 1.

" Where wast thou when I laid *the foundations* of the earth " ?

" Who hath laid the measures thereof (or strata of the earth) if thou knowest? or who hath stretched the line upon it?

" Whereupon are the *foundations* (sockets) thereof fastened? or who laid the corner stone thereof," *Job.* XXXVIII. 4-6.

Hebrew, " Tasad erets al mekoneha al-timoth olam vaed " which is rendered, (God), Who founded the earth on its bases that it should not be moved for ever and aye.', *Ps.* CIV. 5.

" For He hath *founded it* (The Earth) upon the seas and *established it upon* the floods." *Ps.* XXIV. 2.

" O give thanks to the Lord of lords, that by wisdom *made* the heavens, and that stretched out The earth (*land*) *above* the waters." *Ps.* CXXXVI. 6.

"For this they *wilfully* forget, that there were heavens from of old, and *an earth* compacted *out* of water and *amidst* (in) water by *the Word of God*, by which means, *the world* that then was, being overflowed with water, perished." 2 *Peter* III. 5.

" Thou, Lord, in the beginning hast laid the *foundation* of the earth." . . . *Heb.* I. 10.

The foregoing Passages, and those that follow are in direct antagonism with the Suppositions of the whole of Modern Theoretical Science ; *both cannot be true.* Therefore, before ignoring or condemning Bible teaching, it is the duty of thinkers to first *prove which side is right*, as only the unthinking or hypocritical profess to accept both. The Truth is magnified by strict and *unbiassed* investigation, and does not require wrapping-up in crafty-question-begging—Sophistry ; such hypothetical juggling only tends to warp the mind, and often leads people to Doubt, if not avowed Infidelity.

"Thus saith the Lord, if *heaven above* can be measured, and *the foundations of the earth* searched out *beneath*, I will also cast off all the seed of Israel." *Jer.* XXXI. 37.

"For He looketh to the *ends of the earth* and seeth *under* the whole heaven." *Job.* XXVIII. 24.

" Again the devil taketh Him up into an exceeding high mountain and sheweth Him *all* the Kingdoms of that region." . . . *Matt.* IV. 8.

" Under the sun, . . . *under* heaven." . . . *Ecc.* I.

" That *spreadeth abroad* the Earth." (land) *Isa.* XLIV. 24.

" My right hand hath *spread out* the heavens," *Isa.* XLVIII. 13.

" Canst thou with Him *spread out* the sky, which is strong as a molten mirror ?" *Job.* XXXVII. 18.

"The likeness of a firmament . . . *stretched forth over* their heads *above.*" *Eze.* I. 22.

" There is a path which no fowl knoweth, and which the vulture's (or eagle's) eye hath not seen, the lion's whelps have not trodden it, nor the fierce lion past by it," *Job.* XXVIII. 7. 8.

" He that sitteth *above the circle of the earth* . . . that *stretched* out the heavens as a curtain, and *spreadeth* them out as a tent, . . . *Isa.* XL. 22.

" When He established the heavens I was there, when He *set a circle* upon the face of the deep, when He *made firm* the skies *above.*" *Prov.* VIII. 27-28.

" He hath described (*or placed*) a *boundary* upon the face of the waters, unto the *confines* of *light* and Darkness." *Job.* XXVI. 10.

The four preceding passages, clearly describe the SOUTHERN CIRCUMFERENTIAL and IMPASSABLE ICY BARRIER OF THE WORLD, (where unthinking people ASSUME they should fall off or over.) *Vide Vasco di Gama's, Cook's and Sir James Ross' Antarctic Voyages.*

Hebrew. "Nothah tsaphon al-tohu toleh erets al-balgamah." (or belimah) *Job.* XXVI. 7 which is rendered. "He spreadeth out the North over the empty waste, (or desolation) and hangeth (or supporteth) the earth (land) upon no thing but the firmament of His power, (the waters of the Great Deep)." *Vide Adam Clarke and other Commts.*

To hang or support, in this case, completely excludes the idea of motion more especially so, in connection with an ASSUMED Sea-earth-globe of 25,000 miles circumference, ASSUMED to have many motions, one of which is that of flashing through ASSUMED space at the brain-reeling-speed of 19 miles per second. It would be as sane, and consistent to talk of supporting or hanging-up flashes of Lightning, as of such a sea-earth-globe under the SUPPOSED conditions.

"The same day were *all the fountains of The Great Deep broken up.*" *Gen.* VII. 11. (One of the most expressively significant passages in the Bible.)

And *all the high hills, that were under the whole heaven were covered* Fifteen cubits *upward* did the waters prevail, and *the mountains were covered.*" *Gen.* VII. 19. 20.

' The Almighty shall bless thee with the blessing of *heaven above*, and blessings of *the deep that lieth under.*" *Gen.* XLIX. 25.

" Blessed be his land, for the precious things of heaven, for the dew, and *for the deep which croucheth beneath.*" *Deut* XXIII. 13.

" **Or in** the waters *under the earth*," *Exo.* XX. 4.

" Or the likeness of anything that is in the *waters beneath the earth.*" *Deu.* IV. 18.

"Heaven *above* . . . earth *beneath* . . . *waters beneath the earth.*" *Deu.* V. 8.

" As for the earth, out of it cometh bread, and *under* it is **turned up** as it were fire." *Job.* XXVIII. 5.

" In them (the heavens) hath He set a tabernacle for the sun, which is as a bride-groom *coming out* of his chamber, and rejoiceth as a strong man *to run a race*. His *going forth* is from the end of the heaven and his *circuit* unto the ends of it." *Ps.* XIX. 4-6.

" The sun also (zarach) *bursts forth*, and the sun (bu) goes in, (or away) and *hasteth* to his place where he (zarach) *bursts forth,*" *Ecc.* I. 5.

" Let them that love The Lord, be as the sun when **he** *goeth forth* in his might." *Judges* V. 31.

" And he (The Lord) brought the shadow (of the sun) ten degrees *backward*, by which it had gone down in the (sun) dial of Ahaz." II *Kings.* XX. II.

" So the sun *returned* ten degrees." *Is.* XXXVIII. 8.

" Sun, *stand thou still* upon Gibeon! and thou, Moon, in the valley of Ajalon! and *the sun stood still*, and *the moon stayed*, . . *Josh.* X. 12-13

" The Sun and Moon *stood still* in their habitation." *Hab.* III. 11.

" And the stars of heaven fell unto the earth." *Rev.* VI. 13.

This last passage IN THE SO-CALLED LIGHT OF MODERN ASTRONOMY, amounts to downright Balderdash.

We may now glance at a few significant passages from the Old and New Testament Apocryphas, which may be of interest IF NOT AUTHORITY.
Old Testament Apoc:

" In the beginning, when *the earth was made*, before the *borders of the world* stood, or ever the winds blew," II. *Esdras* VI. 1.

" Let the earth be made, and it was made, let the heaven be made and it was created."

" In His word were the stars made."

" He hath shut the sea in the midst of the waters, and with his word hath He hanged *(or supported)* the earth *upon* the waters." II *Esdras* XVI. 55-61.

" He spreadeth out the heavens like a vault ; *upon the waters* hath *He founded it*." (The Earth).

" He *made* man, and put his heart in the midst of the body, and gave him breath, life and understanding." II. *Esdras* XVI. 55-61.

" *Great is the Earth*, high is the heaven, *swift is the sun in its course*, for he *compasseth* the heavens round about, and fetcheth his course again to his own place *in one day*." I. *Esdras* IV. 34.

" For the dumb water, *and without life* (at God's command) brought forth living things." II. *Esdras* VI. 48. also 38-47.

" The mountains also and *foundations of the earth* shall be shaken with trembling, when the Lord looketh upon them." *Ecc.* XVI. 19.

" In his time (Isaiah's) the sun *went backward*." *Ecc.* XLVIII. 23.

" Did not the sun go *back* by his (Joshua's) means ? and was not *one day as long as two ?*" *Ecc.* XLVI. 4.

The whole of Ecc. XLIII. is most expressively antagonistic to Modern Theoretical Astronomy, and the so-called Science which ramifies from it.
New Testament Apoc:

" Unconstant, not knowing the Majesty of God, how great and wonderful He is who *created The World*." II. *Her. Com.* XII. 19.

" Who with the word of His strength *fixed* the heavens, and *founded the earth upon the waters*." I. *Hermas. Vis.* I. 28.

"And even the world itself is upheld (or supported) by the four elements." I. *Her. Vis.* III. 130. *Vide. Zetetic Astrony.* by "*PARALLAX*" *Chap* 13.

Now then consistent and reasonable Christians ! confronted by these passages from your own Text Book, (irrespective of those from the Apocryphas) EXCEPT YOU DELIBERATELY IGNORE THEM, or resort to subterfuges extraordinary, AWAY MUST GO the PRIME HYPOTHESIS of the " GLOBULARITY and mad-whirling-flashing-motions of this World," with the closely related assumption of "Universal Gravitation," and consequently the assumed and unproved Inconceivable Distances and Magnitudes of all things celestial, the Pluralities of Worlds, Myriads of suns and their various systems, the Profundities of ether-filled Limitless-Space. The undefinable-Periods of Geological-Times,

Atomic-Origin of all Things celestial and terrestrial, and other Evolutionary-quagmires; disbelief in "The Creation," and "The Universal Flood," ignorance as to the meaning of the words Up, Down, Above, Below, Under and Over; and all the other BEFOGGING AND UNPROVEABLE ASSUMPTIONS, along with the mystical and sophistical arguments and contradictions of Modern Theoretical Astronomy, and the numerous (so-called) Scientific Hypotheses connected with, and ramifying from It, and without which, It could not possibly exist.

IF you persist in stultifying your senses and reasoning faculties, thus ALLOWING WITHOUT PROOF all the brood of Suppositions and assertions of Modern Theoretical Astronomy, then, The Bible with all Its Facts, can be quietly ignored, or at best considerably whittled-away, as IN THE SO-CALLED LIGHT of the Modern Theoretical Assumptions, It is covertly hinted, or openly stated to be nothing more than a collection of Childish fables and absurd myths, therefore, not worthy even the notice of Agnostic-Infidels, much less the respect and absolute confidence of those CALLING THEMSELVES Christians, no matter of what denomination: BUT, with a knowledge of THE VITAL AND FUNDAMENTAL TRUTH, that THE WORLD IS NOT A GLOBE, the way becomes clear and intelligible, The Biblical Records standing out bright and sharp as unmistakable FACTS and ACTUAL REALITIES, thus proving to the centre the Impregnability of the Rock of Holy Scripture, notwithstanding the assaults made upon It by MODERN-EDUCATIONALLY-BIASSED, and misdirected minds.

"The two beliefs"—Modern Astronomy and The Bible.—
"cannot be held together in the same mind, he who *thinks* he believes *both*, has *thought* very little of either."—Thomas Paine. "*Age of Reason.*"

"If Moses can be shewn to be caught red-handed in ignorance or error, what shall we think of The Christ who quoted and referred to him as an authority." *Present Day Atheist.*

"I had been told so often that The Bible was no authority on scientific questions, that I was lulled almost into a state of lethargy."

"If it shall turn out that Joshua was superior to La Place, that Moses knew more about geology than Humboldt, that Job as a scientist was the superior of Kepler, that Isaiah knew more than Copernicus . . . then I will admit that Infidelity must become speechless for ever." "*Ingersoll's Tilt with Talmage.*"

"In whatever way or manner may have occurred this business (Modern Theoretical Astronomy, and the Modern Elementary Theoretical Science) I must still say that I curse this Modern Theory of Cosmogony, and hope that perchance there may appear in due time some young Scientist of genius who will pick up courage *to upset this universally disseminated delirium of lunatics.*"

J. WOLFGANG VON GOETHE.

BIBLICAL COSMOGRAPHY.

GEOGRAPHY has hitherto consisted of a series of unintelligible and irrational dogmas, by which the memory alone has been taxed, without any attempt to appeal to or encourage the exercise of the reasoning faculties.

It must be understood that we purpose confining ourselves strictly and exclusively to a description of the Universe as a whole, without any special reference to the various Continents apart from their connection with a bird's eye view of the Earth's surface generally.

All that we propose to interfere with and abandon is the Newtonian or Copernican system which identifies the Earth on which we live and move with the planets above our head. There is no possible analogy between them! East and West, North and South, fire and water cannot be more distinct. So that all that we have hitherto learnt upon the subject must be forgotten and laid aside, not partially or here and there, but wholly and entirely, from the first line to the last. "Mercator's Projection" of a square world, with its rectangular meridians and latitudes, is quite as preposterous as the spherical theory, and is a most clumsy and senseless substitute for the impossible " Globe." But one falsehood begets a hundred others; and is only supported by a tissue of fabrications as discreditable as the original lie. We can afford to be very brief, because truth requires no tautology, no special pleading.

The Newtonian or Copernican theory, from the first hour of its invention, has never dared to submit to an appeal to facts! Not a schoolmaster or Scientific professor has ever been known to illustrate its principles or enforce its teaching beyond the pasteboard and paper to be seen in our school-rooms or shop windows. We, on the other hand, are prepared to abide by the severest tests which *practical* science or rational ingenuity could suggest. TRUTH always glories in and courts daylight and discussion; fraud and falsehood can only exist in silence and secrecy and absence of suspicion. The Newtonian imposture would not have lasted as many days as it has years, if its disciples had not resisted and resented all discussion and inquiry. But every system of deception has its limits; and this, the most wide-spread and the most baseless superstition ever imposed upon the ignorance and credulity of childhood, is on the eve of a tremendous revolution.

Defenders it never had; and no threats, no taunts or exposure will ever rouse the energies of a single champion. Cowardice always accompanies conscious guilt; and this vaunted system, "the most exact of all the sciences," will ignominiously perish without a single tongue or pen being moved to uphold its crumbling ruins.

First and foremost, every map of the World with its two hemispheres must be ripped out of our atlases; all the pasteboard models called "Globes" must be stamped into fuel for the furnace, and all our Geography books must be discarded as any authority upon the subject of the shape and surface and size and motion of the Earth and the Oceans which surround it. All previous instruction upon this subject must be unlearnt and forgotten! We must abandon and utterly reject all that we have been told about " Gravitation " and " Attraction," about " Centripetal and

Centrifugal forces," about distances and immensities in space. Jerusalem is still in Asia, Egypt in Africa, New York in America, and the map of Europe is not essentially changed. But this is about all that can be found in our ten thousand schoolbooks which will justify any further reference being made to them! They had much better, however, be all sent to the mill; nothing can be gained by mixing up truth with falsehood. The very sight of them will only serve to embitter the recollection of the baseless frauds they were the means of imposing upon us. They will ever be a disgrace to our intelligence and a scandal to the nations that could be fooled with their teaching.

The very configuration of the several Continents is essentially changed; all our Nautical tables are greatly misleading and deceptive, and involve an amount of calculation and theory which many years' employment of them has never rendered familiar or trustworthy.

On the Plane system all is simple and intelligible; it is natural, and requires no invention or assumptions to illustrate its principles. Its Science is inspired, and its philosophy based upon reason, and can unblushingly invoke an appeal to ten thousand facts to confirm the practical character of the entire system.

The Earth as it came from the hands of its Almighty Creator, is a motionless Plane, based and built upon " foundations " which the Word of God expressly declares cannot be searched out or discovered. All we know or read of is a Hell beneath, a Heaven above us, and the Earth on which we stand; that it is not" hung upon anything," because supported, as *all* material objects *must* be supported, by some material attachment. The *words* " Gravitation " or " Attraction " will not support a material Earth, any more than a magnet will support a single grain of iron or steel or cause it to float in mid air.

The sun, moon, and stars are luminous and imponderous gases only; and, consequently, float upon an atmosphere heavier than themselves. There are no mountains or volcanos or living creatures of any kind in any of the planets. The stars are hardly bigger than the gas jets which light our streets, and if they could be made to change places with them, no astronomer could detect the difference.

The North is the central point of the World. All the compasses that were ever constructed, converge to it; and radiate in all directions, horizontally towards the south.

The South is, therefore, the circumference of the Earth and its waters; East and West are relative terms, according as we move right or left, and always imply a circular direction, as North and South always describe a straight line.

The Circular Charts are more properly divided into 24 meridians, representing the 24 hours in the day; while the concentric circles of latitude can only amount to seven; or three North and three South of the Equator; nine hundred Geographical miles between each parallel, to correspond with the 15 degrees of 60 miles each on the Equator, which is the true scale or standard of all measurements, North, South, East or West. The lines of latitude are severally numbered 15, 30, and 45. There are $12\frac{1}{2}$ beyond the 45th parallel, North and South; but these may be virtually excluded from any calculation and only technically represent unknown and unnavigable regions.

The Summer Solstice reaches about 1,000 or 1,200 Geographical miles North of the Equator, and the Winter Solstice extends as far to the South. Thus the Sun is six months decreasing and six months increasing its horizontal and concentric orbit. Its height above the Earth, is under 3000 miles; but its elevation never varies, although the angle of its altitude of course depends upon its orbit being nearer to or more distant from the observer's standpoint, according to the season of the year.

The Sun makes its circuits, of course, in 24 hours; travelling faster and faster from June to December, and slower and slower in its Northward journey, from December to June. When it is nearing our meridian we enjoy the daylight, and the opposite one is left in darkness. We can never rise above or even approach the Sun's height. To a partial extent, we can keep it in view, by going higher and and higher; but at the seventh or eighth mile we are glad to relinquish the attempt, and it passes beyond our horizon, and the mists and clouds of 7000 miles in density, obscure its rays till the dawn of the following day Its passage or direction, of course, is from East to West.

The Moon is nearer to us than the Sun; one hemisphere is light, (its own light, not a reflected one) and the other side, dark. But it has no solidity, nor is it material in any respect. The Stars also have all their own light; the Sun may shine on them, as it would on a lamp or candle, but their light is quite irrespective and independent of the Sun; otherwise, they would be one side light and the other dark. The surface of all water under the skies is a dead level; at right angles to any one given perpendicular. *All* perpendiculars, wherever situated, are parallel with each other. Neither of these indisputable facts are possible on a spherical surface. These are not questions for argument, but are simple matters of measurement, with which all our children ought to be made familiar.

If we could remove the Continents, the Northern Ice Plains could be circumnavigated in about a 3000 mile circuit. The extreme boundary or circumference of the Southern Ice is nearly 30.000 miles !!!

Two ships, starting either from Cape Horn or Cape of Good Hope, the one sailing in a South-east and the other in a South-west direction, would *never* meet again round a point or "Pole" due South of both ! Why not?

Rivers do not, cannot flow round the spherical surface of a Globe. Attraction does not act upon water; nor is there half an ounce of the atmospheric pressure to the inch; if there was, the said "Globe" would be impeded in its rotation as a narrow wheel is impeded by a "brake" or pressure of very considerably less than half 15 pounds! Atmospheric pressure is as great a delusion as a spherical Earth; a bag of hair or shavings, with all the air pumped out, will not be flattened $\frac{1}{16}$th of an inch, although several cwts. of pressure is *said* to be surrounding it!!

Rivers even do not flow downwards in their progress towards the Sea. In several instances, two rivers are flowing in opposite directions in the same county or district. They are carried forward solely by the impetus and accumulation of the waters behind; and when, as is generally the case, their sources are above the sea level, they descend by sudden "falls" or partial "rapids," till they again reach a level bed. Tidal rivers flow backwards and forwards over the same level channels. The Moon has no sort of effect upon the tides, anymore than upon the ebb and flow of the blood in our veins. *How* they are caused, may be discovered when we grow wiser. The Eclipse of the Moon is *not* caused by the shadow of the Earth, or by the Earth obstructing the light of the Sun. The Moon has its own light, as was before stated. Non-luminous Moons may exist in the firmament and cause all these appearances, or the Moon itself turn its dark side to us. At all events, a flat Earth, with the Sun *always* above it, cannot throw a shadow; that is quite enough for us.

All our present latitudes and longitudes are purely fictitious, invented to suit a surface which only exists in the mind of the mathematician. As was before specified, only 45 degrees or grades of 60 miles each, can be reached, either North or South. ALL meridians diverge; consequently, longitudes increase as they approach the South.

The Continents are of an entirely different shape to what they are represented on a pasteboard Globe. They are much wider, in proportion to their length. This would not apply to those portions *North* of the Equator. Perhaps, the most extraordinary fact connected with the Newtonian imposture is, that there should be no such shape as a curve or curvature on any part of the World's surface! Not even on land, is any such configuration to be found! Whereas, the perfect flatness of *water* can be exhibited and found on every five square miles of the length and breadth of the ocean.

Ships on the water and objects on any flat surface beyond a certain distance, disappear beyond the artificial horizon, which can *only* be seen on a plane or flat surface! No similar prospect could possibly occur on a spherical or globular surface. In the latter case, the intervening obstruction, (it could not be termed an horizon), would nvariably be *below* the tangent line from the observer's standpoint; whereas it is invariably seen as it were above it; and, in *all directions*, the same! Thus, *on a plane*, the observer always fancies himself *below* the horizon; on a Globe or convex surface, the observer would always feel and know himself to be above all his surroundings! Let the Student throughly master these facts.

We have nearly said enough to enable the reader to re-arrange his Maps and Geographical notions. The annexed Skeleton or Linear Chart will furnish a sufficient framework for all purposes. Nothing will ever gainsay or refute it. The *exact* configuration of the Continents we must leave for the results of a more practical Exploration.

The following Tables may possible help some of our readers to detect the fallacy of their own erroneous system, and the tremendous difference between it and the truth. The amount of curvature on a Globe or sphere of 25,000 miles circumference, is 8 inches in the mile, multiplied by the square of the distance. So that in the first mile, the fall or decline from the spot of observation would be 8 inches; in the second mile, 32 inches; in the third mile, 6 feet; in the fourth mile, 10 feet; in the fifth mile, 16 feet; in the sixth mile, 24 feet; in seven miles, 32 feet; in eight miles, 42 feet; in nine miles, 54 feet; in ten miles, 66 feet; in twenty miles, 266 feet; in fifty miles, 1,666 feet, and so on.

The circumference of the Sun's orbit in the June or Summer Solstice, is 16,200 miles, and it travels at the rate of 675 miles per hour. On the Equator, in March and September, its orbit is 21,600 miles, and its speed, per hour, is 900 miles. In the December or Winter Solstice, its orbit is 27,300 miles, and its speed, per hour, is 1,125 miles.

On the globular theory, the arc of the Earth's hemisphere, or the surface distance from the North "Pole" to the South, would amount to 10,800 miles, or 180 degrees of 60 miles each. Whereas, the true measurement, or chord of the supposed arc, is but 6,900 miles, or 115 degrees, if this expression was correct, which it is not. The ½ radius or ¼ diameter is, consequently, 57½ instead of 90. So that all our Nautical Tables must be simplified and corrected. At present, they are an extraordinary jumble of figures; and are made to represent distances which, in reality, do not exist!

The measurement of any five or ten miles of water, would set the whole question at rest. In England during the Summer Solstice, the Sun at *midnight*, is about 3,000 miles away from us. In March and September, about 4,500, and, in the December Solstice, about 6,000 Geographical miles. The *mean* diameter of the Sun's orbit, (in March and September), is rather more than 7,000 Geographical miles; that is, from sunrise to sunset.

The half radius, that is, from the Northern Centre to to the Mean Equator, is 57½ grades of 60 miles each; or 3,450 Geographical miles. Of these, only 2,700 are generally navigable. The radius, from centre to the World's outermost limit, is 6,900 Geographical miles; of these, only 5,400 are practically navigable. The outer circumference of the generally navigable Southern oceans, may be set down as about 30,000 miles. But this would be found to vary considerably, according to the severity of the consecutive seasons; and, whether it is more or less must prove the palpable absurdity of the idea of there being any "South Pole" or antipodean centre. Without a "South Pole" there can be no "Globe," and the Earth *must* be a circular *Plane*.

The Transit of Venus cannot possibly give the height of the Sun, for the best of all reasons; the height of Venus herself is only guessed at, and the method of measuring either one or the other, grossly fallacious. The process of spherical triangulation gives a product of millions, when not so much as 3,000 has to be accounted for. The base, from London to the spot on the African Coast where the Sun is vertical in the June Solstice, would be ample for all purposes. Carefully measure the length of the chosen base, the angle at the point of observation would determine this long-disputed question. No sphericity, of course, allowed for. The base, an horizontal plane. The elevation would be found 100 or 200 under the thousand miles!

Fixed Stars could be measured as to height, in the same manner. The following facts will not be disputed by the mathematician, whether he calls himself a Geographer or not. The quadrant of the circumference of every circle is 90. The diameter is to the circumference as 1 to 3 and some fractions, or say, as 7 to 22. The superficial area of the circle is equal to the product of the circumference and half the diameter. The radius of half an arc or semi-circle is also 90; but half the diameter of the chord of the semi-circle, is not 90 but only $57\frac{1}{2}$. So that there cannot be 90° of latitude, North or South, but only $57\frac{1}{2}$, if taken to the very edge. But, as before stated, the edge or terminus cannot be everywhere approached, within about $12\frac{1}{2}°$ or 750 miles; leaving but 45 as the half radius of our circular plane, and the extent of our latitude. One hour corresponds to 150°, or $1042\frac{1}{2}$ statute miles, or 900 Geographical.

Sir Richard Phillips, who lived and wrote some 60 or 70 years ago, seemed thoroughly familiar with the absurdities of Newton's teaching. We will quote a few of his remarks.

"Newton himself made some feeble attempts to illustrate the mechanical causes of this principle of attraction; but his explanations have never been received or respected by his followers; and so the proximate cause of attraction and gravitation, even assuming their real existence, has continued to be as little known since his time as in any period of antiquity; and the variety of explanations and definitions which his learned followers have resorted to, while hoping to justify the employment of such phrases, has only made confusion worse confounded.

"In spite, however, of all their casuistry, sophistry, and equivocation, it is notorious that the Newtonians still teach the doctrine of a drawing or attractive power existing innately and universally in matter, with a variety of false analogies, new principles, and erroneous reasonings in every branch of this Philosophy, in order to maintain an original mistaken principle, which necessitates an abuse of terms which would be offensive and repugnant in any other branch of Philosophy. It is preposterously illogical to say that an effect is its own cause, that the phenomena produce phenomena,—that attraction causes attraction,—or that weight or gravitation is caused by weight or gravitation.

"The assumption of such palpably equivocal reasoning and phraseology may have been temporarily employed for Geometrical analysis; but after these purposes had been effected, the fictions should have been discarded.

"Newton lived in a superstitious age and district; he was educated among an illiterate peasantry; he was a student in Astrology and of the works of Bœhmen.

"The gratuitous principles of attraction and gravitation for which only an undefinable or metaphysical cause could be assigned, led to a variety of equally baseless assumptions, and Newton's philosophy was, throughout, governed by the bad taste of his age, and grew out of its vulgar and superstitious faith. He himself made use of the unerring tools of Arithmetic and Geometry, but he began with hypotheses and obsolete metaphysics of a dark age, which ought, in our improved state of knowledge, to be exploded and abandoned.

"It is the business of the Philosopher to examine the phenomena of Nature with perfect good faith and with an absolute deference to truth. It is impossible to reason on the operations of Nature, if the bases of our reasonings are incorrect and if they lead to false analogies which inevitably mislead all subsequent inquiries.

"If we would diligently and impartially investigate the Newtonian hypotheses we shall soon be convinced that they are essentially incorrect and imaginary, and unworthy of having any subsequent hypotheses built upon such shallow and visionary foundations."

From the above random extracts may be seen that the Newtonian imposture was detected and exposed at least fifty or sixty years ago; and that its present opponents are only reiterating the strictures which abler men did not flinch from publishing, and who acknowledged that they had "many intelligent disciples, and some in the seats of authority who" (then as now) "had not the courage to acknowledge their heresies."

The Author also refers to the slowness and reluctance with which Newton's theories were accepted by his contemporaries. "Newton," he informs us, "only printed 500 copies of his "*Principia*" in 1687; and, though he was a Professor at Cambridge and a member of the Convention Parliament, yet a second Edition was not required till 1713; which Edition remained on sale till long after his death in 1727, or forty years after its first publication! Even then, his Philosophy had been but very partially adopted in his own University, while to the rest of Europe it was utterly unknown."

So much for the birth and introduction of this groundless and execrable superstition, which had never seen daylight 200 years ago, and was "wholly unknown to the rest of Europe" for much more than a century afterwards!!!*

No wonder that no man of honor or of any scientific reputation can in the present day, be bullied, or taunted or bribed into appearing for its defence! Mr. Proctor is allowed to air his crotchets in the Birkbeck Institute, and Mr. Wallace gets a Sporting Editor to award him £1000 for his trickery at the Bedford Canal; but neither of these men would dare to defend their own principles, much less their conduct, by an appeal to facts; they can but re-echo the same whimsical fallacies which the Authors and Inventors themselves were so thoroughly ashamed of, and repudiated so strongly when they discovered that by some they were actually recorded as truthful realities.

The Rev. W. Jones, who wrote a masterly treatise on "The First Principles of Natural Philosophy, 1762," spoke quite as strongly against the baseless fallacies of Newton as Sir Richard Phillips, and thoroughly exposed many of his leading absurdities. There has not been an interval of ten or fifteen years without some strong, vehement and intelligent protests against the baseless and irrational dogmas of Newton and his modern satellites. As long, however, as credulous dupes will be found to listen to Mr. Proctor's lectures, or be led by Mr. Lockyer's logic, or be hoaxed by Mr. Wallace's still more dishonorable deceptions, we can but hope to influence a few outsiders, and to urge them, with all the earnestness we can express, to rebel

---

*It may not be generally known that there are, at this very day, some five or six of Sir Isaac Newton's last Manuscripts, locked up in the Libraries at Cambridge, which the Authorities dare not publish, for fear of injuring their idol's reputation!

against and reject what is such a scandal and disgrace both to our pretended civilisation and to our very humanity itself.

With one more remark we close this brief summary of the grandest subject that can possibly engage the attention of the Christian student or the man of practical Science. The Arctic crew lately returned from their utterly vain and useless search; as we predicted it would be, from the very moment it left our shores. Captain Nares announced on his arrival that the Expedition had reached the 83º of North latitude, and a little beyond. And, in the good faith of such a representation, Her Majesty conferred upon him the Order of Knighthood. Now we are prepared to meet Captain Sir Wm. Nares, face to face, and show him and all the Navigators in Europe that no such latitude can possibly exist on the surface of the circular Plane we live on! A knighthood may be a very proud distinction for merit honourably earned; but, in this instance, the recipient has obtained the credit of doing what cannot be done, and for going to a spot which only can be found on a pasteboard or brazen Globe. Both English and American Explorers would accomplish a far nobler object if they will only attempt to circumnavigate their pretended " South Pole," and report to the bewildered Philosophers at home the result of their fruitless search.

For a more lengthened exposition of the foregoing statements the reader is referred to a larger pamphlet now preparing. Enough, however, has been said to make all the Geographical and Educational professors of Europe blush their skins into blisters till they can reply to and refute every paragraph and sentence we have written That pretence of " dignified unconcern " which our guilty opponents so persistently assume, may be considered very professional amongst themselves, but it will soon give rise to a feeling and expression of public contempt which will not easily be modified by any subsequent concessions they may make. Messrs. Proctor, Lockyer, Wallace, and Dyer, have each in their turn received an amount of exposure and indignant censure for their repeated fallacies and fictions, which would have made any third or fourth-rate tradesman frantic with shame and confusion. But, unless the British Press is bribed at any cost to overlook and ignore the appearance of this little pamphlet, we do not see that it can avoid challenging every statement we have penned, or else insisting upon a full and complete rejoinder being made by the professors to the crushing attacks upon their honour, their veracity, and their ability as teachers of Science. But it will be seen that these men dare not face any open and honest attacks! No amount of bribe could induce Mr. Wallace to repeat his experiment at the Bedford Canal; no flattery would influence Mr. Proctor to dispute our assertions in the columns of the "English Mechanic." The Royal Geographical Society exists only on the contributions of its deceived and ignorant dupes; what the public get in return is best known to themselves. But, sooner or later, these Institutions will be compelled to show what they have done, and to what extent they have deserved the confidence reposed in them. We may not, perhaps, live to see the indignant surprise with which the detection and exposure of their baseless fallacies and frauds will be received by their insulted and injured disciples, but our readers owe it to themselves to insist upon knowing whether they or their opponents are in the right. If they persist in refusing to answer or reply to the charges we have made, not only in this pamphlet but in various strictures which have from time to time appeared from Mr. Carpenter, on the statements of Messrs. Proctor, Lockyer, Wallace, Dyer, and Co., we do not envy the sense of utter degradation which such exposures must entail. If British science involves such moral serfdom on its disciples, the professors no doubt derive some consolation in the enjoyment of their ill-gotten revenues.

NOTE.—We may now have said enough to enable our readers to form a clear notion of the real configuration of this Universe, as it came from the hands of the Almighty Creator, was recognised and approved by all the inspired historians, as well as by the most learned and practical philosophers of the first 5,500 years, and such as can be confirmed by the indisputable testimony of ten thousand facts, such as no Geographer or Engineer in the United Kingdom or elsewhere will venture to gainsay or deny. Either the whole of the foregoing statements are wholly and provably false, or every lesson that is taught upon the subject is a gross fraud upon the credulity of the public. We challenge our opponents to face the question as men of candour and intelligence and as lovers of truth for truth's sake; and we especially call upon Mr. Professor Alfred Russel Wallace, F.R.G.S., who is still retaining the sum of £775 out of the £1000, forfeited by his own act and deed and handwriting in the year 1870, under the plea that he had proved the convexity of water on the Bedford Canal, which none but himself and his accomplices have ventured to assert. The credit and credibility of the whole Geographical Society is at stake till this dispute is fairly and finally settled. We do not intend the question should rest where it is. Mr. Wallace is either honourably or dishonourably in possession of that £775; and, if he has any self-respect or regard for the reputation of his associates and their science, he would hasten to determine the point at issue, and relieve himself from the discredit and annoyance which our persistent claims must expose him to. The actual shape of water surface is all that we need ascertain to prove the truth or falsehood of the Newtonian theory. Mr. Wallace is the only man in the kingdom who has avowed himself the champion of the convex theory, and his only approver has been the Editor of a sporting newspaper!

Chart & Compass, Sextant & Sun-dial, Latitudes & Longitudes, Plumbline & Pendulum, Globe or Plane?

A LETTER OF REMONSTRANCE

Respectfully addressed to the Officers of

THE NAVAL AND MERCANTILE MARINE OF ENGLAND AND AMERICA—THE FORMER STYLING HERSELF "MISTRESS OF THE SEAS."

Published by the Zetetic Society,

BEING THE ONLY COMPETENT AUTHORITY ON THESE SUBJECTS THROUGHOUT THE WORLD.

CHART AND COMPASS.

A LETTER

RESPECTFULLY ADDRESSED TO THE OFFICERS OF THE NAVAL AND MERCANTILE MARINE OF GREAT BRITAIN AND AMERICA.

GENTLEMEN,

At a time when naval men and shipowners are beginning to inquire whether the science of navigation, with its collateral conditions, is not susceptible of some measure of improvement, may we venture to draw your attention to the undeniable fact that they are still using a projection or Chart of the World which was invented three centuries ago, and which has never been considered otherwise than a make-shift, and retained only because there was no accurate knowledge of the ocean's surface or its configuration. But this ignorance is hardly creditable in the present condition of our commercial interests and the extent of our foreign and colonial traffic.

If those whom it may concern are really in earnest when they propose to take into consideration any means that may be suggested for affording greater security in ocean travelling, and leave them less dependent on systems of navigation which may have been considered efficient for coasting purposes, or before our colonial empire was so extensive as it it now, they will, I trust, be willing to admit that not merely the substantial construction of the vessels themselves should be practically considered, but shipowners and naval men generally should honestly and intelligently ascertain how far their existing charts and projections may be regarded as incapable of improvement, or be practically relied on as affording the most accurate information as to distances and relative position of every principal spot on the navigable ocean.

Without any reference to existing theories, may we venture to ask how many of our most experienced seamen and naval officers could say if they knew what was the actual difference between the northern and southern meridians of longitude? Would they not be obliged to confess that they were navigating their ships on an artificial or purely conjectural system of hydrography, and that the training they had received had never taken into account the possibility of there being a more simple and far more rational method of navigation by a stricter adherence to demonstrable facts instead of to artificial rules and astronomical principles, which can have no more analogy to terrestrial or water surfaces than there could be between the length of a ship's deck and the date of the owner's birth.

If their system was as correct as it professes to be, ought not a fully trained and perfectly educated officer be able to take a ship, say, to Malta and back, on second or third voyage? Instead of which, when his theoretical education is over, he has to put his theories in his sea chest and learn from actual observation, for many long years, how to take a vessel, say, to the Cape of Good Hope, without finding himself, some foggy morning, on the coast of Mexico?

We are afraid these remarks will be received with much disfavour, and be resented as an insult to the services they are humbly intended to serve. But we are sure the issues are too serious to admit of any palliation, even in spite of the undeniable fact that thousands of vessels every year find their way out and home again with the most infallible accuracy. But is this any justification for retaining a system which has to be supplemented with the most voluminous tables, and practically ignored by every experienced sailor?

We are merely insisting on the fact that a comparatively perfect projection is as easy of construction, and a thousand times more intelligible, than one which makes a burlesque of the whole subject, and is as unlike what it is intended to represent as if concocted by a village schoolboy. Whether the earth be a globe or a plane, a *square* chart, with its rectangular meridians, cannot possibly represent its natural divisions, or define its relative distances, or give the true position of its several continents, or the actual bearing of any spot out of sight of the "crow's nest." Any accuracy of detail is utterly impossible under such theoretical conditions.

Surely no one will venture to say that this unnecessary and inexcusable defect in one of the most essential elements of practical navigation is not a slur on two great maritime nations like England and America? If other countries are no wiser, is that any reason why we should persist in setting such an example of barbarous ignorance? Our Admiralty and Boards of Trade are, unfortunately, regarded as oracular on all subjects relating to seamanship and naval science. We are, therefore, the more inexcusable for encouraging others to retain what we dare not defend in our own principles.

We are ashamed to have to insist that the proper construction of our charts of the habitable world is as easily determined, and the distances as accurately laid down, and the latitudes and longitudes as mathematically infallible, as would be the plan of an estate of a few hundred acres. And yet, in this professedly most enlightened age, we are still using a square chart, with every figure and line of it wrong, and as unlike the real earth and its waters as the ingenuity of man could conceive! Is not such ignorance as pitiable in our case as when the Ninevites were said not to know their right hand from their left?

The distances on Mercator's chart are computed on the arc measurement, when no such shape as a curve can anywhere be found! So there are not 90 degrees north or south of the Equator, but only 57½ degrees; barely 45 degrees of which are practically navigable! Meridians of longitude cannot *possibly* be parallel, but *must* be *divergent* from the central north to the southern circumference. Mercator's straight edge is as preposterous as a South Polar centre!

But, without pursuing this subject further, we are quite prepared to show that the topography of the circular plane on which we live, admits of exactly the same infallible accuracy as the compilation of the multiplication table.

By using the present imperfect and most misleading projections, ships must inevitably wander hundreds of miles out of their direct courses; while time and money and valuable cargoes are often risked, merely because the authorities t home think they would forfeit a certain amount of prestige by abandoning their time-honoured fallacies.

If the Royal Geographical Societies of England, Scotland, and Ireland cannot give you better instructions, on what grounds, it may be asked, do these scientific Associations venture to collect incomes of £6,000 or £8,000 a year each, and obtain royal charters for diffusing imposture and fraud and fiction?

Now let us make a few practical remarks on the quadrant and sun-dial. From models or specimens now to be seen in the British Museum, we learn that the quadrant was known and used so early as the end of the 14th century, or several hundred years before sailors ever dreamed they were sailing on the surface of a spherical ocean. The inventor, therefore, never contemplated its employment otherwise than on an horizontal base, at right angles to the sun's vertical rays or to the perpendicular of any other distant object; and it can, by no possibility, give a true elevation, if used on any other conditions! If these absolute conditions are not practically complied with, the observers would have three-fourths of their time occupied in making "allowance for curvature" which it is well known they are never insane enough to think of.

Again, all observers are supposed to be on one and the same horizontal base, so that a greater angle than 90 degrees is never contemplated or provided for or obtainable in any survey that can be made!

Every arctic explorer knows that in the latitude, say, of Behring's Straits, or Hudson's Bay, an observation of the sun can be made at midnight, looking *due north;* which would be physically *impossible* if a large segment of a circle intervened between the observer and the June solstice; looking over the North "Pole!"

How discreditable it is to two such maritime nations as England and America, to say nothing of the other European governments, that such gross and wholly inexcusable ignorance on some of the most important points and the most elementary conditions in navigation should still prevail!

Again, we do not believe that there is one nautical man in ten thousand who knows the true cause of the compass needle's pointing north and south! Unless our naval instructors are no wiser themselves, their sole object seems to be to allow our seamen and officers to be mere machines, and, like the children of landsmen, to be unacquainted with the most rudimentary principles and conditions of practical science. An exposure inevitably will take place some day; and the professors will then learn a lesson from their ill-instructed dupes they are not likely to forget.

The English press, both secular and religious, is unfortunately in the hands of those who are as ignorant of the genuine facts of science as the most thoughtless of their readers. And to hide their shameful incapacity, they toady, like brainless sycophants to the professional authorities, who like abject cowards allow these half-witted journalists to display their wisdom by burking all open discussion, and publishing their vulgar sneers when any honest inquirer ventures to refer to the subject.

We could fill many pages with the description of the manifold fallacies in connection with some of the most familiar objects around us. Phenomena and facts are daily taking place before our eyes, while we go on, year after year, and age after age, as ignorant of their true character as if no such things had ever been seen. Even men of the most undoubted intelligence have, for example, gone on for the last 150 years using a ship's quadrant, and never pausing to inquire or to perceive that the instrument was utterly incapable of being accommodated to the conditions they supposed to exist!

Another instrument, quite as familiar as that of the quadrant or sextant, is the sun-dial. It is the very oldest philosophical instrument ever used or invented by man. And for more than 5,000 years it was used in all its simplicity and perfection of principle; that is, on the motionless disc of a stationary earth, with a revolving sun to mark the shadows in its daily progress, or circular orbit, from its rising to its setting.

But, for the last 200 years, a set of crafty, and fraudulent, and infidel professors have dared to assure their dupes that the sextant and the sun-dial could be as easily used on a round and revolving globe, as on a circular and motionless plane! And, in the ignorance and credulity of our childhood, we have been led to accept as genuine truth some of the most degrading and senseless superstitions ever devised by the arch-fiend himself, in order to persuade mankind that they were wiser than God, and had far sounder views of His creation.

We have scores of other facts, quite as palpable and important, but we forbear, and only ask our readers to weigh well what we have already stated.

Mr. Jordan, of 81, Spring Street, Hull, will send a lithographed sketch of the true and false application of the sextant, which he was the first to detect. The inutility of a sun-dial on a revolving globe will be apparent at a glance, when it is remembered that a body rotating on an axis, in one uniform plane before a fixed sun, *must* cast a *straight* shadow, east and west of any upright staff, placed anywhere or in any latitude between the solstices, so—

W━━━━━━━━━━━●━━━━━━━━━━━E

Does it do so? Why not! Is it hardly credible that millions of intelligent men have used these two instruments during the last two centuries, without detecting the glaring absurdity of adapting them to the conditions of a round and revolving globe? And yet the very professors who have most fraudulently hidden these facts from us, have been year after year pocketing their enormous revenues, while leaving us to believe that all that science could reveal was honestly and instantly proclaimed for the benefit of those who have so unreservedly trusted them. And so far from exhibiting the commonest courtesy towards those who with every mark of respect sought to discuss the matter with them, they have either refused to see them or have given them plainly to understand that no unprofessional assistance could be entertained for a moment.

The Secretary of the Zetetic Society will gladly furnish any further particulars on application.

We have left ourselves no room to refer at any length to the plumbline and pendulum. We have had, however, many opportunities of proving that they are both of them as fatal to the globular thory, as any of the other instruments we have mentioned. The plumbline has never been shown to deviate a hair's breadth from a true perpendicular to any horizontal base. And as for the pendulum, its vagaries have as much consistency as a weathercock! So that, literally, every possible test to which ingenuity can appeal—*all* of them, without one solitary exception, utterly refute the globular theory and fully establish the truth of the motionless plane, which the most learned and godly men maintained and taught for the first five thousand years. And the pagan blasphemy now insisted on in all our secular and religious literature has not the very ghost of a fact on which to rely. Pythagoras, its original inventor, openly asserted that he had "spent six months in hell;" and even if he had not told us so, we should know that there is but one source from whence such a system of lies, and imposture and blasphemy could have emanated. Has it got one defender, except among the very scum of the literary world? God's bitterest curse *must* rest on any country and people enslaved by such heathen superstition. And, in spite of all the extravagant and senseless ecstacy now being displayed on the occasion of her Majesty's Jubilee, we should have thought that after a fifty years' slavery to such pernicious fallacies, and such worse than heathen ignorance on the most elementary subjects, we might have hoped that this year would have been characterised by some less childish and sensuous transports than those we are giving way to.

All communications to be addressed to the Secretary, ZETETIC SOCIETY, *Cosmos House, Balham, Surrey.*

☞ The *only* correct diagram of Latitudes and Longitudes and Solar courses ever published or procurable, with coloured Map of the World as a circular plane, may be had of the Secretary, post free, 2s.

G. W. COOK, Printer, 3, 4 & 5, Swan Buildings, Moorgate Street.

COLLAPSE OF THE GLOBULAR THEORY.

The Surface of ALL Standing Water Proved to be Absolutely LEVEL.

COLLAPSE OF THE GLOBE THEORY.

We give, on the first page, two reproductions of the photograph taken on the Bedford Level.

The second photo is marked with dots, a cross, and the letter A, to indicate to the reader the position of the screen, etc.

The two dots (:) are the screen and its reflection in the water below it—near Bedford Bridge.

Trees near this bridge form a background to the screen, and its reflection rests upon their shadow in the water. The continuation of the canal beyond Bedford Bridge cannot be seen from the direction of Welney Bridge, even quite near it, because there is a junction of canal paths at that point, and their several courses can only be seen by standing on the bridge itself, or proceeding to a point beyond it.

Dallmeyer's latest pattern Photo-Telescopic Camera was used for the experiment. It was placed in position less than two feet above the ground-level by the expert operator from Dallmeyer's, and that gentleman, Mr. Clifton, being a globularist (see his letter, printed below) it cannot be suggested that he would lend himself to unprofessional practice, and were such conduct on his part possible he would have tampered with the instrument, or the *locus in quo*, so as to favour plane-earth teaching; he, however, irrespective of results, acted up to the letter of the test experiment. Mr. Clifton had to lie down in order to manipulate the instrument, close under Welney Bridge, a distance of six miles from Bedford Bridge, the screen being fixed rather to the right of the bridge. The cross at the edge of the photograph marks the position of the camera.

The letter A is intended to draw your attention to a dark chimney, connected with some works near the canal. This chimney is just midway betwixt the two bridges, *i.e.*, it is three miles from Welney Bridge, and three miles from Bedford Bridge.

This experiment was carried out in misty and very unsatisfactory weather, on May 11th, 1904, before Lady Blount and several scientific gentlemen, and proves conclusively that if the world be a globe having a circumference of 25,000 miles, the bottom of the screen should have been certainly over 20 feet below the line of vision in the six miles view. As the whole of the screen, and its reflection in the water beneath were observed and photographed, no curvature can possibly exist; the theoretical scientists are wrong and beaten, and Parallax, John Hampden, Wm. Carpenter, and the army of Zetetics were, and are, right in their contention that the world *is not* a globe!

* * * * * *

To Lady Blount.

Dear Madam,

Referring to the experiments at Salter's Lode, Downham, Norfolk, May 11th, 1904, I have much pleasure in testifying to the fairness of the conditions under which they were conducted. I arrived on the spot with the distinct idea that nothing could be seen of the sheet at a distance of six miles, but on arrival at Welney I was surprised to find that with a telescope, placed two feet above the level of the water, I could watch the fixing of the lower edge of the sheet, and afterwards to focus it upon the ground glass of the camera placed in the same position.

The atmospheric conditions were very unfavourable, a day of sunshine having succeeded several wet days and thereby caused an aqueous shimmering vapour to float unevenly on the surface of the canal and adjoining fields. This prevented the image from being as sharply defined as it would be under better conditions; but the sheet is very plainly visible nevertheless. This trouble is well known to all who have practised telephotography.

With regard to the lens used, I may say that this had an equivalent focal length of between 16 and 17 feet, which ensured an image of appreciable size being obtained at such a distance.

I should not like to abandon the globular theory off-hand, but, as far as this particular test is concerned, I am prepared to maintain that (unless rays of light will travel in a curved path) these six miles of water present a level surface.

Yours faithfully,

For *J. H. DALLMEYER Ltd.,* E. CLIFTON.

(*Chairman: The Rt. Hon. the EARL CRAWFORD, K.T., F.R.S., etc.,*)
25, *Newman Street, London, W.*
The Scientific Department under the control of *T. R. Dallmeyer, F.R.A.S., etc.,*
And
Managing Director: *G. E. St. L. Carson, B.A. (Cantab), B.Sc.*

Thus, by the aid of the latest discoveries and improvements in the art of photography, the earth's unglobularity is proved, and this fact coupled with Proctor's admission that, "*if with the eye a few inches from the surface of the Bedford Canal, an object close to the water, six miles distant from the observer can be seen, there manifestly would be*

Something Wrong in the Accepted Theory,"*

should awaken present-day scientists to the reality that there IS something wrong.

*Myths and Marvels of Astronomy.

Reprinted from

"THE EARTH,"

A Monthly Magazine of Sense and Science, upon a Scriptural Basis;

And of Universal Interest to all Nations and Peoples under the sun.

Edited and Published by E.A.M.B., 11, Gloucester Road, Kingston Hill, Surrey, England.

Copies of this leaflet, price 1/- per 100, may be obtained of Lady BLOUNT, 11, Gloucester Road, Kingston Hill, Surrey.

The Coming Man.

FROM THE "LEICESTER DAILY POST."

LETTERS TO THE EDITOR.

MODERN ASTRONOMY.

SIR,—I was pleased to read in the *Post* of last Friday your sensible remarks on the ambitious pretences of modern astronomy. It is, as you remark, one of the "most fascinating while it is one of the most unsatisfactory of all the sciences;" whilst its professors assume the loftiest tone imaginable, and expect us to receive their mere speculations and fancies as gospel truths. In fact, the teaching of the Bible is entirely ignored in their fascinating speculations, and one is almost scouted in these days for suggesting that possibly the ancients were more correct in their ideas of the universe than are the moderns. If, as you say, "the nebular hypothesis of Laplace really represents the extent of our astronomical knowledge," and this hypothesis should prove an unfounded speculation, how much real knowledge is there in this modern and much-vaunted "science" after all? It would, as the Apostle Paul says, be a "science falsely so-called;" yet many professing Christians, alas! swallow down anything in the name of "science" with open mouths, while the account given by Moses of the creation of the world is pooh-poohed as old-fashioned and out of date.

If all matter were originally nebulous, what, I should like to know, caused its condensation into stars, or hot and flaming bodies, as they are again supposed to be, like our sun? Gravitation? But what is gravitation? I have seen a great deal of astronomical conjuring with this word; but what is the thing itself which is called "gravitation?" Has solar attraction ever been proved, or is it only another "hypothesis" or assumption? If the latter, then the whole theory of modern astronomy rests upon two baseless ideas or speculations, rather than upon the well-founded facts of eternal truth. Now I seriously ask for one single fact proving solar or stellar gravitation. How can one star or sun pull another body said to be millions of miles away? What is the rope, or connecting rod, or coupling, by means of which the "pull" is effected? When we travel by train we find that before the engine can pull its load of following carriages each car has to be hooked on to it, but this mysterious kind of matter called gravitation is a sort of elastic web, which is always hooked on, and which is supposed to pull with the greatest tension when the distance between the two objects is the least! This is contrary to our ordinary experience on the earth, is it not? Perhaps some of your more learned readers will explain it for us, namely, solar gravitation, because it appears to me that this hypothesis of Newton is at the base of all the subsequent hypotheses or speculations of his now numerous admirers. If Dr Huggins and the astronomers cannot tell us whether the heat of our sun is now less or more than it was a hundred years ago, they had better stop at home a little more and not wander so far away amongst the stars. Let them secure their base line, and then sally abroad amongst the stars. Your figure about the household fly having the audacity to suppose it could master the secrets of social science is not a bad one to represent the pride of those titled mortals who not only think they can pierce the heavens, but who dare to impugn the teachings of the Creator through his servant Moses whose writings have been endorsed by the Son of the Most High.—I remain sir, yours, &c., ZETETES.

150 St Saviour's Road, August 25, 1891.

Sir,—There are still a few old-fashioned persons left who profess to believe that the world is flat and that the stars are mere points of light, situated in the heavens solely to illuminate this earth. Probably your correspondent "Zetetes" is one who entertains these curious ideas. He tries to explode the law of gravitation, than which nothing more certain has been proved to exist. If Zetetes had ever carefully studied a handbook on the subject I am inclined to think he would not have written the letter on "Modern Astronomy" which appeared in the *Daily Post* of the 26th inst. Gravitation is no "hypothesis" or assumption. It is an established fact, and its proofs are almost innumerable. Gravitation controls the moon as it revolves around our earth ; our earth, as it revolves around its centre of attraction, the sun ; our sun as it, in its turn, revolves around some other and greater sun. To attempt to prove that gravitation exists in the limited space at my disposal is unnecessary, since explicit proofs are to be found in every work on astronomy. Again, is the account which Moses gives us of the creation of the world to be taken literally ? Are we really to believe that the world was formed in seven days ? Might not a "day" in this case be a period of time extending over countless ages ? Doubtless it was so. That this globe was millions upon millions of years in its formation, from the time that it was a nebulous mass to the time that God created the first man, is acknowledged by the wisest men of our day, and I fail to see that it is opposed to Bible teaching.

Astronomy does not tend to raise man in his own estimation, and make the Creator seem less glorious in the eyes of mankind. The more we learn of this most fascinating science the more we feel our own insignificance and the stupendous greatness and wisdom of the Most High, who ever rules every body which exists in the infinity of space, yet deigns to care for the poorest and meanest mortal upon this earth —I remain, yours, &c.,   GERARD WARNE.

Sir,—Your correspondent Gerard Warne has not even attempted to answer my questions. He refers me to handbooks on astronomy for proof of solar gravitation. He thinks I cannot have read one or I should not question the existence of what is called "gravitation." He is mistaken. It is because I have failed to find any proofs there that I appeal to your readers. It is all assumption. Sir John Herschell says: "We shall take for granted, from the outset, the Copernican system of the world;" and if we turn to Copernicus he admits that his theory of the universe is founded upon hypothesis or assumption rather than actual fact. He says: "It is not necessary that hypotheses should be true, or even probable; it is sufficient that they lead to results of calculation which agree with calculation. . . . Neither let anyone, so far as hypotheses are concerned, expect anything certain from astronomy, since that science can afford nothing of the kind. . . . The hypothesis of terrestrial motion was nothing but an hypothesis, valuable only so far as it explained phenomena and not considered with reference to absolute truth or falsehood." Yet your correspondent says, "Gravitation is no hypothesis or assumption." I am afraid I must ask him to take his own advice and study some handbook on the subject. However, if the proofs of solar attraction are, as he says, "almost innumerable," will he kindly give your readers one? One will be sufficient, if a good one; and out of so many there should surely be one suitable for your pages. He says: "Gravitation controls the moon as it revolves around our earth; our earth, as it revolves around its centre of attraction, the sun; our sun as it, in its turn, revolves around some other and greater sun." But why does he stop here? Why not say this "greater sun" revolves around "some other and greater sun" still; and this around another larger than that, and so on *ad infinitum*, until you get the last sun of the series (pardon my "bull") large enough to fill the universe, and so stick fast.

This would be the logical outcome of this central gravitating theory; but we will, for argument's sake, stop with the first three centres given—moon, earth, and sun. My critic says that gravitation compels the moon to go round the earth, the earth around the sun, and the sun around some sun greater. He does not offer this assertion as proof, but simply sets forth the modern and now popular theory. I ask for proof, not now of the whole system, but of its fundamental and underlying assumption—gravitation. The theory is that every atom of matter in the earth acts on every other atom of matter in the heavenly bodies—sun, moon, planets, and stars. I own that I cannot understand how bodies can act at a distance without some connecting medium, and I want to know what the connecting rod, or coupling, is between the sun and the earth for instance, and between atom and atom. How many hands or "bonds" has each atom to enable it to lay hold of and "pull" every other atom in the universe? And how are all these connecting lines or ropes attached? and do they cross and intersect each other? Yet this tangled mass "is an established fact" forsooth, and its proofs "almost innumerable!"

But let us briefly view the question from another point presented by our astronomical friend. The moon goes circling around the earth: the earth revolves in a greater orbit around the sun; the sun in a vaster orbit still rushes away with both around some greater, say, Sirius; and Sirius—but no! I have promised to stop here. Well, what, on the above assumption, would be the path of the moon? and how if the moon's path be not exactly known, would it be possible to calculate her exact position months beforehand? Let me use a homely illustration. Suppose a gentleman has a dog circling around him at some distance in play ; the owner of the dog is on horseback and galloping at a greater distance around some railway train in motion, and the railway train is rushing along and making [illegible] city ; what would be the curvilinear path of the dog? Would it always have to run at the same speed? and would it be possible for anyone to predict when and where the dog might be seen in a straight line with horse and engine and city? I do not say this problem would be impossible of calculation, but I do say that to calculate it would be mere child's play compared with defining the path of the moon according to the theories of modern astronomy. Yet for thousands of years before these theories were believed or formulated by Newton, astrologers could predict eclipses of the moon with nearly as much precision as astronomers can now.

As I cannot obtain any proof of the theory of solar or stellar gravitation so essential to modern astronomy, I will ask your correspondent to kindly furnish us with some proof that the earth, with all its inhabitants, has the prodigious speed it must have if the popular theory be true? Or to put it more modestly, I will ask him to give us one good proof that the earth has any motion at all. He need not refer me to the text books ; they all assume terrestrial motion as well as solar attraction down to our School Board primers. I ask for proof; and as I only ask for one proof my request cannot be considered unreasonable. The Psalmist said, " He hath founded the earth upon her bases, that it should not be removed for ever." . And Joshua thought that it was the motion of the sun, not that of the earth. which was the cause of day and night ; yet our friend can see no discrepancy between Bible teaching and the theories of the astronomers ! I do not expect Mr Warne to say that Joshua was right : but I shall require, at least, something more than assumption before I believe he was wrong. I am old-fashioned enough yet to believe the Bible to be true, and I think I am sufficiently modernised to know the difference between a fact and a hypothesis, between true science and mere sound, between the teachings of Moses and the theories of Laplace. If, as the latter writer supposes, all matter was originally nebulous, how long was it in this state, and what had gravitation been doing to allow it to get into that loose condition? when did the impulse begin to act, and the truant atoms begin to pull all together? And if they are all still pulling each other to a common centre, how is it they are so long in arriving at it? and what has prevented them from

forming one vast central globe, leaving neither sun, moon stars, nor nebulae to be seen in the surrounding heavens? These baseless speculations are leading men into doubt and infidelity, and it behoves all faithful Christians to withhold their assent to them, at least until some decent proof can be offered on their behalf.   ZETETES.
August 28th.

Sir,--"Zetetes" makes a quotation from your leader on the above subject—"that the nebular hypothesis of Laplace really represents the extent of our astronomical knowledge." The correspondent then refers to gravitation, inquiring what is this power that pulls without chains. Your sentence hardly deals fairly with modern astronomy, as hypothesis and theory are not reckoned in the true science of astronomy, which practically acknowledges only those principles that can be mathematically proved. The nebular theory of Laplace pertains rather to the science of cosmogony—treating of the formation of the universe. Laplace gave his idea of how the world or universe was formed. Your correspondent "Zetetes" prefers to believe in the traditions recorded by Moses. In considering these two accounts do not let modern astronomy be blamed if neither can be proved. The most wonderful part of modern astronomy consists in its exactness. The modern astronomer predicts to a second the movements of bodies that the greater portion of mankind have never seen. The art of navigation depends for its very possibility upon astronomical prediction, the survey of extensive regions of country, the accurate determination of time, and the arrangement of the calendar, these and the laws that can be mathematically proved as correct are the feats of modern astronomy. The law of gravitation is one of these laws—that every particle of matter attracts according to its weight, and inversely as the square of its distance. "Zetetes" would find there is practical proof of this law if he fell out of a balloon, and as he neared the earth his speed would increase, till at the earth's surface he would be travelling at the rate of sixteen feet per second. Newton found that this coincided with the fall of the moon towards the earth, her distance being about sixty times the earth's radius ; therefore the force would be sixty times less, or sixteen feet in a minute, during which time the moon travels in her orbit 38 miles.

This one instance would not prove the law of gravitation, but the same law was found to rule the planets. Some of the stars have been found to be travelling round each other, evidently affected by the same law, but most wonderful was the discovery of the planet Neptune by his having attracted Uranus slightly out of his known orbit. That gravitation is a fact who can doubt? But, like electricity, we cannot explain how it works. It holds bodies in their orbits by invisible chains, far stronger than were ever forged by man.

As regards the nebular *Hypothesis* it has more evidence in its favour than the tradition handed down by Moses. For instance, if we take up a hot stone we naturally conclude that it has been in or near fire ; its outer surface gradually cools. The earth *may* be compared to a hot stone with the crust cooled, and reasoning back we come to a time when the earth was red hot, molten, and gaseous If we scan the heavens we see bodies in all these stages. There are the nebulæ, the sun, and the planet stages. In Saturn we have an example of how nebulæ may condense, leaving a ring of meteoric matter still circling round, until this matter may gather into one body and form another satellite to the parent planet. Both Laplace and Newton were good Christians, and to me the beauty and simplicity of the laws by which the Great Creator rules the Universe, impress one that there is a mind working behind the scenes, the greatness of which the mind of man cannot grasp, and that we are, indeed, conceited beings when we assume that all this great Universe has been designed for our special benefit.—Yours, &c., J. M. B.

———

Sir,—As books on astronomy have failed to convince your correspondent, "Zetetes," that gravitation exists and that the earth annually describes a circle round the sun, I am afraid that my writing will be of little avail. How does "Zetetes" explain the phenomena of day and night, and how does he account for the seasons? How is it that we do not see the same constellations in the heavens all the year round? The fact that we do not is a sure proof of the earth's motion. Does "Zetetes" believe that the heavens are always moving and that we are stationary? There is nothing stationary in the universe. All the *other* planets describe circles around our sun; this is known. Then why should *this* earth alone be fixed and immovable! Again, "Zetetes," questions the law of gravity, yet professes to believe that the earth is a stationary body, supported in space by nothing! How is this? What mystic force prevents it from falling away into space, or being drawn into the sun? The centrifugal and centripetal forces being adjusted the earth retains its place, and ever will while the law of gravitation governs the universe. "Zetetes" will say, this is all conjecture. Perhaps, then, he will tell us what he considers to be the truth of the matter.

GERARD WARNE.

SIR,—As our first friend, Mr Warne, has failed to give us a single proof either of the earth's supposed motion or the theory of solar attraction, the two underlying assumptions of modern astronomy, another writer, J. M. B., has come to the rescue I have no objection to this; and will try briefly to notice both. J. M. B. seems to admit that Moses and Laplace are at variance; while J. W. seemed to think that the teachings of the Bible and those of astronomy are harmonious. Both positions cannot be correct. The latter writer evidently has some respect for the Word of God, though holding teachings which make the Word of none effect; while J. M. B. boldly impugns the Bible account of creation as a "tradition handed down by Moses," and thinks that the "nebular hypothesis has more evidence in its favour." I would not complain if this larger amount of evidence were forthcoming and reliable; but what is it? Fact or assumption again? Let us see. He says: "The earth may be compared to a hot stone with the crust cooled, and reasoning back we come to a time when the earth was red hot, molten, and gaseous." Good. I suppose this is a specimen of what he calls "reasoning back." My friend, you may compare the earth to a hot stone, or to a large elephant, if you like ; but will your comparison make it into one ? If the earth ever was in the condition of a hot stone I will own it must at some time have "been in or near fire" But has it been in that condition ? and into what "fire" was it placed to make it so hot ? Was the order "red hot, molten, and gaseous," or gaseous, molten, and red hot ? And if the latter, how is it that such large mixed bodies contract under heat instead of expanding. Have we to reason "backwards" again ? I suppose that there would be no sea on the "globe" while it was red hot ; and no life of any kind when the crust was in a molten state ? Whence came, then, all the unfathomable waters of the ocean, and all the varied forms of life now existing on the earth ? Evolution ? Another kindred hypothesis. How did the first forms of life appear, and non-living inorganic matter change into living and wonderful organisms ? Evolution, sir, evolution ! And so we go on piling hypothesis upon hypothesis until we deny the Creator of the Bible, and admit that our forefathers were apes or baboons !

But J. M. B. thinks that modern astronomy is free from the charge of being based on assumption, and that it appeals only to the principles of undeniable mathematics. I have already quoted the admissions of two eminent astronomers, which show that their system of the universe is based upon assumption, and Copernicus further says that if any man "should adopt for truth things feigned for another purpose, he may leave his science more foolish than he came to it :" so your correspondent had better be careful. Mathematics can be applied to any theory, whether that theory be true or false ; and in respect of the world, if the relative proportions, distances, and sizes of the heavenly bodies be maintained, the resultant calculations would come out the same whatever those distances might be supposed to be. But we are told that "the most wonderful part of modern astronomy consists in its exactness." Oh ! when I went to school I was taught that the distance of the sun from the earth was ninety-five millions of miles : subsequently I have been gravely informed by an astronomer that this is an error, and that it was only some ninety-two odd millions of miles. Perhaps a difference of two or three millions of miles is a mere nothing in the vast speculations of our astronomical stargazers. But what about Kepler's calculations? Did he not measure the sun's distance at 12 millions. Ricciola 27 millions, Newton 28 millions, Martin 81 millions, and Mayer 104 millions? All doubtless worked out with " mathematical exactness !" It is, indeed, "wonderful " ! Dr B. Rowbotham, of London, by plain triangulation, made out the sun to be a comparatively small body, and something under three thousand miles distant. But then he was only a medical man, not a well paid astronomer.

But your correspondent J. M. B. thinks I should have proof of the law of gravitation if I were to fall out of a balloon. I asked for proof of the Newtonian assumption of solar attraction, and he refers me to the fact that heavy bodies fall to the earth. If a body fall downwards, by its own weight, to the earth, I am asked to believe that the sun can pull a large body upwards by some mysterious force in itself called solar gravitation ! I reply that there is no necessary connection between the premisses and the conclusion. The sun has never been proved, by experiments, to have any attractive force whatever upon earthly bodies ; nor, on the other hand, has the earth ever been shown to have any central power of " pulling " at the heavenly bodies. It is all pure conjecture. Why does not the earth " pull " down the balloon while it is suspended in the air inflated with gas ? Or, if the gas escape, why not pull every particle or atom of it down to the earth ? How is it that a little atom of hydrogen can mount upwards in spite of the combined " pull " of all the atoms of which the world is composed ? But we are informed that Newton " found " that the moon " fell towards the earth." " at the rate of sixteen feet per second." Well, all I have time now to say is, that we had better look out ; there must be a great crash somewhere soon. And if there be mountains in the moon, as they say there are, let us hope that some kindly valley will, at least, fall over Leicester. I was going to ask you, sir, if you could supply me with the " exact " distance of the moon, to divide that distance by sixteen so as to get the "exact" number of seconds when the crash would come; but I will not alarm your readers, and I hope there may yet be time for our friend J.M B. "to explain" the matter so as to ward off the "fall." Perhaps the sun will come to our rescue and give the moon a jerk the other way! But a word in conclusion for my first critic. He cannot see how the phenomena of the seasons can be explained apart from the popular theory. I think I can. But suppose that neither of us could; is our ignorance to be taken as proof of the popular theory? It ought not to be, surely. Let us first see the baseless character of these modern assumptions; and then we shall be in a position to inquire what evidence there is in support of the Bible doctrine that the earth is an "out stretched" and motionless plain, resting upon the fathomless waters of the mighty deep. ZETETES.

September 1st.

Sir,—Your correspondents appear to assume that what they call the nebular theory is antagonistic to the Bible. What is the meaning of the following:—1st chapter of Genesis? Will "Zetetes" explain? "And the earth was (that is, existed) without form and void, and darkness was . . . and the spirit moved . . . and God said, Let there be light." But "Zetetes" wants everything explained. Will he explain all the movements of a spinning top—the motions of procession, of gyration, and of revolution? If he will do so, fully and satisfactorily, your correspondents may be better able to make the other problems clear to his mind.—Yours, W. T.

Sir—Your correspondent "Zetetes," who states that I have "failed to give a single proof either of the earth's supposed motion, or the theory of gravitation," also denies that the earth may be reasonably compared to a "hot stone with the crust cooled." Yet this is evidently its present condition, or from whence comes the fiery matter which our great volcanoes have been ejecting for centuries past? Stromboli has been in a state of constant activity for more than 2,000 years, and still keeps casting out burning rocks and scoria from the bowels of the earth. At the remote period when " the earth was without form and void." it was doubtless in the same condition as is the planet Jupiter now. Owing to the great size of Jupiter as contrasted with our earth, it is evident that ages must elapse before it cools down to a temperature in which life could exist. The moon, owing to its smallness, has long been cold and dead. Our earth, therefore, is in a medium state ; the interior a glowing mass, with a few miles of solid crust upon the surface.

"Zetetes" wants to know "what fire the earth has been in to make it so hot." Laplace believed that the sun, having a revolution on its axis, was surrounded by an atmosphere which extended far beyond the orbits of the planets, which, as yet, were not formed. As the temperature of the sun decreased the rotation increased, and the centrifugal force of the atmosphere overcoming the centripetal, a ring of vapour was separated, which, breaking into pieces, united together and revolved around the sun ; as the cooling process continued they ejected other zones, a series of vaporous planets, and they, in turn, threw off minor satellites. This, surely, is a plausible explanation of the present heat of the interior of the earth. How else can we account for it ?

If gravitation does not exist, "Zetetes." why do we not fall into space ? What holds us to this earth ? We are exactly opposite to our friends in the Antipodes. Yet we both remain secure. It is to the law of gravitation that we owe our safety, day by day.

If "Zetetes" believes that the earth is flat (which has been disproved in hundreds of different ways, and by none better than through observing the circular shadow cast upon the moon during an eclipse) he certainly cannot refer us to the Bible as a proof that it is so, for nowhere in the Bible are we so told—a sure proof, I think, that the Scriptures are inspired.—I remain, yours truly, G. WARNE.

Sir,—After your editorial warning I will only ask you to allow me to make a few concluding remarks in connection with the position assumed by Dr Huggins. The language in his address is much more cautious than that of some of his admirers. He does not affirm that the nebulous theory of Laplace is true: but says that from certain "considerations Kant and Laplace formulated the nebular hypothesis, resting it on gravitation alone." He owns it is only a "supposition"; and a supposition encumbered with the fact that we have still "nebulæ" existing "in a relatively younger state." He, like a true astronomer, hesitating to affirm, suggests the idea that dark suns may have "collided" to make them into hot and bright suns; but he candidly owns that "there is no record of such an event" having taken place. So it is clear, to an unprejudiced mind, that the whole theory of a multiplicity of worlds moving about space, and dragged about by some power called attraction or gravitation, is purely conjectural: or what may be called scientific guesswork. I think therefore that the criticisms of the editor of the *Daily Post* were justifiable, and that Dr Huggins, in his late address, did speak "in the tone of a man who feels that, after all, he is launched on a vast sea of uncertainty." And I for one, decline to give up the ancient cosmogony of Moses, endorsed as his writings were by the great Teacher who came from God, for these modern and baseless speculations. One of your correspondents asks for proof that they are at variance. Allow me to give him one fact. Laplace and modern astronomers require millions and millions of years for the evolution of their universe, while Moses taught that the world, earth, sea, sun, moon, and stars were all made in six days.

Each of my letters has been diverted chiefly against this "nebular hypothesis," and that which Dr Huggins allows was its underlying assumption, namely, "gravitation alone." Newton himself never pretended that the theory of celestial gravitation was founded upon fact ; and some of his friends opposed it as a theory. Yet the hypothesis of Laplace is based upon what one writer says "is but one guess amongst many." Dr Woodhouse, professor of astronomy at Cambridge, about fifty years ago, was candid enough to acknowledge the weak and artificial nature of the Newtonian speculations. He says : "When we consider that the advocates of the earth's stationary and central position can account for and explain the celestial phenomena as accurately, to their own thinking, as we can ours, in addition to which they have the evidence of their senses, and Scripture, and facts in their favour, which we have not ; it is not without some show of reason that they maintain the superiority of their system. . . . . However perfect our theory may appear in our estimation, and however simply and satisfactorily the Newtonian hypothesis may seem to us to account for all the celestial phenomena, yet we are here compelled to admit the astounding truth that, if our premises be disputed and our facts challenged, the whole range of astronomy does not contain the proofs of its own accuracy." Sir, if the writer of the above extract had been editor of a daily paper or only a private individual and investigator like your humble correspondent, he would have had the critics down upon him ; but he was an astronomer, and a little more candid than some. Another writer, the Rev. W. Jones, nearly a hundred years earlier than the above, says : "The attraction of gravity is devoid of all geometrical evidence." Again, the Catholic Church has always rejected these modern theories as "absurd, philosophically false, and formally heretical" A writer in a work entitled "Solar Fictions" asks : "When will men learn to know a fact from a fiction ?" Sir Richard Phillips boldly writes :—" It would be much wiser at once to pull down the whole than to continue the system of patchwork of which the Newtonian theory consists. For I am convinced that such a mass of deformity must, in due time, offend the common sense of mankind, however admired and cherished it may be." And a writer in Chambers' Encyclopedia for the People, in view of certain unanswerable objections to the theory, says: "Such being the state of the case, the reader will consider whether, when Copernicus wrote, he held the doctrine of the earth's motion as a mere hypothesis, and as absolutely in facts true," p. 119.

But I must conclude. One writer asks me the silly question: "If gravitation does not exist why do we not fall into space?" His own theory would supply the answer, because on that theory we are already in "space," and there would be no "gravitation" to pull us anywhere else. But what is "space"? a vacuum? a plenum? or an assumption? If he can tell me I may answer him further. Another writer refers me to the spinning of a top. The case is not a parallel one. The top rests, while it spins, supported on something, while the earth with all its mountains, rivers, and oceans is said to be gyrating unsupported in "space," with half its population standing at the Antipodes head downward! Let my friend spin his top unsupported in mid-air and it will be time enough for him to boast of his analogy. But I must conclude, sir, by thanking you for your impartiality, and hope to remain, yours truly, ZETETES.

September 3rd.

Communications to EDITOR, 204 Dumbarton Road, Glasgow; or "ZETETES," 130 St. Saviour's Road, Leicester.

COMPARE THE
FOLLOWING STARTLING DISCREPANCIES
BETWEEN

The True Facts and the Modern Theories; and then say if is it not a national disgrace that we should prefer such baseless superstitions, for the mere credit's sake of a few infidel scientists.

MODERN ASTRONOMY and its crafty or, most deluded sycophants teach that "the sun is stationary and the centre of the planetary or celestial bodies, and that it is over One Million miles larger than the earth!"

Whereas, it is not one thousandth part of its size,—is constantly moving in spiral orbits, round and over the face of the motionless earth; increasing the latitude of those orbits about ten or eleven miles every day from June to December, and decreasing them in the same proportion from December to June.

The astronomers now teach that the Sun's distance from the Earth is from 92 to 93 M*illion* miles; whereas, it is *proved* to be under 2,400 miles, and the stars and planets even less than that.

So far from the Earth being a planet and "very much inferior in size to all the other planets but Mars," it is no part *at all* of the celestial system; but is wholly distinct and far superior in size and character to all the planets put together, which exhibit very little variety in size among themselves, and no parallax whatever, when viewed under any possible conditions!

The astronomers again assert the Earth to be a revolving and rotating Globe, with an axial speed of over one thousand miles an hour, and an orbital speed of over one thousand miles per *minute!* whereas, it is an absolutely motionless plane, surrounded by a barrier of impassable icebergs, extending into unknown regions, where "day and night come to an end," or, in other words, cease to be distinguished, and will only be removed when "there shall be no more sea;" and a "new firmament and a new earth" shall replace this sin-stained and misused world.

The astronomers teach an atmospheric pressure of fifteen pounds on the square inch; whereas, it does not amount to half as many grains.

Gravitation (central), is the wildest fiction ever invented.

All Attraction,—Solar, lunar, central, and magnetic, is purely conjectural and physically impossible.

The astronomers further declare that this globe has no material support, but is sustained and controlled in its terrific orbits by solar *attraction* only, at a distance of *nearly one hundred million miles;* although Sir Isaac Newton himself declares the idea was "grossly preposterous and discreditable to any rational mind;" meaning, of course, that the "centripetal and centrifugal forces" were hypothetical conditions only, and like *all* the other *data* on which the Copernican theory was founded, were purely geometrical devices, which no one in their sober senses ever thought could be illustrated by an appeal to facts, But such was the ignorance or bigotry of his professional friends, that they would not allow even Sir Isaac Newton himself to publish his repeated disclaimers of being anything more than a mathematical inventor or formulator of a system which was in no degree dependent on the truth of his premises.

The rank absurdity and downright impossibility of any *one* of its most elementary conditions, and the fact that no man of honor or scientific reputation would venture openly to defend it, ought to secure the instant rejection of it by every lover of truth in the world. The whole thing lies in a nutshell. All that is insisted upon is, that results or conclusions drawn from conjectural theories of philosophy, however plausible and apparently simple they may be, are wholly worthless and unworthy of any claim to rank as true science, till they have been *proved* correct by actual experimental demonstration, and can be thus confirmed and illustrated as often as their truth is challenged or denied. Even if a theory be true in the abstract, yet, if incapable of practical demonstration, it is not science, merely to suppose or assert it to be true. Newton himself was too sound a philosopher to attempt to illustrate the truth of the globular theory; he repeatedly discouraged all discussion, and, if left to himself, would never have attached his name or sanction to such a bungling absurdity.

This is the ground on which the ZETETICS take their stand, and all the professors in the kingdom will strive in vain to dislodge them; and, what is more, they are too cowardly to attempt it.

The above is so startling an exposure that the public press is ashamed to announce it; although the globular theory has been shown to be a Pagan superstition, unscriptural, irrational, unscientific, and physically impossible!

The true cosmogony of the Universe should be the primary subject, and one which ought to take precedence of every other on which the mind of a child can possibly be engaged. Ignorance of its true nature is a gross scandal on all connected with any phase of education, either religious or secular; and the imposition of a fraud which the intelligent child soon learns to detect, exerts a positive injury on the mind all through life. At the present moment there is not a teacher in the kingdom who would venture to prove himself familiar with the most elementary principles of natural science; and yet the writer is abused for declaring the world is full of knaves or dunces. Even savages never boast of being wiser than they are, or teach their children what they know to be untrue.

Never before was there such a fuss made about education. Do we really know or only think we know? Or is it learn, learn, learn,—and leave off no wiser than before? If bricks and mortar could make brains— if fine buildings could ensure great thoughts—if learned professors knew all that they profess, Thomas Carlyle would have belied his countrymen when he said we were a nation of "mostly fools."

What is our standard of true knowledge? an imperfect standard is as misleading as defective weights If we deal with false analogies, we are only comparing true metal with an ingenious counterfeit!

What are we aiming at by these remarks? We mean to say and are able to prove that at this moment, we are living in an age of shams and false pretences, and that we are totally unfamilliar with the most elementary conditions of some of the most important subjects that can possibly occupy the attention of a professedly intelligent and commercial nation;— that even our Clergy and Theological professors do not yet understand the teaching of the Mosaic records;—that our Schoolmasters, on some subjects, do not understand or know more than their pupils; that even our nautical charts and tables would discredit a nation of savages; and that our geography and astronomy are just where they were left by our university authorities of two hundred years ago; and that the most eminent of our educational professors would not now venture to defend what our poor children are compelled to carry with them through life.

What shall we say to those "Scientific" Journalists and Reviewers who persistently discourage any desire for discussion or ventilation of this important subject? Is it to hide their own ignorance or to display a mock contempt for any suggestion that runs counter to popular prejudice or opinion; or do they go so far as to imply that they cannot find among their numerous literary connexion, one individual writer who has intelligence enough to maintain his ground in a discussion with an honest opponent? It is a gross fraud on the credulity of the public as well as on the helplessness of childhood to enforce the acceptance of certain teaching which they are too ignorant to explain or too cowardly to defend.

If modern astronomical science could be shewn to be demonstrably true, how glorious would be the victory, and how dire our confusion! But till they cease their insolent taunts and courteously show us our error and the inutility of our opposition, we shall certainly not cease to regard them as vanquished foes, who lack the moral courage to acknowledge their defeat.

And we can but trust that when Prof. This or Prof. That ventures to pose as a Scientist or as an oracle on scientific subjects, he will learn to exhibit a little more modesty and confess that for once in his life, he had to bow to the superior influence of common sense, and, above all, to the omnipotent authority of the inspired records.

As it is, Science is irredeemably prejudiced in the eyes of all honest men and can never again assume those insolent airs which have so frequently disfigured its encounters with the less dogmatic searchers after truth.

The following Extract was taken from the "Birmingham Weekly Mercury" of February 15th, 1890, and must have been first published very soon after the experiment on the Bedford Canal, when that Professor fraudulently appropriated the sum of £1,000, on the grossly false plea that he had proved a curvature on six miles of the surface-water of that Canal.

"An Engineer of Thirty Years Standing" writes to a magazine in 1874 quoting the following sentences as the result of his experience in the construction of railways, more especially :—"I am thoroughly acquainted both with the theory and practice of civil engineering. However bigoted some of our professors may be in the theory of surveying according to the prescribed rules, yet it is well known amongst us that such theoretical measurements are incapable of any practical illustration. All our locomotives are designed to run on what may be regarded as true levels or flats. There are, of course, partial inclines or gradients here and there, but they are always accurately defined, and must be carefully traversed. But anything approaching to 'eight inches in the mile, increasing as the square of the distance,' could not be worked by any engine that was ever yet constructed. Taking one station with another all over England and Scotland, it may be positively stated that all the platforms are on the same relative level. The distance between the eastern and western coasts of England may be set down as three hundred miles. If the prescribed curvature was indeed, as represented, the central stations say at Rugby or Warwick, ought to be close upon three miles higher than a cord drawn from the two extremities. If such was the case, there is not a driver or stoker within the kingdom that would be found to take charge of the train. As long as they know the pretended curve to be mere theory, they do not trouble themselves about what may be stated in the tables of the geographers. But we can only laugh at those of your readers and others who seriously give us credit for such venturesome exploits, as running trains round spherical surfaces. Horizontal curves on levels are dangerous enough; vertical ones would be a thousand times worse, and, with our rolling stock constructed as at present, physically impossible. There are several other reasons why such locomotion on iron rails would be as impracticable as carrying the trains through the air."—SURVEYOR.

☞ JUST PUBLISHED, IN MONOCROME, A SKETCH OF THE WORLD AS A CIRCULAR PLANE ; SHOWING SOUTHERN ICEBERGS, Size, 15 by 15. Price, 2½d.

A Shilling's worth of Papers (various) sent on receipt of Seven Stamps, and a larger assortment for Thirteen.

All Communications may be addressed to the Secretary, 3, Park Street, Croydon, Surrey.

THE DAUNTLESS ASTRONOMY.

A SHORT ADDRESS.

"*Many shall investigate, and knowledge shall increase.*"

JUNE 4TH, 1894.

LADIES AND GENTLEMEN,

Truth is great and *must prevail*, so that all who are taught, led, or persuaded to contend for the impossible, must sooner or later knuckle down to the possible, the inevitable, and the true. Man's nature, since the fall, is so constituted that imposture has more hold on his intellectual faculties than truth; and that which is false can attain much more ready and universal acceptance than that which is true; thus a lie, once made to fit, can get round the world while truth is putting its boots on.

It has been thus with modern astronomy, which is considered by all the educated and learned, to be such a certain, settled and intellectual science, that they are disgusted with the least hint that this boasted science is founded on a scanty and baseless foundation; but the truth will out, and if any man wishes to be wise, even in the things of this world, he must be willing to commence fool, that he may be wise. Prince Bismarck, said, "I have always endeavoured to learn new things, and when I have had, as a consequence, to correct an earlier opinion, I have done it at once, and I am very proud to have done so. A Portsmouth tradesman, said, when spoken to on the subject of astronomy, "I am quite satisfied about it, and if I am wrong, I have no desire to be put right." An American writer says, "The man who does not care to learn if his decision is right or wrong, is not half a man." This is a lamentable state of mind on any matter, but on the subject of astronomy is very inconsiderate. Suppose Dr. Cousins, for instance, were to tell you that your heart was the breathing apparatus of your system, and the lungs, the organ that propelled the blood through your veins, you would think he was a long way from being an eminent Portsmouth physician; and

DAUNTLESS ASTRONOMY.

rightly too, for no medical man could understand the human frame if his knowledge of physiology held such a baseless theory. It is precisely the same with the science of astronomy; no astronomer can be correct with the science, unless he understands fully the relative and active positions of the sun and the earth; the sun being the *anima mundi*, the soul and heart of the universe. It matters not how many lines, angles, pretty wheels, and cat sticks they may draw, nor how wonderfully they may seem to magnify, calculate and exaggerate the distances and magnitudes of the heavenly bodies; if they reckon on a revolving earth, and simply a staring sun, all is wrong. They make their boasted knowledge the mere tool of astonishingness, not the natural or sublime science of astronomy. Much will be the surprise to most readers to learn that all modern measurements of the heavenly bodies, is based upon the results of experiments, that really mean the peculiar position, and toss up of a halfpenny. Mr. Richard Proctor—the great modern astronomer, tells us, "Anyone can tell how many times its own diameter the sun is removed from us. Take a circular disc an inch in diameter, a halfpenny for instance, and see how far it must be placed to exactly hide the sun, the distance will be found to be rather more than 107 inches, so that the sun like the halfpenny which hides his face, must be rather more than 107 its own diameter from us; so that the supposed distance of 95,000,000 miles rests, *probatum est*, upon the peculiar position of one halfpenny. Whatever the halfpenny should happen to reveal, that decides and regulates the magnitudes, distances and calculations of all the rest of the heavenly bodies, *ad infinitum*, to the world's end. For he says, "We are so constituted as to seek after knowledge; and knowledge about the celestial orbs is interesting to us, quite apart from the use of such knowledge in navigation and surveying. It is easy to show that the determination of the sun's distance is a matter full of interest, for on *our estimate of the sun's distance, depend our ideas as to the scale, not only of the solar system, but the whole visible universe.* The size of the sun, its neaps, and therefore its might, the scale of those wonderful operations which we know to be taking place upon, within, and around the sun, *all these revelations,* as well as our estimate of the earth's relation and importance in the solar system, depend absolutely and directly on the estimate WE form of the sun's distance. Such being the case, this being in point of fact the cardinal problem of dimensional astronomy, it cannot but be thought that, great as were the trouble and expense —sometimes reaching a quarter of a million of money—of the expeditions sent out to observe the transit of Venus of 1874, they were devoted to an altogether worthy cause—To establish and vindicate the gigantic fabrications given out for an educated and trained youth as well as the public to swallow as the pure and undeniable results of a toss up of a single halfpenny; some have ridiculed the bun, what about the halfpenny! Well, sure enough, astronomical falsities are like the proverbial lawyer's lies, they are official, therefore current.

DAUNTLESS ASTRONOMY.

As long as they are official, cannonised, stamped by government, and printed by authority, they may be used by the bushel; but you must not indulge in them personally or singly; if children do, their parents will be angry, their schoolmaster will cane them, their governess will stand them on the form, their minister will denounce them, if adults, from the pulpit as black sheep, for others to be aware of; and some ministers will thunder against Jacob's deception, yet coolly allow their school children to be taught the greatest deception and lies of the age, in geography.

Science ought to be a collection of truths, but the astronomers and geologists have made it a collection of lies. We cannot verify as truth that which is not of itself indisputably true. Nor are the scriptures a ladder we can kick away with regard to the science of nature, any more than the science of the soul. Everyone is considered now-a-days behind the times, and the age, who does not believe in the results of the halfpenny. But there are some phenomena of nature which suggest false ideas, there are manifold ways in which our senses may deceive us unless their evidence is carefully cross-examined; it strikes us that *this halfpenny* requires a great deal of cross-examination. We tried it ourselves and could hide the sun at 37 inches, that would make a lot of difference all round, from Greenwich to Venus, and from the transit of Venus to the sun who never budges an inch for the greatest astronomer, or the most costly coin of the realm. Not even Sir William Pink's Christmas calculations would make the slightest difference to Phœbus, who simply avenges all mistakes by still travelling and shining. It would be difficult to say which was the most ridiculous problem; the Hindoo notion of the earth resting on the back of an elephant, or the British notion of the distance of the sun resting on the top side of a halfpenny, but doubtless the Hindoo's is the most beautiful and natural; but Sir William Pink places undoubted reliance on results from the top of a halfpenny, as witness the first paragraph in his Christmas address of 1893, which commences, "Ladies and Gentlemen, some astronomers a few weeks since, held a meeting, with a profound lover of the heavenly bodies in the chair, for the purpose of demonstrating that the earth is not round, and does not revolve, but is somewhat of the shape of a wedding cake, with figures on the top and sugar representing the everlasting snows, glaciers and frozen arctic regions. These learned men had the temerity to attest that the sun revolves round the earth, so that the sun must travel 270 millions of miles in 24 hours, or over 11 millions of miles each hour. Well if this assertion was true, we are afraid we would not get any seasons, and Christmas addresses would not be required; or perhaps, all the inhabitants may want to burrow in the huge plum cake. However, as the season has arrived, notwithstanding all these opinions, we again resume the pleasing duty of addressing you."

The sun is a concentrated body of light, heat, and attraction;

DAUNTLESS ASTRONOMY.

not an expanded substance at all; its actual distance less than 6,000 miles.

The worthy Alderman makes a mistake when he says, "that we held a meeting for the purpose of demonstrating that the earth is not *round*," we thoroughly believe it *is round*, like a table, or like a circular wedding cake, but not globular, like an orange, nor a spheroid. Then astronomers could not afford such a luxury as sugar to represent the everlasting snows, pure white wadding did that; the iceberg represented by grocer's soda, which "Betsy Mullins" was unaware was to be found at the North Pole. Then Sir William says we had the temerity to assert that "*the sun revolves round the earth*;" we only had the temerity to assert that which the glorious sun has had the temerity to perform for the last 6,000 years. And even Mr. Richard Proctor, the late great astronomer, details very graphically "the rathway of the sun." If the sun has its daily, monthly, and yearly pathway round the heavens, the earth can have *no pathway round the sun*, that phenomena would be entirely obnoxious to the economy of nature, where nothing is arranged but on the basis of philosophical necessity. Consequently He who made both heaven and earth always proclaims the sun the traveller—never the earth; so we render unto all their dues. We can assure the knight of Shire that it is only necessary for the sun to travel in a spiral path, 15 miles per minute, 15 degrees every hour from east to west, altering its position northward or southward about one degree a day and all is done by its constant journeys through the twelve signs of the zodiac, the belt of the sky, which if he refers to Keith, on the Globes, page 4, he will find mapped out for each month in the year, which movement *alone*, and *constantly*, regulates the seasons, which produces in regular order the Christmas fruits he secures for his customers and which he acknowledges in his next paragraph was produced by the extraordinary warmth of the sun.

Perhaps the address was commenced in the middle of the day when it was time for the afternoon nap, and we imagine, therefore the pen dropping on the desk and a respite is had in dreamland, and lo! one of the Alderman's fields may produce next year a million of pumpkins, that will produce a million of threepenny-bits from his millions of customers, who surround him millions of times during the year, from all parts of the known-world; and lo! in the middle of the field, one bumper pumpkin grows to the representation of the sun, that may go to the Drill Shed in November, and win a first-class prize; won't that be good luck! He is astonished at the dream. "It does not matter what these fellows said at the Albert Hall, I've seen it all in a dream; talk about the sun not being larger than the earth, why it had to travel 270 millions of miles in 24 hours, it must burst asunder soon, and if my sunny pumpkin burst, all the juice would run out, then my prize would be lost at the show." "Oh, oh, oh, I will never have that!" But there, we will believe the whole-

DAUNTLESS ASTRONOMY.

sale grocer understands the evolutions and revolutions of tea, cheese, and bacon, much better than the sun, moon and stars, so grant a free pardon for all astronomical blunders.

If the earth, therefore, remains in the same situation while the sun revolves round it, its mass must be much greater than that of the sun, for it is contrary to the laws of nature for a heavy body to revolve round a light one as its centre of motion. The fact is, there is no doubt that our Creator has worked upon the number ten, in the creation; ten divides the allotted period of the world's present era, 6,000; ten divides the Millenium, 1,000 years, to be spent by the Redeemed of all nations in heaven, not on earth, as is usually reported; ten divides the jubilee and the captivity; the life of man; the plagues of Egypt; the kings, and the horns of Antichrist; also the ten days of persecution. Now the earth is at least, ten thousand miles in diameter, not eight thousand; enough has not been allowed for the south pole, the circumference of which no ship approaches within 2,000 miles; while the centre of the earth *is the north pole*, immediately under the north star, the centre of the heavens, which never varies its position. The President of the Geographical Society stated at the Portland Hall, on one occasion, that, "Tropical vegetation had been found as far north as Disco; if found any farther north, they would have to change the present accepted theory of the shape of the earth." The sun is not more than one tenth the diameter of the earth, nor more than 6,000 miles distant; everything else in proportion; and *everything* is within ten thousand miles of us, so that the redeemed often are permitted to hear the songs of heaven before they reach there. "What beautiful singing of birds I hear," said a doctor as he was dying in St. George's Square. "Open the windows and doors, and let me hear that beautiful music," said another dying saint in Sultan Road, Landport.

Eclipses are never occasioned by the earth's intervention, or they would not be recorded three times out of four "invisible at Greenwich." They are entirely associated with, ruled by, and dependent on the heavenly bodies. When a planet crosses the sun, it is a black disc; eclipses are easily and correctly predicted by the Metonic system, not at Greenwich, but those able to compile the Nautical Almanack for Greenwich. The Chaldeans had the celebrated Metonic style of 19 years, and could calculate eclipses for hundreds of years in advance. An eclipse has been known to occur while both luminaries have been above the horizon, so the earth was entirely left out.

The moon has also a spiral motion 12 times faster than the sun through zodiac stars; all agree it is the nearest of heavenly bodies. Then how ridiculous to state that the nearest star is 100 times more distant than the sun.—*Gen. Drayson.*

DAUNTLESS ASTRONOMY.

With regard to the ocean it is most certainly *a level*; several gentlemen that have been to sea all their lives, attest the same. "As level as the road," said one to me; also experienced by the authorities that have had years of experience at the Dockyard Semaphore. A globular ocean is absurd in theory, ridiculous in imagination, and never had an existence. The same phenomena that occurs with a ship outward or homeward bound, would occur on the plains of the Orinico, which are level for 1.000 miles. Would anyone declare the level plain to be globular because the pole of a van might be seen first, owing entirely to laws of perspective and angular vision. This has been tested in our own experience with an opera-glass. If the earth revolved 120 times swifter than a cannon ball, no engine would keep the metals, as nothing will remain on the drum of the shaft, unless it is spliced on. The life of no living creature would be worth five minutes' purchase after it had started one revolution. The laws of gravitation is a myth, a vain imagination, and a scientific toadstool. A light is never made larger than the place to be enlightened. A room is never taken round the light, but the light round the room. A light is never placed far distant, but as near as possible. All matter is inert and motionless, so we are not a set of leap frogs, leaping about on Newton's merry-go-round, but stately men and women walking on a stable and fixed earth. Thus we cannot do without railways. by reckoning so many jumps to each place, as we want it, and it comes round, or we might be able to reckon—

> Three jumps to London, four jumps to Kent,
> Eight jumps to Birmingham, and back again in Tent.

With regard to navigation, an artificial globe is never taken to sea to sail by, but always superficial charts as flat as the surface of the sea; one of the greatest lighthouses just opened is fully visible at the distance of 300 miles. Navigators always take the sun, never the earth; and they use tables compiled 200 years ago, and handed to them by pagans 200 years before that. How much more simple would navigation be if the dictates of nature were accurately and fully followed. Captain Parry and several of his officers on ascending high land in the vicinity of the north pole, repeatedly saw for 24 hours together the sun describing a circle upon the southern horizon, a spiral motion. And the moon, which has charge of the tides, Captain Parry says, "had the appearance of following the sun round the horizon." Nature can be watched and tested, and no baseless theories can long survive attack. "Truth is great and must prevail."

N.B.—It cannot be too emphatically impressed on the reader's attention that the earth has nothing whatever to do with regulating or causing the seasons, no more than a guard's van has to do with supplying the locomotive power for a railway train. The mandate at the creation was, let THEM—the heavenly bodies—be for seasons, &c. Not let IT—the earth—be for seasons. When Job was asked by his Creator if he could bind the sweet influences of the Pleiades,

DAUNTLESS ASTRONOMY.

or loose the bands of Orion, what was meant was, canst thou prevent the return of either spring or winter? For the term Pleiades referred to certain stars or constellations denoting spring, and Orion to certain others denoting winter, as was well understood by the people in the East. Mazzaroth is the Chaldean name for Zodiac. So the Creator acknowledged the direct cause of the seasons, and all His works praise Him and are sought out by all them that have pleasure therein.

   See the rising hills 'neath yonder azure sky,
   And under spreading woods the sloping valleys lie.

     Yours faithfully,

Fratton, Portsmouth.   EBENEZER BREACH.

ECLIPSE IN THE MOON.

It is said the diameter of the moon is about 2,160 miles, and that of the earth about 8,000. It is also stated that the moon's motion round the earth works out at about thirty-seven miles per minute, while in its journey round the sun the earth travels along at about 1,080 miles per minute. Now supposing the shadow cast by the earth on the moon is equal to half its (the earth's) diameter—4,000 miles is an outside estimate, as the shadow would tend to converge. And if these figures, given by astronomers of the earth's and moon's motion are correct, readers will see it is *impossible* for an eclipse to last in the moon for more than *seven* minutes, although eclipses have been *known* to last for over *four hours*, so that this shows the eclipse cannot possibly be caused by the shadow of the earth's rotation.

It is known that there are dark bodies in the heavens, and eclipses may be caused by the periodical motion of one of these. In any case, eclipses are no proof of the earth's rotundity.

It is argued that the shadow cast on the moon is generally circular. A circular shadow is not necessarily cast by a moving globe. It can be demonstrated in a room with one light. Take a ball or orange and place a *flat* ruler, so that its shadow is cast upon the ball, and it will be seen that its shadow is curved or circular, it could not be otherwise.

Again, where there is a gaslight with an ordinary mantel over the kitchen table and with an ordinary plate or plates underneath, although flat it throws a circular shade on the table-cloth.

Just a word as to the shape of the earth which I consider flat but possibly round like an ordinary plate,—as the seven seas run into each other this is surely agreeable to a fixed, flat earth.

The earth can be circumnavigated, but it is no proof whatever that it is a globe or rotating.

HOW WE GET OUR SEASONS.

Astronomers tell us that we get our seasons by a daily rotating earth to a fixed sun, but I maintain that this is absolutely impossible.

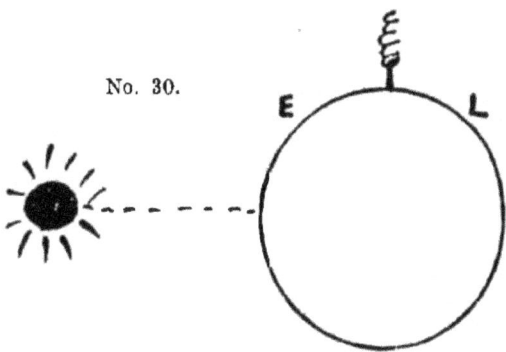

No. 30.

E on the sketch is England, **L**, Labrador, North America.

On a 12-in. globe I possess, I find Labrador is on the same circle as England, and assuming that the earth rotates daily to a fixed sun, both countries should receive exactly the same heat as each other daily, but do they? No!

It should be precisely the same as a joint of meat turning before the fire; that is, where it turns to the fire in the same circle, such as the position of **E** and **L** in the illustration. The joint of meat would of course receive the same heat at *each revolution* to the fire, and certainly **L** could not freeze for weeks at a stretch, while **E** was quite warm for the same period.

Readers, you can readily see that my theory is the correct one, and that we get winter and summer by the sun travelling over our fixed earth, and certainly not by the earth rotating to a fixed sun.

Astronomers inform us that the sun is at a greater distance from us in summer than it is in winter, as may be seen in orbit illustration in school book. Can anyone imagine a more unreasonable proposition? There might have been some reason in the theory, if the earth had made one revolution in twelve months, for then if the sun had been a fixture and it took the earth six months to rotate half the way, and this half of the earth turning before the sun would have given some reason for concluding that the other half of the earth, that had not come into position, or faced the sun for six months, would be having its winter months; and the other half that had kept turning before the sun for six months would be having its summer months.

If, however, our earth makes one revolution *each day*, then Labrador, which is in North America, and in the same circle as England, or same latitude north, which readers will see by cunsulting the rotating globe, that represent the earth as used at school, should have the *same climate* and each day of same length. This should be the same for each place of the same latitude. We are, however, aware that the weather is very severe at Labrador, when it is midsummer in England, and also that the day is of short duration, and these *facts* alone should convince readers, without a shadow of doubt, that I am correct in saying that it is utterly impossible, for all of us in this world, to get our seasons *as we now get them*, except by the sun passing nearer to us in summer over our fixed earth and then farther away during the winter months.

This is undoubtedly the cause of winter and summer and our seasons. I might mention that a great part of Russia in addition to Labrador, referred to, comes in the same circle as England.

If readers will only carefully examine a 12in. globe (this size will enable you to see the various places in the world printed on it) and if they will rotate the globe, in front of a lamp to represent the sun shown in illustration, it will be found that it would have been impossible to get these places in position when rotating the globe, to work correctly with the astronomers' theory, and to me the whole rotation theory is undoubtedly a false one and will be found upon examination to be ridiculous.

I might mention that it cannot be said that the water is the cause of the climate being much more severe in Russia than that of England, for it will be seen, that the region I refer to in Russia is on the same circle on the globe as England and is hundreds of miles distant from any sea or canal. It is entirely all land, and ought to be warmer.

I therefore maintain that as the *climate* and seasons at Labrador in North America, Warsaw and Moscow in Russia, Kamchalka in Siberia, and *numerous other places in Russia*, are quite different to the climate in England each day, although it should be *precisely the same*, please refer to globe, this fact alone must be a *positive* proof that the earth does not rotate daily before the sun as illustration.

Readers must remember too, that from December 21st the sun begins to travel nearer to us each day till June 21st. The heat from the sun accumulates and gradually warms the atmosphere until the end of March, when then we begin to feel it quite dry. Then as you are aware, each day from March, the sun gets nearer and nearer, until at mid-day on June 21st the sun has arrived overhead, and seeing that the air in March was dry and the heat still accumulating till June, with the sun getting nearer and more overhead, this causes the earth and atmosphere to be very warm and hence summer is upon us to grow our crops.

Now as readers are aware, it is then midwinter in New Zealand. Why is it midwinter in New Zealand when midsummer in England ? Why ? Because the sun has kept away from that country ; it has taken a direction over England. Why is it midwinter with us six months hence ? Why ? It is because the sun from June 21st begins to travel nearer and nearer to New Zealand every day and night for six months till it finally arrives at a position overhead, and then it is of course summer at that place. Often the weather is very warm in England at the end of August and September, and often as warm as it is in summer, although the sun has been travelling farther away from us then for several weeks. The reason for this is, that the heat had accummulated so fast, that we found the weather very warm, even after the sun had left the position of being overhead as is the case in June. Take the 21st of September till the 21st of December and note how gradually day after day, we begin to find winter weather coming upon us, as day by day the sun gradually moves further from us through the heavens. There can be nothing more simple for one to understand how we get our seasons with a fixed earth and a travelling sun.

The astronomers inform us that in June my garden and also your garden slope different to what they do in December, and this is the cause of summer.

As my garden is a part of the earth, and my house is built upon a part of the garden, which was levelled when built, by the aid of a spirit level, and I find it is of the same level to-day as it was twenty years ago, when built, and the same level now while I am writing, which is in the month of June, as it was last December, surely there can be nothing more ridiculous than to imagine and say that each garden and building throughout the world slopes towards the sun in June of each year while it does *not* in December. What would really happen with the *very high* buildings, high chimney stacks, church towers, etc., if the earth where they are built upon, altered its level every six months, would they be perpendicularly upright.

May I ask is it worth while assuring pupils at school that this kind of thing takes place in June and that this is the cause of our summer?

In seriously thinking over the question one cannot but come to the conclusion, that the Creator caused the sun to travel over our fixed earth, in the various directions, to give warmth to grow our crops in each country. If the Creator had caused the sun to take one direction only all the year round some countries would get no heat to grow their crops and if the Creator had made the sun a fixture and caused the earth to rotate *daily*, surely it is easy to see, that we should not get the seasons as we now get them, it would be quite impossible, utterly impossible.

PART I.

...THE...
EARTH A PLANE.

BY
JOHN EDWARD QUINLAN

(Commissioned Land Surveyor of St. Lucia and St. Vincent. British West Indies),

7, CHARLWOOD PLACE, PIMLICO, LONDON, S.W.

PART I

THE EARTH A PLANE.

By JOHN E. QUINLAN, 7, Charlwood Place, Pimlico, London, S.W.

It is an unquestionable fact that the earth is an extended plane with an irregular land surface, and not of spherical or globular shape with two flattened ends as scientists and astronomers assert in their speculations, and most people believe. Neither is it pear-shaped, as Professor W. J. Sollas suggested as recently as the 24th May, 1906, at the Royal Institution, Albemarle Street, London.

That the earth is flat can be verified by our senses from every point of view. Practical demonstrations can be given of this shape which should satisfy every unprejudiced and reasonable mind; and there are numerous references in God's Holy Word —the Bible—to an earth of this shape.

I am aware that one of the first questions that globites will put to me is: How can ships sail round the world if it is not a globe? In answer to this query, please note that ships cannot sail round the world on a uniform course, whether the world be a globe or a plane. This is impossible except in a latitude which is south of Cape Horn. Land would intervene on every uniform course that lies between this extreme southerly latitude and the Arctic regions. But when airships shall have been brought to a high state of perfection, a globe-trotter will be able to sail round the world on a uniform course by getting into one from a sailing ship when land was in the way, and out of it, back into another sailing ship when water was reached.

The magnetic compass, that would help the globe-trotter to keep on his course, always points to the magnetic north when there are no local attractions. The magnetic north is near the North Pole. The North Pole is the centre of the extended plane —the earth—and not one of the two flattened ends of a globular world as is so often asserted.

It was not by white scientists and astronomers that the magnetic attraction towards the North Pole was discovered, but by learned Chinese who knew then, as their offspring know now that the earth was flat.

Take a piece of cardboard and trace a circle on it with a pair of dividers. The centre point of that circle would represent the position of the North Pole on the flat earth, the circular line the southern extremity of that earth, and not the South Pole. There is no such point, and there never can be, as the alleged South Pole. Draw another circle from the same centre, so as to be midway between that point and the extreme circle, and this middle circle would represent the Equator.

Place a magnet near the centre of the circles and a sensitive needle anywhere within the outer circle, and the needle would be bound to point towards the centre. This must be so. A right angle on the right side of the needle points east, a right angle on the left side west, and the opposite end of the needle points to the south.

Place one leg of your divider in the centre point and the other leg near the sensitive needle on its right side ; then, by making a circular sweep with that leg of the divider till it touched the needle on its left side, an easterly course would be traced round the flat earth. Reverse the operation. From the left side of the needle make a circular sweep back to its right side, and a westerly course round the flat earth would be traced.

It is therefore possible to outline and eventually traverse a uniform course round a flat earth by means of a seagoing ship and a navigable airship.

When the petrol launch sails round the islands that stand in the centre of the lakes in the parks of the London County Council, English children get a practical demonstration of the possibility of sailing round a flat earth—the islands representing the earth—in water that is always in its natural state—a dead level.

Let scientists and astronomers, who assert that the earth is a globe, give us a practical demonstration of this by placing a magnet near the North Pole of an artificial globe and a sensitive needle anywhere on its convex surface. If this be done it will be found a matter of impossibility for the needle to point towards this North Pole as a needle does on the natural world, and would do on my cardboard. Let them also give us as practical an illustration of the sailing of a ship on a globular ocean as I have given of an ocean with a flat surface in my illustration of the sailing of a petrol launch on level water round the islands in the parks of the London County Council.

A sea-going ship could sail round the navigable world on portions of a uniform course, say like the Equator, in the following manner. It could start from the east coast of Africa at the Equator, and keep along the easterly course till it reached the many islands of the East Indies. It could sail round each of the islands till it got back to the Equator on their east side; it could then proceed on its uniform course across the Pacific till the west coast of South America was reached. Then by sailing southerly and around Cape Horn, and northerly till it reached the Equator near the mouth of the Amazon, it could pursue its uniform course across the Atlantic Ocean and the Gulf of Guinea (where it would cross the meridian of Greenwich), to the west coast of Africa, and by again sailing southerly and around the Cape of Good Hope and then northerly it would get back to the point from which it started; but the ship would have been sailing all the time along a flat ocean, and never along a convex or globular one, and its captain would have been consulting a flat chart, and never a globe to sail it along the desired course.

As it is impossible for any reasonable man to imagine such a feat as looping the loop from the outside of the circular ring, so must it be equally impossible for him to accept the ridiculous dogma of a ship sailing round a globular earth. He cannot accept this without being false to his reason.

If the ship to which I have referred had been sailing along the Equator of a globular earth, every six hours it would have been in the following different positions: At midday it would be in the horizontal position in which all ships appear to our view; six hours later it would be in a perpendicular position, with its bow pointing downwards; at midnight it would be topsy turvey upside down; at six a.m. it would be once more perpendicular, but this time with its bow pointing upward; and at noon, twenty-four hours later, it would be back in its former horizontal position.

If, on the other hand, the ship were sailing northerly from the Equator along a meridian: at six p.m. it would be lying on its right side with its masts in a horizontal position; at midnight it would be upside down; at six a.m. it would be lying on its left side with its masts again in a horizontal position; and only at noon, as at the previous noon, would it be in the position in which all ships appear to our view. It would be highly amusing to a reader if I were to describe the hourly positions of the ship, but unfortunately very limited space disallows of my doing so.

A ship lying in a dock fully laden has only a few feet of water below its keel. If it draws at noon 18ft. 3in., the minutest observation fails to discover the difference in its drought which ought to take place at midnight when it is upside down. Will scientists explain?

Having mentioned the meridian of Greenwich, I must point out, in connection with the alleged interference of the London

County Council electric generating station at Greenwich with the delicate instruments of the Royal Observatory, that it is the Royal Observatory that should be removed and not the generating station. An observatory ought to be erected on a spot at the level of a large sheet of water, which would serve as a natural horizon. Great Britain is surrounded by such water ; but Great Britain is otherwise unsuited for a proper observatory for reasons which I give in the next paragraph, and also because every meridian that passes through Great Britain crosses the Equator in mid ocean, where a sister observatory cannot be erected.

All are agreed that when we have equal day and equal night twice a year over the earth, the sun is vertical to the earth at the Equator, but it is not generally known that when the sun is in this position—called its equinox—observers at 45° north latitude and 45° south latitude must record 45° as the angle of elevation from the horizon to the sun's centre at noon, and that the spot north or south of the Equator from where the sun is observed at this angle at noon is the spot that marks the exact distance to the Equator as the sun is above the Equator. This spot—on the north side of the Equator—is exactly midway between the Equator and the North Pole. It therefore follows, as a matter of course, that the exact distance of the sun above the Equator at its equinox is exactly half the distance between the Equator and the North Pole.

Two observatories are therefore indispensable, one at the Equator and the other 45° north or south of the Equator. There are no such observatories in existence, and I venture to say, without fear of contradiction, that there are only two spots in the world on the same meridian—one at the Equator, and the other 45° north or south of the Equator—where the angle of elevation to the sun could be measured from a natural horizon. The Astronomer Royal and his fellow-scientists may have this opportunity of saving their faces by naming the positions of these two spots before I publish them later on.

With observatories at these two spots the distance of the sun from the earth, which is the very foundation of all astronomical researches, could be known exactly. Scientists and astronomers say it is about or nearly 93,000,000 of miles. How did they measure it ?

Here is a practical illustration of the infallible method I have given for finding out the distance of the sun accurately. Take a square piece of paper ; each of the four corners forms an angle of 90°, and together a total of 360° ; the same as the degrees of a circle. The four sides of the paper are of equal length. Fold the paper diagonally, and it becomes of triangular shape. The triangle that it forms is not only a right-angled triangle, but one that has its diagonal at an angle of 45° or one half of 90°. As the four sides were of equal length, the two sides that are at the right angles remain of equal length.

Hold up the triangular paper before you with one prolonged side of the right angle pointing towards the centre of the heavens, and the other towards the North Pole, and imagine the corner pointing upward to be the sun at its equinox at noon : that pointing northward to be the spot on the earth which is 45° north of the Equator; and the third corner to be the spot at the Equator which is perpendicularly under the sun at noon when at its equinox. This clearly demonstrates that the spot from where the sun can be seen at noon at an angle of 45° above the horizon, is at exactly the same distance from the Equator as the sun is above the Equator, and not a foot more or less. And whether the earth be globular or flat the same method answers for ascertaining the exact distance of the sun.

Scientists and astronomers who hold to the globular world theory have then no excuse for saying the sun is about or nearly so many millions of miles away when there are two spots with natural horizons even on a globular world for ascertaining the exact distance. And it is with their uncertain distance of the sun that they measure the distances of other heavenly bodies. How much better to have an accurate and reliable base with which to make these measurements. Are scientists and astronomers ignorant of this method?

THE EARTH NOT A GLOBE,

BUT

POSITIVELY A PLANE.

PREFACE.

THERE is an agreement between the word and works of God. The allusions of the word to the works are strictly accurate, and the facts of nature attest that accuracy.

The works of God are magnificent; but their magnificence is diminished and misrepresented by fanciful theories at variance with His word. To maintain the connection is to hold fast an important truth; which is a delightful duty, and is one of those *few things* by being faithful to which we may be "made rulers over many things."

This work is certainly of God, for it is a vindication of the inspired accuracy of Scripture cosmogony.

I have a grateful remembrance of the late John Hampden, Esq., who kindly sent me many useful papers, one of which was:—"The Geometry of the Circular-plane and the Harmony of the Solar Courses." I am also indebted to Albert Smith, of Leicester, England, for his suggestive paper—"The Sun-dial, a Strange Fact and a Forgotten Truth." May God's blessing go with the book.

JOHN LAWSON.

THE EARTH NOT A GLOBE, BUT POSITIVELY A PLANE.

CHAPTER I.

The Earth and Ocean not a Globe; but a circular-plane, according to the two great books of God, Revelation and Nature.

According to the book of Revelation, the word "world" is used more than 260 times in the Bible, and the word "earth" 350 times. The words "round" or "globe" or "sphere" are never once applied to it Not a single expression is used from the beginning of the inspired book to the end, suggestive of the idea that the earth is a planet, or, suggesting that the earth is anything else than a stationary plane.

1. Passages speaking of the heavens above as "stretched out," and the earth spread forth and stretched out upon the waters, so that the line of the heavens is parallel with that of the earth.—Isaiah 42: 5; Psalm 136: 6. "To Him that stretched out the earth above the waters."—Psalm 24: 2. "He hath founded the earth upon the seas and established it upon the floods." Genesis 1: 10, informs us that the waters were gathered together unto one place, and called seas, and upon those seas the earth was founded, consequently the waters are beneath and around the earth. The waters sustain the earth, sustain the earth as a whole.—2nd Peter 3: 5. "For this they willingly are ignorant of, that by the word of God the heavens were of old, and the earth standing or consisting out of the water and in the water." R. V. reads: "For this they wilfully forget, that there were heavens from of old, and an earth compacted out of water and amidst (or through) water, by the word of God." Compacted—held together—leagued with—united. Now look! The earth is one—the continents appear to us who are on the surface as being a great distance apart,

yet these continents are joined together; they are connected in the water and together they constitute one earth. These connections deep down in the water of continent with continent, are themselves sustained by the waters underneath. "The waters under the earth." Those waters under the earth, sustain the earth as a whole; sustain the earth in its entirety. "The earth compacted out of water and amidst water by the word of God." The dry land above, or out of the water compacted, united with, the earth in the water and that again compacted, united with, the other continents amidst or through water.

The continent of America is compacted, united with the continents of Europe and Asia, amidst or through the water of the Atlantic. The continent of Africa, though connected with Asia by the Isthmus of Suez, yet is compacted, united with Asia and Europe amidst or through the water of the Mediterranean Sea, Red Sea, and probably Arabian Sea. Australia compacted, united with Asia amidst or through the water of the Indian Ocean.

As the heavens are spread out above, and the earth stretched out upon the waters, so the line of the heavens is parallel with that of the earth. That is the reason why the heavens seem to close or touch the earth at the horizon. They are parallel. Parallel lines appear to converge in the distance. In a long tunnel, the floor appears to rise to a level with the eye at the entrance, and the roof to come down to a level with the eye, they are parallel; so with the earth and sky, they are parallel. The distant horizon appears on a level with the eye, and the sky in the distance seems to descend to a level with the eye. This has been particularly remarked by balloonists. At their greatest elevation, the horizon seemed to be on a level with the eye, and the sky in the distance seemed to close with the horizon. This is according to the law of perspective, parallel lines appear to converge. In Isaiah 40: 22, we read: "It is He that sittteh upon (or over) the circle of the earth, and the inhabitants are as grasshoppers." It is not said here " circle of a globe," no, but "circle of the earth." He sitteth over the arctic circle and beholdeth to the ends of the earth. The earth being stretched out upon the waters, and therefore in its general configuration it is a plane, it is flat; the centre of the earth is north, the circumference is south. The ends of the earth, the extremities of the earth are out towards the southern circumference. Cape Horn is an end, New Zealand is an end, Tasmania and Cape of Good Hope are ends. The great God sitteth over the circle of the earth—over the north centre and seeth to the ends of the earth. Isa. 45: 22. "Look unto me and be ye saved all the ends of the earth." We could not live on a globe—a globe revolving on its axis and shooting away through space in its orbit round the sun. It would be impossible to live upon it, constituted as we are. In Isa. 45: 18; "Thus saith the Lord that created the heavens; God Himself that formed the earth and made it; He hath established it, He created it not in vain, He formed it to be inhabited." He says: "I am the Lord and there is none else." A revolving globe—a planet earth would not be habitable. A planet earth, such as the modern astronomers speak of, is a thing of the imagination; it does not exist in nature. it is a vain thing; it is not the habitable earth. What a cutting rebuke is here in these few words in Isaiah to the modern astronomers, the followers of Copernicus and Sir Isaac Newton! What a contrast to their theory. God formed the earth and made it; it became a reality; He established it; He created it not in vain; He formed it to be inhabited. Can you apply these words to the planet earth of Copernicus and Sir Isaac Newton? "Formed the earth and made it, established it." Is it so with the planet earth? No. It is not formed, not made, not established; it has no real, material existence. It is just a thing of the imagination, supposition, hypothesis. God created the earth not in vain, He formed it to be inhabited.

God said, let the dry land appear, the dry land did appear. Let the waters be gathered together unto one place, and the waters flowed into the place appointed for them, beneath and around the earth. Here we are on this earth—living, moving, working, trading, buying and selling. This earth is a real earth, it is inhabited as God meant it to be, and some of our fellow beings go down to the sea in ships and see the wonders of God on the great waters. This earth is a real, material earth, and this material earth, really, unmistakably and without doubt, is compacted out of the water and amidst the water. But the planet earth of Copernicus and Sir Isaac Newton is not real—it has no material existence, it is fictitious and visionary.

2. There are also passages showing that the waters surrounding the earth have their bounds on the great southern circumference. Job 26: 10, "He hath compassed the waters with bounds, until the day and night come to an end." In Margin, "until the end of light with darkness." In the R. V. it is "He hath described a boundary on the face of the waters, unto the confines of light and darkness." Job 38: 8, 9, 10, "Who shut up the sea with doors when it brake forth." 9— "When I made the cloud, the garment thereof and thick darkness a swaddling band for it." 10th and 11th, "And prescribed for it my decree (or boundary) and set bars and doors,

and said, hitherto shalt thou come, but no farther, and here shall thy proud waves be stayed." And far out there on the southern circumfereuce, are solid and impassable ramparts of ice, barriers—cliffs of ice—forbidding the further progress of daring navigators. It was so with Captain Wilks and Jas. C. Ross. Well, you ask: Do the Scriptures mention these solid walls and barriers of ice out there on the southern circumference? Yes, they do. Job 38: 30; "The face of the deep is frozen." The daring navigators on the southern seas, who have told us of the solid walls of ice—the barriers and cliffs of ice— disclose to us the meaning, the sublime meaning of such passages as Job 38: 30; Ps. 33: 7.

II. The earth and ocean together constitute an immense circular plane, according to the other book of God, *Nature*. There are phenomenal proofs that the earth is not a globe, with north and south poles, but that the earth is a plane, having the central region for its north, and, the southern circumference for its south.

1. Long periods of light and darkness, regularly alternating, is a phenomenal peculiarity of the north, but not of the south, and proves that the north is the central region, and the south is the circumference.

During the summer solstice, the northern or central region of the earth is illuminated for several months together, during those months it is a long day without a night. This is a phenomenal characteristic of the north. This being the central region, the diameter of the sun's orbit in June is much smaller than that of its December or winter solstice, its speed is not so great or rapid as it is in December when on its outer path, or orbit on the Tropic of Capricorn, consequently its rays continue over the northern centre for several months. But in the south this is not the case, though it would be if the earth were a planet. In the south, on the contrary, the day closes abruptly in summer, they have little or no twilight. In the south seas beyond the 50th parallel, the sun will be shining brightly, and, in a very short time, the sailor who happens to be aloft, will be in pitch darkness. The sun seems to drop below the sea. At Auckland, New Zealand, there is little or no twilight. At Nelson, it is light till about 8 o'clock, then in a few minutes it becomes too dark to see anything, and the change comes over in almost no time. Twilight lasts but a short time in so low a latitude as 28 degrees south, according to Captain Basil Hall, so that from 28 degrees south, to beyond 50 degrees south, there is little or no twilight. But, in the corresponding latitudes north, the twilight continues for hours after visible sunset. In the northa t midsummer, for many nights in succession, the sky is scarcely darkened.

2. The differences between north and south with regard to organic life, vegetable and animal, show that the earth and ocean is a circular-plane. The long periods of sunlight in the north, develop with great rapidity numerous forms of vegetable life, and furnish subsistence for multitudes of living creatures. But in the south, where the region is circumferential (not central as in the north, the sunlight cannot linger, but sweeps quickly over that greater southern circle, completing it in the same time as the shorter circle of the north, viz., 24 hours, and so has not time to excite the surface, has not time to aid and stimulate animal and vegetable life to the same extent as in the north, consequently in comparatively low southern latitudes, everything wears an aspect of desolation.

The South Georgia's latitude 54 and 55 degrees in the very height of summer, is covered deeply with frozen snow; but in the farthest north, nature is adorned with summer beauty; flowers and grasses bloom during a brief and rapid summer. Kerguelan, 49 degrees south, boasts 18 species of plants, only one being useful in cases of scurvy, it is a peculiar kind of cabbage; but Iceland, 65 degrees north, 15 degrees nearer the pole in the north, boasts 870 species. Kerguelan's land, or, Desolation island, was discovered in 1772 by M. de Kerguelan, a French Navigator. Here December corresponds to our June. According to Captain Morrell Kerguelan is situated in latitude 48 degrees, 40" south, longitude 69 degrees 6" east. Many of the hills on this island, though of moderate height, were covered with snow, nothwithstanding that the season was midsummer. January corresponding to our July. There is not the appearance of a tree or shrub on the whole island. Captain Morrell, 1822 to 1831, in latitude 62 degrees 27" south, longitude 94 degrees 11" East, met with extensive fields of ice, one of which would have measured 150 miles, east and west.

The bones of musk oxen, killed by Esquimaux, were found on the 79th parallel north, while in the south, man is not found above the 56th parallel of latitude.

These differences between north and south could not exist, if the earth were a globe, turning upon axis and moving in an orbit round the sun. The latitudes corresponding north and south, would have the same degree of light and heat and the same general phenomena. The distance round a globe would be the same at 50 degrees south as at 50 degrees north, and the surface at the two places would pass under the sun with the same velocity, and the light would approach in the morning and recede in the evening in exactly the same manner. There would be a sameness of phenomena north and south, if the earth were a globe; but the differences are in harmony with the doctrine of the circular-plane of the earth and ocean.

3. The meridian lines diverge southwards, and the degrees of longitude increase accordingly; but if the earth were a globe, the degrees of longitude northward or southward from the equator would diminish.

From the known distance between two places in the south on or about the same latitude, and the difference of solar time (or difference in longitude) we can calculate the length of a degree at that latitude.

JOHN T. LAWSON,
KEARNEY,
PARRY SOUND DISTRICT,
ONTARIO.

CHAPTER II.

The circular-plane of the earth and ocean an immense sun-dial, witnessing to its own level and immobility.

I. The position of the sun in the firmament in relation to his diurnal course indicates the time of day. The meridians are straight lines from the north centre and diverge more and more southwards, are 24 in number, corresponding with the 24 hours of the day. The terrestrial surface is the dial plate. The time of day advances according to the progress of the sun in the firmament. The sun moves from east to west and comes round to the same point in 24 hours, thus completing a circular path above a stationary, planary earth.

A man looks, yes actually sees the sun move in an arc of a circle, and in so watching the progress of the sun his eye-line is something like the finger on the earth-dial. During the summer solstice he sees the sun rise a little north of east, then passes on to east, south-east, then to the meridian, then south by west, south-west, west, next sets a little north of west. A man watching the sun in its daily course, measures very nearly two-thirds of the circle, both of the earth plane and of the firmamental path of the sun.

The motion of the shadow on the sun dial in some part of a circle or curve around the column is caused by the motion of the sun in an arc of a circle, in the same way as the shadow of a narrow bottle in a kind of curve on a table is caused by moving a light in a circle around the bottle. The motion of the shadow on one side the bottle corresponds with the motion of the light on the other. The light moves in both cases and the surface on which the shadow is cast is stationary in both cases. Our own body may serve as a column to cast the shadow. That the earth is a plane was believed by the ancients. Yes, was believed by men for 5,500 years. Narrien in his history of astronomy says: "The accounts collected from the most ancient authors concerning the nature of the universe coincide nearly with each other in representing the earth as a plane, bounded on its whole circumference by an ocean of vast extent.

We say that the position of the sun in the firmament indicates the time of day. The sun completes a circular path in the firmament in 24 hours. Rising in the east, then advancing to the noon-day position over the southern horizon at 12 o'clock, setting in the west. Then during our night passing from west along the other side the north centre to the east, where we see him rise in the morning. If the sun was stationary and the earth revolved our day would only be six hours long and our night 18 hours long, and the sun instead of moving in an arc of a circle, or completing a circular path, it would rise, pass over head and set in the plane of our position. At 12 o'clock the sun would not be over our southern horizon, but would be setting. The fact that the sun is over our southern horizon at 12 o'clock and that it is noon along the whole meridian proves that the earth is a plane and stationary and that it is the sun that moves. To place the matter of the sun's moving above the earth beyond a doubt, the observations of arctic travellers may be quoted. Captain Parry and several of his officers on ascending high land near the arctic circle repeatedly saw for 24 hours the sun describing an arc of a circle upon the northern horizon.

During the summer solstice the sun is above the horizon for 15 hours 26 minutes when he is 1035 miles nearer to us than the equator. When on the equator he is above the horizon for 12 hours. The sun, during the summer solstice, being vertical at the tropic of cancer, 1035 miles north of the equator, is the cause of the day being then 3 hours 26 minutes longer than at the equinox.

Captain Beechy says: "Very few of us had ever seen the sun at midnight, and this night being particularly clear, we saw him sweeping majestically along the northern horizon."

In July, 1865, when the sun was at the summer solstice, Mr. Campbell, United States Minister to Norway, with a party of gentlemen went far enough north to see the sun at midnight. They were 69° North latitude and they ascended a cliff 1000 feet above the arctic sea. It was late and the sun swung along the northern horizon from west to east. We all stood silently looking at our watches. When both hands stood together at 12 midnight the full, round orb hung triumphantly above the wave a bridge of gold spangled the waters between us and him. There he shone in silent majesty which knew no setting. We involuntarily took off our hats, no word was said. During the summer solstice at our latitude there are only 8 hours and 24 minutes out of the 24 that we do not see the sun. Going as far north as these gentlemen went would just be extending our horizon 8 hours and 24 minutes, so that we would see the sun describe that part of his circle from west to east, that we do not see in this latitude.

II. The position of the sun in the firmament in relation to his monthly courses indicates the season. The sun's revolutions from solstice to solstice are eccentric or spiral. It is summer during his 90 eccentric revolutions from June to September. It is autumn during his 90 revolutions from September to the winter solstice. Winter during those from the winter solstice to the March equinox, and spring during his 90 spiral revolutions from March to the summer solstice.

The sun's speed per hour on the equator from east to west is 1035 miles. his speed north or south in 90 days is 1035 miles and this distance embraces 15 degrees latitude. When the sun is moving from the Tropic of Cancer towards the equator it is summer in the north and the days are 14 and 15 hours long, it is then winter in the south where the days are 11 and 12 hours long. When the sun is moving from the Tropic of Capricorn northward towards the equator, it is summer in the south and the days there are 13 and 12 hours long.

III. The circular plane of the earth and ocean as an immense sun-dial witnesses to its own level and immobility. The shadow on a sun-dial in some part of a circle or ellipse is caused by the motion of the sun in an arc of a circle, and if the sun moves then the earth is stationary. If the earth moved then the end of the shadow would not describe a circle, but would describe a straight line.

When the sun is on the meridian then it is 12 o'clock along the whole meridian line. This would not be if the meridian line were a semi-circle as on a globe. The mariner's compass points north and south at the same time, but it could not do so if north and south were at the centre of opposite hemispheres. This coincides with the meridian which is a straight line north and south. The north is the one fixed point, the centre; the south is a vast circumference, a circular boundary; to all parts of this circular boundary the south point of the compass shifts around as it is carried around the central north. There is therefore no south point or pole, but an infinity of points forming a vast circumference.

The meridians are straight lines north and south, and latitude is distance along the meridian line. The degrees of latitude are $57\frac{1}{2}$ not 90 as upon the globular theory where the meridians are semi-circles. $57\frac{1}{2}$ is the proportion of radius to circumference, and the degree of latitude is a definite, unvarying quantity as measured throughout, upon the total meridian length. It is $69\frac{1}{4}$ miles upon a plane surface and this agrees with the most exact measurements ever made on the face of the earth by men of greatest skill and by the best instruments.

The Swedish Government in latitude 66° 20′ 10″ makes a degree of latitude 265,782 feet, that is more than 69 miles.
The Russian Government, 58° 17′ 37″ = 265,368 ft., more than 69 miles again.
The English Government, 52° 35′ 45″ = 364,971 ft., 69 miles.
The French Government, 46° 52′ 2″ = 364,872 ft., } 69 miles.
" " 44° 51′ 2″ = 364,535 ft., }
The Roman Government, 39° 12′ 0″ = 363,786 ft., 69 miles.
The American Govern't, 1° 31′ 0″ = 362,808 ft., 68½ miles.
The Indian Government, 16° 8′ 22″ = 363,044 ft., 68½ miles.
" " 12° 32′ 21″ = 363,013 ft.
The African. Cape of } 35° 43′ 20″ = 364,059 ft., 69 miles.
Good Hope

If the earth were really flattened at the poles the degrees would shorten in going from the equator towards the north, and yet men of the greatest skill, using the most perfect instruments, making the most exact measurements ever made on the face of the earth have found results the very reverse of the Newtonian theory. Well, then, $57\frac{1}{2}$ degrees of latitude from the north centre to equator being proportion of radius to circumference (on a level surface of course) give us $69\frac{1}{4}$ miles to a degree, and this agrees very nearly with the beforementioned exact measurements, the most exact measurements ever made on the face of the earth, by men of the greatest skill, using the most perfect instruments.

Parallels of latitude are circles concentric with the northern centre. A degree of latitude is a definite and unvarying quantity as measured throughout upon the total meridian length. But a degree of longitude is a varying quantity, according to the radial distance. On the equator, the degree of latitude will equal the degree of longitude. North of the equator, latitude exceeds longitude; south of that, longitude exceeds latitude.

The Meridians are straight lines from the centre to the circumference 24 in number to correspond with the 24 hours of the day. There are six parallels of latitude—three North and three South of the Equator. The distance between these parallels from each other and from the Equator, is precisely the same as between any two meridians on the equatorial circle. This distance on the equator is 1,035 miles, a 1-24th part of 25,000 miles. The parallels of latitude show sections of 15° or 1,035 miles each. The sun's speed per hour on the equator is 1,035 miles, and the distance northward from the equator to the summer solstice, which the sun makes in 90 days; or in 90 eccentric or spiral revolutions is also 1,035 miles. This is the sun's northward journey, and it decreases its orbit and its speed in proportion to this distance. The distance southward from the equator to the winter solstice, which the sun makes also in 90 days, 90 eccentric or spiral revolutions, from September to December is again 1,044 miles. This is the sun's southward journey, and it increases its orbit and its speed in precisely the same ratio. Just as the revolution of the sun in his 24 hour path, from East to East again gives us alternate day and night, just so does the increase southward or decrease northward of its orbit, provide for the change of seasons.

The sun remains at each solstice 64 hours—two and a half days = (and 8 hours over), before renewing its spiral courses northwards or southwards, so that its orbit for those extra five days (and eight hours) is concentric ; at the other 360 revolutions it is eccentric or spiral, giving the two solstitial months June and December $32\frac{1}{2}$ days each; the other ten just 30 days each. This doctrine of the Earth plane presents educational advantages—would greatly facilitate the progress of the scholar, boy, or girl in his or her physical geography. By adding pleasure to the study it would make their progress easy and rapid.

The works of God in nature are a counterpart of His word, and when studied attentively ; when spelled out carefully, give emphasis, often a startling emphasis to His word.

There are many sentences with great depth, and breadth, and height of meaning. " Encompassed the waters with bounds until day and night come to an end," or " until the end of light with darkness."

"The face of the deep is frozen," referring undoubtedly to the far South—the solidly frozen region. The barriers of ice—the cliffs of ice—the solid, impassable ramparts of ice

JOHN T. LAWSON,
Kearney P.O.,
Parry Sound District,
Ontario.

EXPERIMENTAL PROOFS

(WITH ILLUSTRATIVE ENGRAVINGS)

THAT THE

SURFACE OF STANDING WATER

IS

NOT CONVEX

BUT

HORIZONTAL.

WITH A CRITICAL EXAMINATION OF THE RECENT ATTEMPT TO DECIDE THE QUESTION—"IS THE EARTH A GLOBE OR A PLANE?" FOR A WAGER OF ONE THOUSAND POUNDS, MADE BETWEEN JOHN HAMPDEN, ESQ., SWINDON, WILTS; AND ALFRED WALLACE, ESQ., F.R.G.S., REGENT'S PARK, LONDON; WITH MR. W. CARPENTER AND DR. COULCHER AS REFEREES, AND J. H. WALSH, ESQ., LONDON, EDITOR OF "THE FIELD," AS UMPIRE.

BY

"Parallax,"

(Author of "Zetetic Astronomy," "Patriarchal Longevity," & other works.)

LONDON:
WILLIAM MACKINTOSH, 24, PATERNOSTER ROW,
AND ALL BOOKSELLERS.
1870.

ENTERED AT STATIONERS' HALL.

EXPERIMENTAL PROOFS, ETC.

In the year 1865 I published a work called "Zetetic Astronomy,"* the object of which was to prove that the Earth is not a revolving Globe, but an irregular Plane without orbital or axial motion, and the only known material world in the Universe. For many years previously I had delivered lectures upon the subject in most of the principal towns of Great Britain and Ireland.

In October of last year (1869), whilst lecturing in the Westbourne Hall, Bayswater, near London, I received a letter dated Swindon, and signed "John Hampden." In it the writer stated that he had then recently obtained a copy of the work above alluded to (Zetetic Astronomy); had been greatly interested in its perusal, and was perfectly satisfied from the evidence it contained that the doctrine of the Earth's rotundity and the Newtonian system of Astronomy altogether was fallacious. An almost daily correspondence seeking and giving information upon various points connected with the subject was afterwards maintained for several weeks. The questions were incidentally asked "how long had I held such convictions, and what steps had I taken to make them known to the world?" on replying that I had been labouring in various ways—lecturing, debating, writing, &c., &c., for upwards of thirty years, he expressed himself as greatly surprised that he had never heard of the matter until lately, and stating that he was so completely satisfied of the truth of what I had written and published, that he would at once begin to do his utmost for its diffusion and establishment.

Arrangements were made that he should make extracts from and reprint a given section of my work in the pamphlet form; and he thenceforth laboured in every way which lay in his power,

---

* Simpkin, Marshall & Co., London.

sparing neither expense nor effort. He wrote in local papers fiercely denouncing the Newtonian system and all who held it to be true. Epithets were used and charges made which no man has a right to employ or make against those who simply differ in opinion or conviction. I wrote complainingly of the style of advocacy he had adopted; and endeavoured to show him that as the world had been educated to believe in the Newtonian philosophy as true and satisfactory, it was our duty to treat such educated conditions with respect and consideration. That we must seek to uneducate or educate afresh and not to denounce and abuse. And I here take the opportunity of earnestly advising *all the converts to the Zetetic Philosophy to treat their opponents as at least equally sincere and honorable as themselves*. It is the first duty of an advocate to be respectful, patient, free from all special pleading and calmly reliant upon the force of truth plainly and solemnly presented. It should ever be borne in mind that all men wish to be right in their convictions. They do not wilfully cling to error. If they appear to do so it will be found that, however false their opinions may be in reality, they at least appear to be true, or they could not be conscientiously defended. Men are often stupid enough in refusing to listen to evidence for opinions contrary to their own, and many there are who are incapable of strict logical reasoning, who cannot trace effects to their legitimate causes, and who are unable to follow out the sequences of the evidence presented. Such people are often more troublesome and obstinate in discussion than those who are gifted with higher degrees of mental power. Evidence appears to have little weight with them, and any change of conviction seems to be the result of some accidental impression, rather than the direct effect of a reasoning process : but until that evidence has penetrated and changed their minds they are to be considered as equal to ourselves in every worthy characteristic. Any other course is persecutive, unjust, and injurious to the cause it is intended to serve

I deeply regret that, as in many other instances, the advice I gave was not regarded ; an unmistakeable and unfriendly defiance arose. For several weeks our correspondence was suspended, and I was entirely ignorant during that time of what was being done ; but at length I was startled by reading in the daily and weekly newspapers the following announcement:—

"500*l*. has been offered and accepted on the result of a scientific investigation as to whether the surface of the earth and water is level or convex. The challenge was made by Mr. Hampden, of Swindon, and has been accepted by a fellow of the Royal Geographical Society of London. The 1,000*l*. has been lodged at Coutts's, and the survey is to be made before the 15th of March, in the county of Cambridge. The editor of an old-established London paper has been chosen umpire; each party names a referee. Much interest in the decision is felt by the innumerable advocates of the Newtonian and Copernican theory of the rotundy and revolution of the earth, which Mr. Hampden affirms to be a downright fiction and a fraud, in the face of all the philosophy and science of the United Kingdom." Shortly afterwards my attention was drawn to an article in "The Field" of March 5th, 1870, from which the following is an extract:—

"EXPERIMENTAL PROOF OF THE ROUTUNDITY OF OUR EARTH.

" For some years a correspondent of a provincial journal, signing himself " Parallax," has attempted to revive the long-exploded theory that the earth on which we live is a plane, and that, while the North Pole is in the centre of this great flat, the South is not a point, but a margin of ice, which is the sole obstacle to an exploring party reaching the edge. The theory is so opposed to numberless facts well known to scientific men, that no member of the latter class has until now, as far as we know, thought it worth confuting; but—whether from this cause, or from its novelty, or from its inherent truth, it matters not — the fact remains that "Parallax" has obtained a numerous following, and among others a gentleman of the name of Hampden, residing at Swindon. So convinced is he of the existence of this plane, that he has for some time offered to test it experimentally, and to risk £500 on the result, on condition that a similar sum is also deposited by the opposite side. For a time no one thought of taking up the cudgels, but at length Mr. A. Wallace, a fellow of the Royal Geographical Society, thinking it desirable to disabuse the minds of the diciples of "Parallax" of this fallacy, as he assumes it to be, offered to comply with Mr. Hampden's conditions, by proving the convexity of the surface of a length of water (six miles) by ocular demonstration; and for this purpose £500 aside have been deposited in our hands, the whole sum to be handed over either to Mr. Hampden or Mr. Wallace, according to the success or failure of the latter in proving the disputed convexity."

I wrote enquiring as to the nature of the experiments to be made and the place and time and persons concerned in the matter: but could get no information. I was kept in entire ignorance of the whole affair until it was over. I could not but feel that this was altogether injudicious on the part of Mr. Hampden and his referee, Mr. Carpenter; and very unfair both to myself and to the public. Common justice ought to have suggested to them that no such attempt to settle so important a matter should have been made without an invitation to the author to be present. More especially should this have been done when it is known that both Mr. Hampden and Mr. Carpenter were literary, and not scientific gentlemen. They knew little or nothing of the nature of the instruments employed in the experiments, and became literally the helpless victims of their more philosophical and practical opponents. What could be more unwise than for Mr. Hampden to deposit the sum of £500 against the same amount from Mr. Wallace, and then to allow Mr. Wallace to dictate his own experiment and to use and manipulate his own instruments? In such a procedure common sense and practical justice were ignored. The only proper plan would have been for both gentlemen to stand aside, and allow two distinctly and separately engaged Surveyors to take the level of the water; the referees noting the result, and the particulars afterwards given to the umpire. But even then it was the duty of these gentlemen to first repeat the experiments which I had made and published in my work, in which at pages 10 to 13 the following account occurs:—

"If the earth is a globe there cannot be a question that, however irregular in form the *land* may be, the *water* must have a *convex surface*; and as the difference between the true and apparent level, or the degree of curvature would be 8 inches in one mile (statute measure), and in every succeeding mile 8 inches multiplied by the square of the distance, there can be no difficulty in detecting either its actual existence or its proportion. Experiments made upon the sea have been objected to on account of its constantly changing tidal altitude, and the existence of banks and channels which produce currents, 'crowding' of waters, and other irregularities. Standing water has therefore been selected, and many important experiments have been made, the most simple of which is the following:—In the County of Cambridge there is an artificial river or canal, called the 'Old Bedford.' It is upwards of twenty miles in length, and passes in a straight line through that part of the fens called the 'Bedford Level.' The water is nearly stationary, often entirely so, and throughout its entire length has no interruption from locks or water gates; so that it is in every respect well adapted for ascertaining whether any and what amount of convexity really exists. A boat with a flag standing five feet above the water was directed to sail from a place called 'Welche's Dam' (a well known ferry passage), to another place called 'Welney Bridge.' These two points are six statute miles apart. The observer, with a good telescope, was seated in the water, with the eye not exceeding eight inches above the surface. The flag and the boat were *clearly visible throughout the whole distance!* as shown in the following diagram.

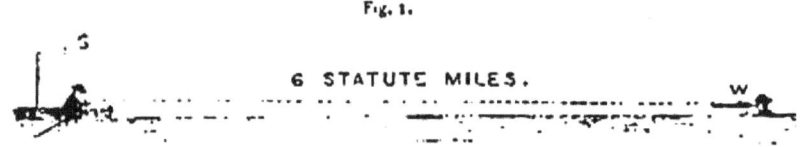

Fig. 1.

6 STATUTE MILES.

" From this experiment it was concluded that the water *did not decline from the line of sight!* As the altitude of the eye of the observer was only eight inches, the highest point, or the horizon, or summit of the arc, would be at one mile from the place of observation; from which point the surface of the water would curvate downwards, and at the end of the remaining five miles would be 16 feet 8 inches *below the horizon!* The top of the flag, being 5 feet high, would have sunk gradually out of sight, and at the end of the six miles would have been 11 feet 8 inches *below the eye line!*"

This will be rendered clear by the diagram.

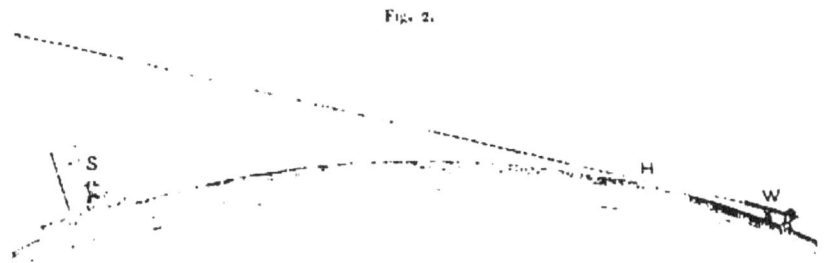

Fig. 2.

W the position of the observer, S the flag-staff six miles away, and H the intervening horizon.

"From this observation it follows that the surface of standing water is not convex, and therefore that the *Earth is not a Globe!* On the contrary this simple experiment is all sufficient to demonstrate that the surface of the water is parallel to the line of sight and is therefore *Horizontal;* and that the earth cannot possibly be other than A PLANE!"

After such an account as the above had been published, and as both Mr. Hampden and his referee Mr. Carpenter were, in a practical sense, perfectly ignorant of the whole matter, having never tried such experiments but relied entirely on the statements made in my work* to agree to make any other kind of experiment without having first tested the truth of my statement as above given, was, to say the least, unfriendly, very foolish, unjust, and logically irregular. The most simple and decisive should always be first employed, and then made use of to test and rectify the more complicated. My own long experience as an experimental investigator has proved to me that however complicated and conflicting may be the mere systems and opinions of men, the great principles and truths of nature are always simple and consistent. It is therefore imperative that in every enquiry after

* "Mr. Carpenter was engaged to decide a disputed question, of which he and his principal (Mr. Hampden) professed to be practically ignorant."— *Field, March* 26, 1870.

truth, the simplest possible means should first be adopted. Had Messrs. Hampden and Wallace conferred with, or invited me to take part in their operations, I could have shown and satisfied them that what they proposed to do and the instruments they were about to employ as well as their mode of application, were in every sense unsuitable for the object they had in view, and could not lead to definite and satisfactory results. Was there any justice in pretending to test the truth of my teachings by any other method than that o repeating the experiments which I had made, and the particulars of which I had published to the world? Were Mr. Wallace and his referee Dr. Coulcher in preparing and carrying out their peculiar and special operations, and neglecting to test my statements, doing that which they could approve if done by others towards themselves? Was it not the duty of Mr. Hampden and his referee Mr. Carpenter to insist upon the experiments described in my work being repeated? Before agreeing to any other course were they not bound in honor, as gentlemen, saying nothing of the friendly feeling which might have been expected from their recent conversion to the "Zetetic Astronomy," to have informed me of their intentions and to have invited me to take part in their proceedings? Their not having done so was to myself individually a needless insult; towards the public an unwarrantable deception—a mockery—a make-believe of a sincere desire to settle an important question; and to the cause of truth and progress an ill-conceived, injurious retardation. There never was an instance where in deed and in truth it could have been more justly said "save me from my friends." For their folly and injustice they have had to forfeit five hundred pounds, but the opposite party have not fairly received it. All concerned were the victims of self-imposed instrumental deception. All were evidently ignorant of the possible behaviour or reading of the telescope and the spirit level when applied in the way they had mutually agreed upon. This will be seen by a very brief examination of the report of their proceedings which appeared in the "*Field*" *of March* 26, 1870, and referred to in a leading article in the same number by the Editor who was also umpire in the case. The Editor says:—

" In the remarks which we ventured to make on the 5th inst., we endeavoured as far as possible to state exactly what Mr. Wallace engaged to prove, namely, that by fixing three discs at equal distances, 12 feet from the surface of this level, one being at *each end* and the third in the middle; according to the received theory, the middle disc ought to range 5 feet (in round numbers) above the level of the terminal discs; while Mr. Hampden risks his £500 on the assumption that the three will range in a straight line."

From the above it will be seen that *three discs* were to have been used, one *immediately near* to the telescope, one in the middle, or three miles from the telescope, and one at the end of six miles. Now clearly the conditions of the experiment were *not carried out!* At the end of six miles instead of *a disc* an oblong flag was placed; a disc was erected in the middle position only! and *nothing whatever* was fixed *close to the telescope!* Thus the agreement was completely violated. Those who suggested and those who were idly and carelessly present, and agreed to such a procedure, were alike greatly to blame. The very life of a great cause was at stake. The wager of twice five hundred pounds, even if the puzzled heads of the wagerers had been thrown in, were as nothing in value compared with the importance of the question to be decided: and the least that ought to be said is, that every man connected with the operations was entirely unfit for the duty. Had a disc 12 feet, or any other altitude, been placed *close to the telescope;* another of exactly the same make at the distance of three miles, and a third of the same character and altitude at the end of the six miles, and the telescope placed immediately *close to* and the line-of-sight directed *over* the first disc, the conditions of the agreement would have been properly fulfilled, and the result would have proved the water surface to have been equidistant from the line-of-sight throughout the whole distance of six miles. Let any one select a long row of lamps of equal altitude, and on truly horizontal ground. Let him elevate himself until he is able to place a telescope just *above* the first lamp; on directing it along the whole row he will find that the line-of-sight will pass at the same distance just over each lamp to the end of the series; but let him ignore one half the lamps immediately nearest to him, and so place the telescope that he *must* observe the *last* or farthest lamp, and he will see all the intervening lamps *standing above* the line-of-sight—the *nearest* to him apparently the highest, and the degree of elevation more or less according to the power of the telescope. This is precisely the case of the first observation made by the gentlemen who so strangely risked their money and their scientific reputation on the result! From the careless and logically dishonest manner in which this experiment was conducted, the agreement broken, the most essential condition neglected, the whole matter falsified as compared with their published programme, it is clear that the recipient of the prize of £500 is not honestly in its possession. The money was neither won nor lost. The race was not run; the experiment agreed upon was not tried, and therefore the stakes should be returned. So far nothing was proved except the childish carelessness of the operators.

The second experiment, that with the spirit-level, was equally valueless in determining the form of the surface of the water; as will be seen from the following representation and report taken from "The Field," of March 26th, 1870.

"Fig. 3. Diagram showing what was seen in the telescope of a sixteen-inch Troughton level, accurately adjusted and placed in the same position and height above the water as the large achromatic."—

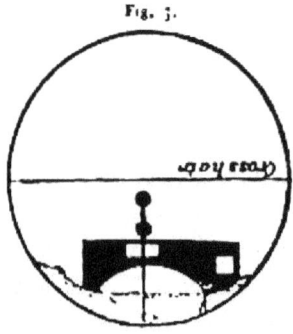

Fig. 3.

In this experiment there was the omission of an important element, as in the observation made with the large telescope, viz. a disc close to the spirit level. It had been agreed that there should be three signals, one at each end of the six miles of canal, and one in the middle. But *only two* were employed! What could be the motive of Mr. Wallace in thus leaving out one of the three signals and that one the most important? It is useless to say that the third signal was the spirit-level or the telescope, because it had been distinctly decided that *three objects* should be fixed at the same altitude above the water, and three miles apart; and that they were to be observed through the telescope of the spirit-level, whether they ranged in a straight line or gradually declined from the cross-hair of the instrument; therefore the instrument (the spirit-level) was intended to have been used in *addition* to the *three signals*. It was absolutely necessary that three signals should be fixed. It had been seen and admitted to be necessary. It was agreed that it should be done; and yet it was not done! Only two signals were fixed; and it is right that those who suggested the omission, or who neglected to carry out the conditions previously agreed upon, should be made to see that by so doing they caused the whole of their proceedings to be u'terly worthless; besides laying themselves open to disagreeable charges of improbity. The third but omitted signal properly placed would have given a certain determination to the line-of-sight which could have been used as a test of declination or curvature both in fact and in amount, from the point of observation.

But let us examine the case as it stands, and as it is represented in fig. 3. First, the spirit-level was accurately adjusted, that is, it was "levelled:" which means that, if the earth is a globe, the axis of the telescope was at right angles to the direction of gravity or the immediate radius of the earth. The line-of-sight indicated by the cross-hair (seen in the diagram) was therefore a tangent. From this tangent, or from the cross-hair, the top of the upper disc should have sunk six feet. It is known (by previous measurement) that the two discs on the first signal pole, three miles away, were *four feet apart*. Now if we take a pair of compasses, or a scale of equal parts, and measure the apparent space between the centres of the two discs as they appear in the diagram, we shall find them to be two-sixteenths of an inch apart. Therefore two-sixteenths represent the previously known space of *four feet*. Now measure from the top of the upper disc to the cross-hair and it will be found to be one-sixteenth of an inch, which only represents a fall of *two feet;* but if the earth is a sphere of 25,000 statute miles in circumference, the fall in three such miles would be *six feet.* Here then, if the top of the upper disc appearing to be below the cross-hair is taken to represent the downward curvature of the water there is an error in the reading and appearance of *four feet.*

Again, if we measure the distance between the centre of the lower disc and the centre of the white flag or signal on the bridge, we find it to be one-and-a-half-sixteenth of an inch, representing *three feet.* Now, as previously demonstrated, from the known distance apart of the two discs, one-sixteenth of an inch represents the actual space of two feet, and as the centre of the white flag, or farthest signal, on the bridge, is not more than five-sixteenths of an inch below the crosshair, it is thus represented as being only ten feet below the tangent, or line-of-sight; but the curvature in six statute miles would be *twenty-four feet.* So that if the appearance in the field of view, as given in diagram fig. 3, is taken to represent the downward curvature of the water in the canal, there is a demonstrable error or deficiency of seven-sixteenths of an inch, representing in practice *fourteen feet!* The only alternative is that the earth, if a globe at all, is very much larger than has hitherto been affirmed! Demonstrably then the appearances in the telescope of the spirit-level cannot be taken to represent declination of the surface of the water; and if the observers will only be wise enough to gather experience from failure, and wisdom from experience, they will never again attempt to decide so important a question as that of the earth's convexity or non-convexity by the use of such an instrument as a "Troughton's spirit-level." If they had employed a good Theodolite it would have been better, because the appearances could have been somewhat tested and rectified by taking the "dip" or angle subtended by the apparent depression to the several signals. As it is we must seek some other explanation of the appearances given in the diagram. Every scientific surveyor of large experience knows that the very best theodolites and spirit-levels require very careful adjustment, that they are all liable to error from various causes, viz. *collimation, parallax, refraction, aberration, spherical confusion,* and *chromatic dispersion;* and that when adjusted in the most perfect manner possible, there will still be minute errors in distances of a few hundred yards. In a work entitled "A Treatise on Mathematical Instruments," by J. F. Heather, M.A., of the Royal Military College, Woolwich, published by Weale, High Holborn, elaborate directions are given for examining, correcting and adjusting the collimation &c., and at page 103 these directions are concluded by the following words, "the instrument will now be in complete practical adjustment, for any distance not exceeding ten chains, the maximum error being only $\frac{1}{1000}$th of a foot."

The principal instrumental error is that of collimation, or slight divergence from the true axis of the eye: and as this might easily amount to $\frac{1}{1000}$th of a foot in ten chains (220 yards,) in the most perfect instrument when manipulated by the most experienced surveyors, we see at once the cause of the appearance in the diagram fig. 3. The top of the disc on the signal pole, three miles away, appears to be one-sixteenth of an inch below the cross-hair. The centre of the signal flag placed against the bridge, six miles away, is five-sixteenths of an inch below the cross-hair. It has already been shown that these distances below the cross-hair are not such as could have appeared from downward curvature in the water. It is now demonstrable, that they were the result of inevitable collimation or unavoidable divergence of the pencils of light passing through the glasses of the instrument: and as we have seen by the quotation from Mr. Heather's work, this divergence cannot be prevented, and might amount in the most perfect instruments to $\frac{1}{1000}$th of a foot in 220 yards, how much greater would it be in distances of three and six miles? Now as the collimation or optical divergence was only $\frac{1}{16}$th of an inch in three miles, and $\frac{5}{16}$ths in six miles, it necessarily follows that the instrument used was a very good one, and that the *utmost care had been exercised in its adjustment.* The unavoidable instrumental errors were indeed so minute that if a hundred additional observations had been made, so far as appearances were concerned, they might never again have given such perfect results.

The folly and injustice with which all the parties connected with these observations have here been charged, consisted in their having agreed to rely upon such appearances, without knowing their actual cause, or not being able to give them their proper interpretation; and in so persistently and unaccountably leaving out one of the three signal poles in both experiments. It is painful to hear the remarks which are made, on every hand, respecting the conduct of Mr. Wallace and Dr. Coulcher in acting so very suspiciously. Again and again have the expressions been heard, " they knew their game," " they determined to win the money, and ' cooked their case' accordingly." Their having thus acted has certainly rendered it difficult to defend them. The thought of such scientific gentlemen having been actuated by any other than the most honorable motives ought not to be tolerated: yet from their strangely unscientific procedure the suspicion is not un-natural; and there is *only one way* of successfully destroying it: viz. their acceptance of the invitation given at page 19, the use of their large Achromatic Telescope in the manner there indicated and their agreeing to stand by the consequences. The charge rests upon Mr. Hampden and Mr. Carpenter for not insisting upon a repetition of the telescope-and-boat experiment described at page 5, fig. 1, and neglecting to invite or confer with the author as the originator of all such observations, and the Founder of the " Zetetic Astronomy " erected upon the results. Had this been done all the confusion and ill feeling which have arisen, as well as the misappropriation of the wager, and the very awkward position in which the Editor of the *Field* must have found himself when called upon to act as umpire in the case might have been prevented. He was required to decide an important question, not from unquestionable results, but from optical appearances only; and as these were not understood, and therefore neither challenged nor allowed for, he could not have done otherwise than hand over the one thousand pounds to Mr. Wallace. But as the principal points in the agreement were not carried out—both parties being at fault; and as the positions of the signals were only apparent and had really no connection with the question at issue, the five hundred pounds ought to be returned to Mr. Hampden, and the whole affair looked upon as a " drawn battle "—as a contest, the conditions of which were not fulfilled.

As long ago as 1838 I made a number of observations on the old "Bedford Canal," and soon found the necessity of specially studying the structure and peculiarities of all the different kinds of levelling instruments; and many times since that period have taken part in levelling experiments with some of the first surveyors and engineers of the day. In all the experiments in which I have been thus engaged I have been able to state before-hand—to predicate, what appearances would be observed in the field of view, and in almost every instance have satisfied the operators that what they saw was simply the admitted unavoidable peculiarities of the instruments, and not indications of the earth's rotundity. So important is this explanation that I deem it right to offer to the reader a simple demonstration. Let him find a piece of ground—a terrace, promenade, line of railway, or embankment, which shall be perfectly horizontal for, say, five hundred yards. Let a signal staff five feet high be erected at one end, and a theodolite or spirit-level fixed and carefully adjusted to exactly the same altitude at the other end. The top of the signal will then be seen a little *below* the cross-hair, although it has the *same actual altitude* and stands upon the same horizontal foundation. If the position of the signal staff and the spirit-level be then reversed the same result will follow. Another proof will be found in the following experiment: Select any promontory, pier, lighthouse-gallery, or small island; and, at a considerable altitude, place a smooth block of wood or stone of any magnitude. Let this be "levelled." If then the observer will place his eye close to the block and look along its surface towards the sea he will find that the line-of-sight will touch the distant horizon. Now let any number of spirit-levels, or theodolites be properly placed and accurately adjusted; and it will be found that in every one of them the same sea-horizon will appear in the field of view considerably *below* the cross-hair. Thus proving that the telescopic or instrumental readings are not the same as those of the naked eye.

The above illustration will be still more striking if a strong tube, without lenses or glasses of any kind, be "levelled" and directed towards the sea-horizon. On looking through it the surface of the water will appear to ascend to nearly the centre of the open end or "field of view," as shown in fig. 4; H H the sea-horizon

Fig. 4.

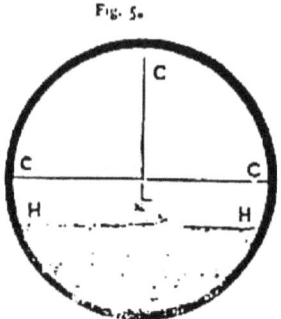
Fig. 5.

On looking in the same direction through the telescope of the spirit-level, when "levelled" the sea-horizon will appear considerably *below* the cross-hair or centre of the tube, as represented in fig. 5:—

C C C C the horizontal and vertical cross-hairs, H H the sea-horizon, some distance below C C. This depression below the cross-hair is found to be greater as the altitude of the observer and therefore the distance of the horizon increases: which is precisely the phenomenon observed in the field-of-view of the spirit-level represented in fig. 3, page 9, where the flag signal on the bridge, being the farthest away, appears lowest, and lower than the disc on the signal pole—which is only half the distance. The top of the bridge in fig. 3, may be compared to the horizon H H in fig. 5.

Thus the apparent depression of the horizon through the influence of the lenses which constitute a telescopic levelling instrument is visibly the same, and arises from the same cause as the apparent depression of the signals observed in the experiment on the Bedford canal.

On repeating the above-named experiments and seeing with his own eyes the actual results, the practical observer cannot fail to be satisfied that when distant objects are seen below the cross-hair of an optical instrument like the spirit-level, the cause is simply *aberrance of light* or "wandering" of the eye-line from the true line, or axis of vision, in passing through the glasses or lenses of the telescope: and not because really depressed in consequence of declination or curvature from the line-of-sight. Hence all such appearances are entirely out of place and valueless in connection with the subject of the earth's true form and magnitude.

These very simple experiments will satisfy the observer that there is in every such instrument more or less divergence of the line-of-sight; and that however small the amount—perhaps inappreciable in short distances—it is necessarily considerable in several miles. He will then be fully satisfied that what was seen on the Bedford Canal by Messrs. Hampden, Wallace, Carpenter and Coulcher, was not the convexity of standing water, but telescopic aberration, and instrumental "error of collimation." This will again be rendered clear and certain by such an experiment as going into the water of the canal with a telescope and observing a receding boat with a flag affixed, for the distance of six miles, as represented in fig. 1, page 5. In such an observation instrumental error will be neutralized and the surface of the water proved beyond all doubt or cavil to be horizontal.

On seeing the reports of the referees in the "Field" of March 26th, and the editorial article in the same number, I determined to visit the scene of their operations and to make some experiment or observation so simple in character that no possible doubt as to its value in deciding the question at issue could be raised. I left London on Tuesday Morning last, April 5th, 1870, and arrived at the old Bedford Sluice Bridge at twelve o'clock. The atmosphere was remarkably clear and the sun was shining brightly upon the bridge and the various objects around it. I immediately made the following measurements: —

		ft.	in.
Height of Arch		12	8
,, ,, Sluice Gate		5	8
,, ,, Abutments		3	8
,, ,, Bottom of notice board (or table of rules for navigating the river)		6	6
Length of ditto		7	2
Width of ditto		5	2
Height of a turf boat moored close to Abutment		2	6

I then obtained assistance and had the turf boat worked into the middle of the canal. A good telescope was then directed along the water and immediately the archway of Welney Bridge came distinctly into view. I saw through the arch, and for a considerable distance beyond it. If the earth is a Globe of 25,000 miles circumference, the convexity of the water between Bedford and Welney Bridges, the distance being 6 statute miles, would be such, that allowing 2 miles for the altitude of the observer's eye (30 inches), the remaining 4 miles would curvate from the summit of the arc of water 10 feet 8 inches. The highest part of the arch of Welney Bridge is 7 feet, so that the top of the arch should have been 3 feet 10 inches *below* the line-of-sight. Whereas not only the top of the arch, but the springs and abutments were distinctly visible. Therefore the surface of the water was *not convex*, but *perfectly horizontal*.

A train of several empty turf boats had just previously entered the canal from the river Ouse; and was about proceeding to Ramsey in Huntingdonshire. I arranged with the captain to place the lowest or shallowest boat the last in the train, and to take me on to Welney Bridge. The telescope was placed on the lowest part of the stern; and was exactly 18 inches above the water. The Sluice gate 5 feet 8 inches high; the turf boat from which I had made the observation to Welney Bridge, 2 feet 6 inches high, and the white notice board, 6 feet 6 inches high, were all before me. The sun was shining strongly upon them; the air was exceedingly still and clear; and the surface of the water "smooth as a molten mirror;" so that everything was extremely favourable for observation. At 1·15 p.m the train of boats started for Welney, the objects above named were plainly visible as the boats receded, and were kept in view during the whole distance; as represented in fig 6, T the telescope and B the notice board.

Fig. 6.

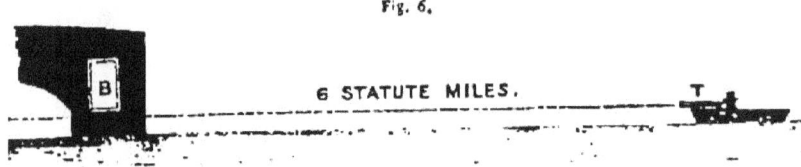

6 STATUTE MILES.

d

On reaching Welney Bridge I made very careful and repeated observations and finding several men upon the banks of the canal I called them to look through the telescope. They all saw distinctly the white board and the sluice gate, and the black turf boat moored near them. Now the telescope being 18 inches above the water the line-of-sight would touch the horizon at 1 mile and a half away—if the water is convex; the curvature of the remaining 4 miles and a half would be 13 feet 6 inches: hence the turf boat would have been 11 feet; the top of the sluice gate 7 feet 10 inches, and the bottom of the white notice board 7 feet *below the horizon*, as shewn in fig. 7, T the telescope, H the horizon, and B the notice board.

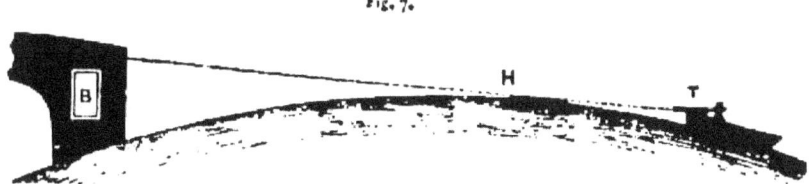

Fig. 7.

It was *not* so; and the unavoidable conclusion is that *the surface of standing water is not convex*, but *horizontal*.

Previous to leaving London it had been suggested to me that the names and addresses of several persons should be obtained certifying to the observations which had been made. This I did; but recollecting that many years previously I had obtained formal certificates from respectable persons in the locality; and that when these documents were referred to in proof of my statements, it was said that no evidence existed that they were genuine—" they might be forgeries" &c., &c. I resolved not to publish them. I considered it would give more dignity and certainty to the cause of truth, to make and describe the simplest possible experiments: to refer to fixed well-known objects, which were selected for observation, and to formally challenge the scientific world to visit the locality and to make observations for themselves. I most earnestly and solemnly invite all those who take a serious interest and who feel the great importance of the subject to make a special journey to the Bedford Canal; and, above all things, to repeat the simple experiments which

I have here described. The whole matter ought to be taken out of the pale of controversy. Believing and disbelieving should have no place in connection with it. It is capable of demonstration to the eyes and judgment of every one who may desire to be satisfied, and who will take the trouble to visit the place and observe for himself. Why not establish a scientific pilgrimage to the Fens of Cambridgeshire? Every man who sees the magnitude and the logical consequences of the question "is the surface of standing water horizontal?" should make it an important part of his education to visit at least once in his life the old Bedford Sluice or Canal, and make such experiments as will for ever satisfy him that the surface of the water is not convex, and that therefore, and of mathematical necessity the EARTH IS A PLANE. *

As many have expressed a degree of doubt that the curvature upon the surface of a Globe 25,000 statute miles in circumference amounts to 8 inches in the first mile, and increases as the square of the number of miles multiplied by 8 inches, the following quotation will be useful :—

"If a line which crosses the plumb-line at right angles be continued for any considerable length it will rise above the earth's surface (the earth being globular), and this rising will be as the square of the distance to which the said right line is produced; that is to say, it is raised 8 inches, very nearly, above the earth's surface at 1 mile's distance; four times as much, or 32 inches, at the distance of 2 miles; nine times as much, or 72 inches, at the distance of 3 miles, &c. &c."

"The preceding remarks suppose the visual ray to be a straight line, whereas on account of the unequal densities of the air at different distances from the earth, the rays of light are incurvated by refraction. The effect of this is to lessen the difference between the true and apparent levels, but in such an extremely variable and uncertain manner that if any constant or fixed allowance is made for it in formula or tables, it will often lead to a greater error than

---

* The Author is willing to accompany any specially arranged party, on receiving due notice—addressed " Parallax," care of the Publisher, (Wm. Mackintosh), 24, Paternoster Row, London.

what it was intended to obviate. For though the refraction may at a mean compensate for about a seventh of the curvature of the earth, it sometimes exceeds a fifth, and at other times does not amount to a fifteenth. We have therefore made no allowance for refraction in the forgone formulæ."—*Encyclopædia Brittanica*, article "*Levelling*."

It will be seen from the above that, in practice, refraction need not be allowed for. Indeed it can only exist when the line-of-sight passes from one medium into another of different density; or where the same medium differs at the point of observation and the point observed. The Ordnance surveyors of England have found that 1-12th of the altitude of an object may be allowed for refraction. Taking this amount from the different altitudes referred to in the several experiments made upon the old Bedford Canal, it will make very little difference in the actual results. For instance, in the experiment represented by fig. 2, page 6, the top of the flag would be reduced 1-12th, leaving it 10 feet 8 inches instead of 11 feet 8 inches as there given.

Others not being able to deny the fact that the surface of the water in the old Bedford and other canals is horizontal, have thought that a solution of the difficulty was to be found in supposing the canal to be a kind of "trough" cut into the surface of the earth: and have considered that although the earth altogether is a globe, yet a canal or "trough" might exist as a chord of the arc terminating at each end. This however could only be possible if the earth were motionless. But the theory which demands rotundity in the earth also requires rotatory motion: and this produces centrifugal force. Therefore the centrifugal action of the revolving earth would of necessity tend to throw the waters of the surface away from the centre. This action being equal at equal distances; and being retarded by the attraction of gravitation, which is also equal at equal distances; the surface of every distinct and entire portion of water must stand equidistant from the earth's centre; and therefore must be convex, or an arc of a circle. Equidistance from a centre means, in a scientific sense, "level" or convex. Hence the necessity for using the term horizontal to distinguish between "level" and "straight."

In addition to the evidence already advanced, that, as the surface of standing water is not convex but horizontal, and that therefore it is impossible that the earth can be a globe, the appearance of the horizon at sea may be referred to. The sea horizon, to whatever distance it may extend to the right and left of an observer on land, always appears as a perfectly straight line, as represented by H H in fig 8.

Fig. 8.

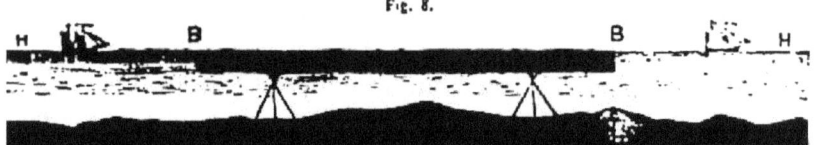

Not only does the sea horizon *appear* to be straight as far as it extends, but it may be *proved* to be so by the following simple experiment :—At any altitude above the sea level and opposite to the sea horizon fix a long board (say from six to twelve or more feet in length) edgways upon tripods or any other kind of stand. Let the upper edge of the board be perfectly smooth and truly " levelled." On placing the eye behind this upper edge and looking over it towards the sea, the distant horizon will be observed to run perfectly parallel with it throughout its whole length ! If the eye be now taken backwards to some distance, so that in looking to the right and to the left at considerable angles over the ends of the board, there will be no difficulty in observing a length of from ten to twenty miles, according to the altitude of the position; and this whole distance of twenty miles of sea horizon will be seen as a perfectly straight line ! This would be impossible if the earth were a globe, and the water of the sea "level" or convex. In twenty miles there would be a curvature on each side from the centre of the distance of 66 feet; and instead of the horizon touching the board along its whole length it would be seen considerably below the two extremities, as shown in the following diagram, fig. 9.

Fig. 9.

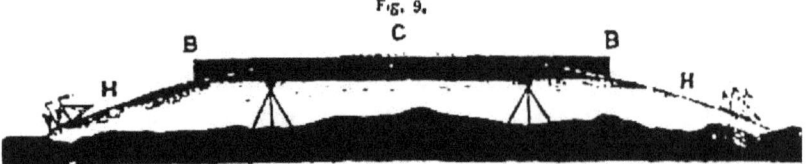

B B the upper edge of the board; and H H the horizon depressed below the centre C 66 feet in 10 statute miles ($10^2 \times 8$ inches $= 66$ feet 8 inches.)

If H were really 66 feet below B, what influence could possibly operate to make it appear at B?

A very striking illustration of the horizontal character of the sea horizon may be observed from the high land at the head of Portsmouth Harbour. Looking along the Harbour across Spithead to the Isle-of-Wight, the base or margin of the land from the extreme east to the "Needles" in the west appears and may be proved by the means just described to be a perfectly straight line: as shown in fig. 10.

Fig. 10.

As the Island is 22 statute miles from east to west it is evident that the two extremities E and W would be the square of half that distance (or 11 miles) times 8 inches or 80 feet below the centre! As 80 feet 8 inches is the amount of declination, which would exist if the earth were a globe such as the Newtonian philosophy affirms it to be, and as no such declination, but the very reverse can be proved to exist, it follows necessarily that in this important particular that philosophy is fallacious—false absolutely! If it be said that upon a globe of such magnitude as the earth a declination of 80 feet could not be recognised or detected, in the distance between the Isle-of-Wight and the head of Portsmouth Harbour, the following experiment will give the answer. Let a long rod or string, S S, in fig. 10, be tightly stretched and "levelled" before the observer, and raised a little above the horizon, so that it cuts the tops of the high lands, or hills, of the Island. It will then be seen that the slightest alterations of altitude—the ascents and descents of the various mountain peaks, and the gradual declinations of the land at its extremities can all be followed and read with the utmost precision. If such gradual and trifling alterations or differences of altitude can thus be read in relation to the line

S S, what can prevent a declination of 80 feet in the horizon E W being detected by the same means? But as such declination cannot be detected, the natural conclusion is that it does not exist! As it cannot be found or proved to exist, then the doctrine of rotundity is a fallacy in idea, and an impossibility in fact. That the earth is A PLANE is thus experimentally, logically, and formally demonstrated.

A few words may here be useful respecting the suggested pilgrimage to the scene of the previously described experiments. For ages past our philosophers, with very few if any exceptions, have indulged in and been quite content with the practice of theorizing or forming hypotheses for the purpose of explaining phenomena. This has necessitated special experimentation. They could not, so long as they desired to maintain their theories, be fearless and impartial in their search for evidence. Experiments and observations specially suggested were all that could be tolerated. To the very last degree is this the case, in our own day, with too many of our otherwise extraordinarily gifted scientific men. What praiseworthy efforts have been made by our Royal and other learned Societies, as well as by individual philosophers, to arrange and amply provide for the most difficult expeditions to various parts of the world; and what immeasurable daring and noble sacrifice have been shewn by those who have been commissioned to carry them out! After such immense sums have been expended, such risks incurred, such great and fearful sacrifices, and the most prolonged physical and mental sufferings and deplorable deaths—often cheerfully borne for the sake of science and philosophy, but, unfortunately, too often, for the purpose of giving additional importance, or, if possible, evidence, in support of some prevailing and favourite hypothesis, will they join in a well-concerted expedition to that comparatively unknown region, called the "Bedford level?" Will they see that their acceptance of such an invitation is only what society has a solemn right to expect? They are looked up to as the very *foci* of human learning: and they surely would not feel themselves undignified in becoming the Judges and Umpires in so momentous a problem as that of whether the Earth is a Globe or a Plane—involved as it is in the previous question, is the surface of standing water convex or horizontal? Let them take up the subject with a full determination to do it justice, and to settle the question once and for ever. Their means are abundant; their time is at their own command; the proper locality is within a few hours from London; the problem to be solved is deeply important, and the Author (who may at any time be found through his publisher), is ready and anxious to join them and to stand or fall by the results.

For the long period of thirty-one years he has laboured single-handed to bring this important subject before the world: not simply by recording and publishing his convictions, but by constant efforts in lectures, discussions—both on the platform and in local journals, and travelling from place to place—never resting longer than a few months in one locality, but like, as it may be said, a scientific or philosophic gypsy breaking up his tent and pitching it "here, there and everywhere" in order to debate this great question, and draw to it the attention of all classes and degrees of intelligence (and as a matter of course has had to bear every possible form of opposition, the bitterest denunciations—often amounting to threats of violence and personal danger, the foulest misrepresentations, the most reckless calumny, and the wildest and most desperate efforts to stay his career and counteract his teachings), but only recently and indirectly has the challenge received public and formal attention. A perfect stranger, a gentleman whom the Author has never yet seen, was so deeply impressed with the truth of what he had read in "Zetetic Astronomy," and its vast importance, that he determined to do his utmost to bring it to something like a practical culmination. He soon found however that upon the breast of Modern Astronomy lay a terrible incubus,—a dead-weight which no amount of argument could lift; a mass of gravitating cohesion which all the truth-love in the world acting conjointly and even centrifugally could not dissever; truth, reason, consistency and magnitude of consequence were all powerless in its presence: but, as in all other human combinations, there was an element of weakness—one little point in the structure was vulnerable. A sum of money was offered to any one who could prove the convexity of water, and when all other means had failed to draw the attention of the scientific to the subject, this one little obtrusive element, self, began to operate and an attempt was made to win the prize. Pseudo and meagre and improper as it was, by its influence the philosophic world has been disturbed. It has drawn and fixed the attention of thousands who otherwise would have remained in permanent opposition or indifference; and the hope may reasonably be held that ere long the Royal Astronomical and Geographical or other Societies will feel it their duty to step forward and give that aid and attention which the subject most undoubtedly demands.

The gentlemen who so courageously advanced the thousand pounds in order to bring the matter to an issue, although their proceedings were altogether faulty and insufficient for the purpose, deserve the warmest thanks of all those who feel that every error is injurious, and that truth alone is the real and abiding friend of humanity.

The Terrestrial Plane:
—OR—
The True Figure of the Earth.

Scripturally and Scientifically
Demonstrated by . . . . . .

FREDK. H. COOK.

INTRODUCTION.

The desire of the author of this small treatise is, not so much to explain in harmony with a plane earth, all the wonderful phenomena in Nature, as to incite a more critical investigation into Natural Science. Individuality of observation is a sure method of ascertaining the truth, therefore, he desires to help beginners over the threshold of this particular branch of science. His object is to contribute towards helping thoughtful and truth-seeking men to solve the problem that true science and true religion are not antagonistic; but that the God of the Bible, the Creator of all things, is harmonious in all His works and words.

"I have made the earth and created man upon it. I, even my hands, have stretched out the heavens, and all their host have I commanded . . . . I am God, there is none else."

Such a Being is worthy of our reverence and worship.

Astronomy took its rise in the East. Since those days when the earth was young, many, indeed, have been the various "world systems." In spite of opposition, the Newtonian-Copernican system has prevailed. "But a reasonable motion of the sun through space, discovered and established by Sir W. Herschel, and others, tends to deprive 'the system' of those pretty pictures in concentric circles." On many sides one hears rumours of a proposed change, in fact, some scientists have already changed. Sir Richard Phillips goes so far as to call Sir I. Newton's ideas "execrable superstitions," and he also says: "Woe to him who for another century shall oppose them." Well, from the great interest taken in the author's lectures upon this subject he feels sure that this book will meet with a good reception. It is a large and interesting subject. The study of it can do nothing but good.

The following words from a work on "Liberty," by John Stuart Mill, should secure an impartial reading of the work.

"If there are any persons who contest a received opinion, or will do so if law or opinion will let them, let us thank them for it, open our minds to listen to them, and rejoice that there is someone to do for us what we otherwise ought, if we have any regard for either the certainty or vitality of our convictions, to do with much greater labour for ourselves."

FREDERICK HENRY COOK.

35a, Lymington Avenue,
    Noel Park, London, N.
        England.

May, 1908.

NEBULA PHILOSOPHY AND GRAVITY.

Professor Haeckel informs us that "the world is nothing else than an eternal 'evolution of substance,'" and that this "periodical process of evolution" is really caused by "the inherent primitive properties of substance—feeling and inclination "—- which he says, are "active causes"—What does he mean? He tells us in plain and unmistakable language in the edition of the "Riddle of the Universe" of 1902, page 92, therein he says:—

"No philosopher has done more than Immanuel Kant in defining the profound distinction between efficient and final causes, with relation to the interpretation of the whole Cosmos. In his well known earlier work on 'The General Natural History and Theory of the Heavens' he made a bold attempt 'to treat the constitution and the mechanical origin of the entire fabric of the universe according to the Newtonian laws.' This 'cosmological nebular theory' was based entirely on the mechanical phenomena of gravitation. It was expanded and mathematically established later on by La-Place. When the famous French astronomer was asked by Napoleon I. where God, the Creator and sustainer of all things, came in in his system, he clearly and honestly replied: 'SIRE, I HAVE MANAGED WITHOUT THAT HYPOTHESIS.' That indicated the atheistic character which this mechanical cosmogony shares with all other inorganic sciences. This is the more noteworthy because the theory of Kant and La-Place is now almost universally accepted; every attempt to supersede it has failed. When atheism is denounced as a grave reproach, as it often is, it is well to remember that the reproach extends to the whole of modern science, in so far as it gives a purely mechanical interpretation of the inorganic world."

Haeckel, in common with others of his school of thought, denies the existence of the Creator, in fact, he goes further, and says that the notion has gone for ever, and that the "eternal iron laws of nature" have taken the place of God; and Haeckel arrives at this conclusion though—philosophy. He candidly admits that "the greatest triumphs of modern science—the cellular theory, the dynamic theory of heat, the theory of evolution, and the law of substance—are PHILOSOPHIC ACHIEVEMENTS." The Apostle Paul says:

"Beware lest any man spoil you through PHILOSOPHY." He also says: "Keep that which is committed to thy trust . . . . avoiding the oppositions of science falsely so called."

Where should we look for Truth? In the Holy Scriptures that are "able to make us wise unto salvation." They warn us against man's "philosophy," and man's "vain deceit." Thus guiding us amid the conflicting and ever-changing theories of men who know not God and yet consider themselves wise! Not only do they consider themselves wise, but far above a faithful follower of the lowly Nazarene! Haeckel informs us that, "Christ Himself had no knowledge whatever of astronomy—indeed, He looked out upon heaven and earth, Nature and man, from the very narrowest geocentric and anthropocentric point of view." Considering the sublime teaching of Christ as to the duty of man to man, and man to God, his comprehensive view of man's life here and hereafter, it is but a step from the sublime to the ridiculous to further consider such a criticism on the Son of God, who knew, as Christ did, the will and ways of God to fit Him to be the heir of all things.

We will continue our investigations further as to when, and how, the world began to evolve *itself*.

In the beginning there was gas, or a "nebulous cloud," according to scientists and evolutionists. This is rather a difficult subject to deal with, because, as we have already read, there was scientifically, no beginning—just an "eternal evolution of substance." Anyway there *was a time* when this "nebulous cloud" arose—never mind where it came from, for no scientist has yet even attempted an explanation on this point, although its existence requires SOME accounting for, considering that it was inorganic matter, and it possessed the powers of "feeling and inclination." According to La-Place, "the particles forming the cloud were very hot," he was not there to see, but I only mention this because some scientists, like Herbert Spencer, state that the "embryo universe" was cold. Anyway, hot or cold, the particles by universal suffrage, or by some other method, unknown to scientists, took upon themselves to form the "solar system;" therefore, it was necessary that this "diffused fire mist" should condense a little, and move its particles a little closer together, "according to Newtonian laws." As the Newtonian laws of attraction, or gravitation, formed the basis of this "world building nebular theory," let us consider these laws.

Sir Robert Ball tells us, that "every body in the universe attracts every other body." He also says that "the law of gravitation underlies the whole of astronomy." But when we read in a "Million of Facts," by Sir Richard Phillips, that: "Universal gravitation . . . is an utterly impossible mode of action," I think it time we consulted Sir I. Newton on the matter. I find, according to a letter he sent to Dr. Bentley, February, 1692, that he expressed the opinion, "that attraction should be innate and inherent in matter so that one body can act upon another at a distance—is to me so great an absurdity, that I believe no man, who has, in philosophical matters, a competent faculty of thinking can ever fall into it." I shall never fall into it, especially considering Sir I Newton's words, that:—

"Gravity must be caused by an agent acting according to certain laws, but whether this agent be material or immaterial I have left to the consideration of my readers." Professor Bernstein's consideration is, that:—

"The theory that motions are produced through material attraction is absurd." Perhaps Sir I. Newton agrees, for he says:—

"What I call attraction may be performed by impulse, or by some other means unknown to me." Well, if Sir I. Newton does not know we must not be surprised that, C. V. Boys, F.R.S., etc., says:

"It is a mysterious power which no man can explain; of its propagation through space all men are ignorant." I quite believe this, and also the following, written by Professors Singer and Berens:

"A body on earth falls to the ground, this is observation, body and earth attract each other, this is an obvious (?) and necessary inference and *inference* only." Dear me! I shall believe as Professor W. B. Carpenter says:—

"We have no certain experience at all . . . the doctrine of universal gravitation then is a pure *assumption*."

The fact is that "gravity" is not required, there is not the slightest evidence in the universe around us of the existence of such a "mysterious power."

In Joyce's "Scientific Dialogues," we read:—

"It seems very surprising that philosophers, who have discovered so many things, have not been able to find out the cause of gravity. Had Sir I. Newton been asked why a marble, dropped from the hand, falls to the ground, could he not have assigned a reason? That great man, probably the greatest man that ever adorned the world, was as modest as he was great, and he would have told you he knew not the cause."

This is valuable evidence, coming from believers in the theory of gravitation.

The learned Dr. Price asks:—

"Who does not remember a time when he would have wondered at the question, WHY DOES WATER RUN DOWN HILL? What ignorant man is there who is not persuaded that he understands this perfectly? But every IMPROVED man knows it to be a question he cannot answer. For the descent of water, like that of other heavy bodies, depends on the attraction of gravitation, the cause of which is still involved in darkness."

Well! It is astounding! Newton invents a theory, which admittedly has no known foundation in Nature; a pulling, a pushing power called "gravitation." Nobody understands its working, no one knows anything about its cause, it has never been seen, tested, or felt, yet such a person as Pope wrote:—

> "Nature and Nature's laws lay hid in night;
> God said: 'Let Newton be,' and all was light."

Where is the light? A question is asked of an "improved man," and he cannot answer! Why does water run down hill? Why does it not run up hill? If the earth is a globe it does both! Fancy! There are, so scientists say, 21,923,200 cubic miles of land, and 323,722,000 cubic miles of water in the "globe." Whatever keeps this preponderance of water underneath, and on the top, and on the sides, and all round *the outside* of the comparatively small portion of "land"? Water DOES run down hill. Why does it not run down the globe hill—and fall off? Light is coming!—"Gravity is a theoretical power necessary to the theory that the solar system made itself into numerous rotating whirling globes, each one that has been, is, or will be capable, perhaps, of supporting life, as we understand it, upon its surface." Apart from this theory, in all its ramifications, gravity can find no place in Nature.

Leave paper astronomy, and come out in the light of Nature. Why does a balloon ascend? Because bulk for bulk it is lighter than air. It will rise to a position; at that elevation it will stay because it will have found its equilibrium. When it loses its bulk by an escape of gas, it will collapse and descend to earth again, for the simple reason that its weight is greater than that of the air it displaces. Wood floats in water; a piece of solid iron sinks; why? Because bulk for bulk the wood is lighter than the water it displaces, whereas bulk for bulk the iron is heavier than the water. The denser a body the greater its weight; recognising this truth scientists say—gravity is another name for weight. They may call it "gravitation" if they choose to do so; but when an apple falls to the earth, it falls by its own weight when released from the stalk on which it grew; not because the apple has been pulled by the earth, or the earth pulled by the apple.

Considering all the contradictions and uncertainties of the "scientific world," as to what is, or is not gravity,—its very existence being questioned, the following words, by Professor T. H. Huxley, are highly significant:—

"If the law of gravitation ever failed to be true even to the smallest extent, for that period the calculations of the astronomer have no application."

DISCOVERY OF NEPTUNE.

From the foregoing chapter it is obvious that "science" can supply no information when we ask for the origin of matter or motion. In fact, when we ask about origins, "science" is dumb! The "world building" scientists who build on atoms—or little somethings—cannot prove the atomic theory upon which they build; or even tell us the origin of atoms; or how they came to be *diffused through space*, or by what law diffused matter did aggregate.

Camille Flammarion, a popular astronomer, says:—

"The most probable hypothesis, the most scientific theory, is that which represents the sun as a condensed nebula. This carries us back to an unknown epoch, when this nebula occupied the present place of the solar system. . . . Let us imagine, then, an immense gaseous mass placed in space. Attraction is a force inherent in every atom of matter. The denser portion of this mass will insensibly attract toward it the other parts, and, in the slow fall of the more distant molecules toward this more attractive region, a general motion is produced, incompletely directed toward this centre and soon involving the whole mass in the same motion of rotation. . . It has begun to turn so quickly as to develop, at the exterior circumference, a centrifugal force superior to the general attraction of the mass, as when we whirl a sling; the inevitable consequence of this excess is a rupture of the equilibrium, which detaches an external ring. This gaseous ring will continue to rotate in the same time and with the same velocity; but the nebulous matter will be henceforth detached, and will continue to undergo progressive condensation and acceleration of motion. This same feat will be reproduced as often as the velocity of rotation surpasses that by which the centrifugal force remains inferior to the attraction."

According to this "scientific theory,"—this "most probable hypothesis," the "planets" were detached from the condensed sun mass, we are to *imagine* how.

Lord Salisbury when President of the British Association for the Advancement of Science, asked the following question, that has not yet been answered:—

"If the earth is a detached bit, whirled off the mass of the sun, how comes it that, in leaving him, we cleaned him out so completely of his nitrogen and oxygen that not a trace of these gases remains to be discovered, even by the sensitive vision of the spectroscope?"

Sir Robert Ball informs us that, "some of the elements which are of the greatest importance on the earth would appear to be missing from the sun. Sulphur, phosphorus, mercury, gold, nitrogen may be mentioned among the elements which have hitherto given no indication of their being solar constituents."

But there are many objections to the probability of the nebular theory being true, even supposing the world to be a globe. It is well known that the planets revolve around the sun from west to east; but, totally ignoring the nebular hypothesis, it was stated a short time back by Professor Lankester, that "one of the satellites of Saturn went round that planet the wrong way!—thus calling for a fundamental revision of our ideas of the origin of the solar system." This is not the only instance. The "moons" of Uranus instead of rotating from west to east rotate from east to west! while the planes of their revolution are nearly at right angles to the orbit of their "parent," Uranus! Sir Robert Ball says that "we are not in a position to give any satisfactory explanation of this circumstance."

I am about to describe now, what Sir Robert Ball calls "a discovery so extraordinary that the whole annals of science may be searched in vain for a parallel." I must be as brief as possible and yet "develop the account of this striking epoch in the history of science with the fulness of detail which is commensurate with its importance."

It is supposed that the supreme controlling power in the solar system is the attraction of the sun, and that every planet in the system revolves around the sun in an elliptic path. Newton's laws of gravitation, of course, underlies all this supposition. According to this law every body in the universe attracts every other body. The planet Uranus was observed to have "perturbations." Le Verrier, a great French astronomer, set himself to investigate the cause of this disturbance. The influences of older planets were found to be inadequate to account for the perturbations, so Le Verrier commenced a search, by the aid of mathematical investigation, for an unknown planet. It also appears that another astronomer, Mr. Adams, had undertaken the same task as Le Verrier, each being ignorant of the others labour. Now for the "discovery."

On the night of the 23rd of September, the sky being clear, a telescope was pointed in accordance with Le Verrier's instructions. The field of view showed a multitude of stars. One of these was really the planet Neptune. The next night the object was again observed. "It had moved, and when its motion was measured it was found to accord precisely with what Le Verrier had foretold. Indeed, as if no circumstance in the confirmation should be wanting, the diameter of the planet, as measured by the micrometers at Berlin, proved to be practically coincident with that anticipated by Le Verrier."

"The world speedily rang with the news of this splendid achievement. Instantly the name of Le Verrier rose to a pinnacle hardly surpassed by that of any astronomer of any age or country. The circumstances of the discovery were highly dramatic. We picture the great astronomer buried in profound meditation for many months; his eyes are bent, not on the stars, but on his calculations.

No telescope is in his hand; the human intellect is the instrument he alone uses. With patient labour, guided by consummate mathematical skill, he manipulates his columns of figures. He attempts one solution after another. In each he learns something to avoid; by each he obtains some light to guide him in his future labours. At length he begins to see harmony in those results, when before there was discord. Gradually the clouds disperse, and he discerns with a certainty little short of actual vision, the planet glittering in the far depths of space. He rises from his desk and invokes the aid of a practical astronomer; and lo! there is the planet in the indicated spot. The annals of science present no such spectacle as this. It was the most triumphant proof of the law of universal gravitation."—Sir R. Ball.

The joyful bells of the scientific world, however, soon stopped ringing. The above "splendid achievement," "the most triumphant proof of the law of universal gravitation," has been weighed in the balances and found wanting.

Mr. Babinet, September 15th, 1848, read a paper before the French Academy of Sciences, as follows:—

"The only sittings of the Academy of late in which there was anything worth recording, and even this was not of a practical character, were those of the 29th, and the 11th. On the former day M. Babinet made a communication respecting the planet Neptune, which has been generally called M. Le Verrier's planet, the discovery of it having, as it was said, been made by him from theoretical deductions which astonished and delighted the scientific public. What M. Le Verrier had inferred from the action on other planets of some body which ought to exist was verified—at least, so it was thought at the time—by actual vision. Neptune was actually seen by other astronomers, and the honour of the theorist obtained additional lustre. But it appears from a communication of M. Babinet, that this is not the planet of Le Verrier. He had placed his planet at a distance from the sun equal to thirty-six times the limit of the terrestrial orbit. Neptune revolves at a distance equal to thirty times of these limits, which makes a difference of nearly TWO HUNDRED MILLIONS OF LEAGUES! Le Verrier had assigned to his planet a body equal to thirty-eight times that of the earth; Neptune has only ONE THIRD of this volume! M. Le Verrier had stated the revolution of his planet round the sun to take place in two hundred and seventeen years; Neptune performs its revolutions in one hundred and sixty-six years! Thus, then, Neptune is not M. Le Verrier's planet, and all his theory as regards that planet falls to the ground! M. Le Verrier may find another planet, but it will not answer the calculations which he made for Neptune.

"In the sitting of the 14th, M. Le Verrier noticed the communication of M. Babinet, and to a great extent admitted his own error. He complained, indeed, that much of what he said was taken in too absolute a sense, but he evinces much more candour than might have been expected from a disappointed explorer. M. Le

Verrier may console himself with the reflection that if he has not been so successful as he thought he had been, others might have been equally unsuccessful; and as he has still before him an immense field for the exercise of observation and calculation, we may hope that he will soon make some discovery which will remove the vexation of his present disappointment."—" Times " Newspaper, Sept. 18th, 1848. "Cosmos," by Humboldt; and "Earth not a Globe," by "Parallax."

It must not be supposed that Neptune was never observed until the time of the above recorded "discovery." Several instances have been discovered of Neptune being noted, and marked as a star on the catalogues of earlier astronomers. On May 8th and 10th, 1795, Lalande observed the same star.

Even supposing that Le Verrier had fully proved his case, it would neither have proved the theory of gravitation true, nor the tremendous distances of the stars and their gigantic sizes as postulated. The perturbations of Uranus were more likely caused by the known powers of magnetism and electricity, for we must not lose sight of the fact that there is every reason to suppose and believe that the sun is the seat of electrical phenomena. As to the distances and size of the stars we shall have more to say later on, but with assurance we say now that there is not an astronomer who knows the distance or size of any one of them. One is led to believe that the star distances and magnitudes are calculated according to the method John Wesley suggested was employed—"distance proves the magnitude, and the great magnitude proves the tremendous distance."

ECLIPSES.

It is often asserted that "the globular theory must be true because astronomers can predict eclipses most accurately." If the capability of predicting eclipses is going to determine the truth or otherwise of any world system, we should get a confused medley of "true systems!" for all theories regarding the order of the universe claim the power to foretell eclipses, one as accurately as the other. It should be recognised that practical astronomy—a science of observation, for the study and development of which the Greenwich Observatory was established—is independent of any, and every theory. Eclipses are not timed by any calculation concerning the rate or distance at which the earth be supposed to fly round the sun and the moon round the earth, or by the rate at which the moon and the sun travel over the earth. The calculations necessary to locate future eclipses are based upon the records of past observances of these periodically recurring—phenomena.

Eclipses occur in cycles. An eclipse of the moon occurs again after a cycle of, practically, 18 years 10⅓ days. If all the eclipses are observed in this period it would be possible to foretell all future ones; a certain amount of mathematical skill, of course, is necessary. It was by this rule that ancient astronomers accurately predicted eclipses.

Thales, who lived 600 years before the birth of Christ, predicted eclipses. Ptolemy also foretold eclipses for hundreds of years to come. Egyptian, Hindoo, and Chinese astronomers of ancient times foretold eclipses. A. McInnes, in his work on "Pagan Astronomy," says:—

"More than 2,000 years ago the Chaldeans presented to Alexander the Great, at Babylon, tables of eclipses for 1,993 years; and the ancient Greeks made use of the cycle of 18 years 11 days, the interval between two consecutive eclipses of the same dimensions . . . . . Mere theorising about the sun and moon—the great unerring clocks of time—has thrown chronology and the calendar into confusion, and hence scientists cannot agree as to the world's age."

If the facts already given are not sufficient to convince the reader that the "globular theory" has nothing whatever to do with the accuracy of eclipse occurrences, the following, from "New Principia," by Morrison, F.A.S.L., R.N., may effectually convince the student:—

"Eclipses, occultations, the position of the planets, the motion of the 'fixed stars,' the whole of practical navigation, the grand phenomena of the course of the sun, and the return of comets, may all and every one of them be as accurately, nay, more accurately, known without the farrago of mystery the mathematicians have adopted to throw dust in the eyes of the people, and to claim honours to which they have no just title .... The public, generally, believe that the longitudes of the heavenly bodies are calculated on the principles of Newton's laws. Nothing could be more false."

How are eclipses of the moon caused? Our astronomers of the " globular theory " school tell us that a times comes when the earth lies directly between the moon and the sun; the moon is thus plunged into " the shadow " of the earth, the light from the sun that the moon is supposed to reflect is intercepted, and the moon is eclipsed. This is very remarkable, and I doubt its possibility, considering that even in the depth of a total eclipse the moon remains visible, and actually glows with a bright copper coloured hue; but there are even greater objections than this against " the shadow " theory.

Now, according to the " globular theory," a lunar eclipse occurs when the sun, earth, and moon are in a direct line; but it is on record that since about the 15th century over 50 eclipses have occurred while both sun and moon have been visible above the horizon. The accompanying illustration will show how utterly impossible it is to harmonise this fact with even the globularists' own theory.

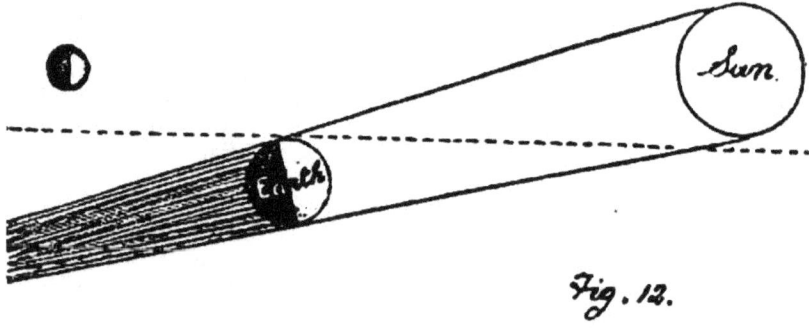

Fig. 12.

The horizon to an observer on the earth would be at right angles from a perpendicular line where he stood, and above this horizon— overhead—was the sun and moon visible—the moon eclipsed.

One may read carefully a whole host of " scientific " books before finding the information that there have been several instances of lunar eclipses being seen with both sun and moon above the horizon —why this silence?

Writing to the Astronomer Royal on this subject I was informed that the above phenomenon was caused by " refraction," which caused the sun and moon to appear " above the horizon when wholly below it." Ah! of course! Professor Airy once said: " One of the most troublesome things an astronomer has to deal with is refraction." But it seems a bit convenient at times! Is the phenomenon of a lunar eclipse with both sun and moon visible above the horizon due to refraction? Let us consider the position.

First, the globularists admit that the facts observed in Nature relative to this case do not agree with their globular theory; that is a proper admission to make. A theory concerning the operation of "refraction," instead of clearing up the difficulty, really adds to the dilemma.

What has "refraction" to do in the matter? The moon has visibly risen, and the sun has not yet set, in accordance with accurate almanac time, and an eclipse of the moon is due, and takes place through "refraction." It must be "refraction," astronomers say so. We will now deal with "refraction."

Refraction only operates when our line of sight, or a ray of light, passes from one medium into another of different density. Get a basin; place it where a light causes part of the rim to cast a shadow into the bowl; place, say, a penholder obliquely in the basin and then pour in some water, and you will see that refraction will apparently raise the immersed part of the penholder, while the shadow will go back and down. Now apply the experimental knowledge thus gained to the theory before us. If refraction did throw up the sun and moon, then refraction would throw "the shadow" further down away from the moon—and there could not be an eclipse. And so it is impossible for astronomers to prove our earth—*terra firma*—to be a heavenly body, whirling and spinning between the two luminaries, the sun and moon.

They say: "The shadow of the earth on the moon proves the world a globe." Oh! How are we to know that it is the shadow of the earth? Is there any special way of identifying it? Might it not be "the shadow" of some other moving dark body? It is supposed that the earth is a globe because the shadow on the moon is curved; but it is not only a globe that can cast a circular shadow on a sphere; experiment with an orange, a cube, and a lighted candle, in a dark room, and whatever "shadow" that is cast on the orange by the cube will be curved—how could it be otherwise?

Eclipses of the moon may be caused in several ways. I do not profess to know how they are produced, for I believe, as it says in Ecclesiastes, 8th chap., that "a man cannot find out all the work that is done under the sun: because though a man labour to seek it out, yet shall he not find it; yea, further, though a wise man think to know it, yet shall he not be able to find it." We cannot know all the works of God; we know that there are "dark bodies" in the sky; the moon may be eclipsed by the periodical motion of one of these.

Eclipses of the sun and moon, or any celestial phenomena, cannot prove the earth to be a globe, or even flat. It is most illogical to search the sky for "proof" as to the shape of the earth. It is quite possible to determine the figure of the earth while we are on it; having done this, all that occurs in the sky must be explained—if explained at all—in harmony with the ascertained fact that the earth is a plane.

THE SUN'S DISTANCE.

Sir Richard Proctor, in his work on "The Sun," informs us that "the determination of the sun's distance is not only an important problem of general astronomy, but it may be regarded AS THE VERY FOUNDATION OF ALL OUR RESEARCHES." So it is the foundation of all their researches, in fact, THE ASSUMED DISTANCE OF THE SUN FROM THE EARTH IS THE "MEASURING ROD," USED BY THE ASTRONOMER TO DETERMINE ALL OTHER DISTANCES. What is the length of this measuring rod? What kind of "foundation" are the researches of the astronomer built upon?—let us see.

Sir Robert Ball informs us that "the dimensions of our luminary are commensurate with his importance. Astronomers have succeeded in the difficult task of ascertaining the exact figures, but they are so gigantic that the results are hard to realise." He says: "The *actual* distance of the sun from the earth is about 92,900,000 miles." Fancy the "actual distance" being "about"! No doubt it is advisable to have a saving clause "about" in the "exact science" of astronomy, for Professors Airy and Stone gave the distance of the sun from the earth as 91,400,000 miles. Evidently THEY made a slight mistake in a few millions, because Encke knew the distance to be 95,000,000 miles —in winter, and a few millions less in summer. This is not an exact scientific statement, considering that it is summer in one part of the earth, and that it is winter in another part of the earth at the same time. But this is a scientific trifle; for after all, what is a matter of 2,000,000 miles in 95,000,000? When we come to Copernicus we find him stating that the distance of the sun from the earth to be "3,000,000"—what!—your book says, "5,000,000 miles"—Oh, yes, it is all right, the 3,000,000 miles was an earlier guess—I beg pardon—calculation.

The ideas of ancient astronomers as to the distance of the sun from the earth were not quite so great as the ideas of modern astronomers, although, no doubt, they considered themselves quite as accurate as do modern astronomers in their statements of "actual distance." Pythagoras gave as his estimate of the sun's distance from the earth a matter of "44,000 miles." However, he was wrong, right enough, for Tycho Brahe, and others, knew the distance of the sun to be about 13,000,000 miles above the earth." Some time afterward it was shown that even Tycho Brahe was a few millions of miles out, and his "observations" must have led Kepler millions of miles astray, for in 1670, Cassini demonstrated, in the usual way of astronomers, that the distance of the sun from the earth was "85,000,000 miles." No doubt he did his best; but of what avail were his efforts when Sir I. Newton afterwards gave the distance as "28,000,000 miles," or "54,000,000"; no need to be particular, for Sir I. Newton said, "either distance would do very well." I am sorry to say it, but I am afraid Newton was forgetful or ungrateful, as the basis of his labours were the laws of Kepler; but he totally ignored the distance of the sun from the earth according to Kepler's law—" 12,376,880 miles."

Mayer gives the sun's distance as over 104,000,000 miles. One of the latest globular theories, "Koreshan Astronomy," which claims to interpret all ancient legends, and mythologies, and to furnish the basis of all reason and science, emphatically states that the distance of the sun from the earth is about "4,000 miles."

Some say it is 96,000,000 of miles. I do not give all the authorities with their "actual" and "about" distances; but, according to the "globular theory," the distance of the sun from the earth may be ANYTHING BETWEEN "4,000 MILES" AND "104,000,000," this represents the astronomical "measuring rod"—"the foundation of all our researches!"

With the above futile results of attempts to ascertain the distance of the sun before us, a thoughtful consideration of the following Scripture may serve a useful purpose.

"Thus saith Jehovah, which giveth the sun for a light by day, and the ordinances of the moon and of the stars for a light by night, which divideth the sea when the waves thereof roar; the Jehovah of Hosts is His name: If these ordinances depart from before me, saith the Lord, then the seed of Israel also shall cease from being a Nation before me for ever. Thus saith Jehovah:—"IF HEAVEN ABOVE CAN BE MEASURED, AND THE FOUNDATION OF THE EARTH SEARCHED OUT BENEATH. I will also cast off all the seed of Israel for all that they have done."

"The heaven for height, and the earth for depth . . . . there is no searching."

Astronomers, with their various ideas concerning the sun's distance, speak concerning millions of miles as though they were but inches. In fact, it is as Joyce, in his "Scientific Dialogues," says:—

"We talk of millions, with as much ease as of hundreds or tens, but it is not, perhaps, possible for the mind to form any adequate conception of such high numbers. Several methods have been adopted to assist the mind in comprehending these vast distances." You have some idea of the swiftness with which a cannon ball proceeds from the mouth of a gun—at the rate of about 8 miles in a minute. The numbers of minutes in a year is 525,600, so it would take a cannon ball travelling at the rate of 8 miles a minute, 22 years to reach the sun from the earth!

The "exact figures" concerning the size of the sun are as various and as unreliable as the distances given. We are informed that the sun is "more than a million of times larger than the earth;" with a diameter variously estimated by modern astronomers in harmony with their different ideas as to its distance. According to Russell, the diameter of the sun is 882,000 miles; but Giberne says it has a diameter of 850,000 miles and Sir Robert Ball, of "exact" figure fame, has found the diameter to be 866,000 miles.

When this gigantic sun is considered one really must wonder where the supply of fuel is obtained from, to maintain the great heat it must have. It is also curious to note (supposing, according to modern astronomy, that all the heat we get actually comes from the sun) that the nearer we get to the sun the greater is the cold. On lofty mountains, even under the equator, are to be found never melting snows; and at sea level the hottest parts of the earth are not under the equator (which, of course, is supposed to be nearest the sun), but are places some degrees—about 10—north and south of the equator. Our astronomers seem to think that their explanations concerning the earth and celestial phenomena are almost complete. They may yet learn that "there is more in heaven and in earth than is dreamt of in their philosophy."

Job said, in enumerating what he knew to be idolatry and sin against God:—

"If I beheld the sun when it shined, or the moon walking in brightness, and my heart had been secretly enticed, or my mouth had kissed my hand: This were also an iniquity to be punished by the Judge: for I should have denied the God that is above."

It is well-known that sun worship was practised at the time in which Job lived; and many years after his time we find God's prophets lamenting the fact that those who by their knowledge ought to have done better, were worshipping the sun, moon, and all the host of heaven. It will no doubt come as a revelation to many to learn that heliolatry, or sun worship, is still practised, and in this country. It is not really to be wondered at, when such men as Sir Robert Ball say:—

"For the power to live and move, for the plenty with which we are surrounded, for the beauty with which Nature is adorned, we are immediately indebted to one body in the countless host of space, and that body is the sun."

How different is this from the words of St. Paul:—

"God that made the world and all things therein, seeing that He is Lord of heaven and earth . . . . . giveth to all life and breath, and all things . . . . . He is not far from everyone of us, for in Him we live, and move and have our being."—Acts. 17.

But are we indebted so much to the sun as Sir Robert Ball supposes? I think not. All the heavenly bodies have their God-given functions. In Deuteronomy, 33rd chap., we read that there are "precious fruits put forth by the sun, and precious things put forth by the moon." In Job we read:—

"Canst thou bind the sweet influences of the Pleiades (the seven stars): Or loose the bands of Orion? Canst thou lead forth the Mazzaroth (the signs of the Zodiac) in their season, or canst thou guide the Bear with her train?" I know that very few believe that the moon, stars, and "planets" have much influence in the affairs of this earth, but Holy Scripture teaches that they do have their *role* to fulfil, as in the 19th Psalm:—

"The heavens declare the glory of God; and the firmament showeth His handiwork. Day unto day uttereth speech, and night unto night showeth knowledge. There is no speech or language where their voice is not heard. THEIR RULE (or direction) IS GONE OUT THROUGH ALL THE EARTH, and their words to the ends of the world."

For the benefit of those who have "faith" in the opinion of scientific men in preference to the Bible, I may mention that Kepler, one of the "greatest astronomers," believed in the influence of the stars; and so did Flamsteed, our first Astronomer Royal.

"CUI BONO."

"Of all terrors to the generous soul, that *cui bono* is the one to be most zealously avoided. Whether it be proposed to find the magnetic point, or seek a north sea passage impossible to be utilised if discovered; or a race of men of no good to any human institution extant, and of no good to themselves! or to seek the unicorn in Madagascar, and when we have found him, not to be able to make use of him; or the Great Central Plateau of Australia, where no one could live for centuries to come; or the great African lake, for all the good it would do us English folk, might as well be in the moon; or the source of the Nile, the triumphant discovery of which would neither lower the rents nor take the taxes off anywhere—whatever it is, the *cui bono* is always a weak and cowardly (?) argument; essentially shortsighted, too; seeing that, according to the law of the past, by which we may always safely predicate the future, so much falls into the hands of the seeker for which he was not looking, and of which he never even knew the existence. The area of the possible is still very wide, and very insignificant and minute is the angle we have staked out and marked impossible. What do we know of the powers that Nature has yet in store; of the secrets she has yet awaiting discovery, and the wealth concealed? Quixotism is a folly when the energy which might have achieved conquests over misery and wrong, if rightly applied, is wasted in fighting windmills; but to forego any great enterprise for fear of ridicule and the dangers attending it; or to check a grand endeavour by the *cui bono* of ignorance, stupidity, and moral scepticism, is worse than a folly—it is baseness, and cowardliness."

A well-known infidel has written:—

"In every Christian country the masses of the people are taught in childhood that God created the universe in six days and rested on the seventh. Yet every student knows this is utterly false, every man of science regards it as absurd, and the more educated clergy are beginning to explain it away."

Though this is not exactly true it is rather near to it. Now the truth of Christianity is called in question to-day, as in old time, by the wisdom of this world—"science." I would ask my readers after having read through this work, before drawing any hasty conclusion for or against the arguments herein, to carefully consider the fact that they have been trained, perhaps from early childhood, to believe the "globular theory" true. It is a compulsory subject at school, and the plane earth teaching is never referred to, except perhaps in derogatory terms by teachers who could not give a lucid exposition of our standpoint. Considering also that such men as our titled astronomers go out of their way to inform the public that "it is only the untutored mind that believes the earth to be flat," it is not to be wondered that so many people, consciously or unconsciously, are prejudiced against any teaching not in harmony with the "globular theory"

Infidels and astronomers, who say that no scientific men believe the Story of Creation, narrated in Genesis, to be accurate, make a great mistake. However, the only appeal I make is, that one and all will judge this little book on its merits, using their natural sense in its study; also doing as the Scripture commands:—

"PROVE ALL THINGS, HOLD FAST THAT WHICH IS GOOD."

SECOND EDITION

THE ENLIGHTENMENT
of
THE WORLD

By JOHN G. ABIZAID

This book contains proof that the earth is flat and stationary, while the sun, moon and stars are in constant motion.
Also letters showing the testimony of Scripture on the subject.

1912

SECOND EDITION

REVISED AND ENLARGED

THE ENLIGHTENMENT

— OF —

THE WORLD

BY

JOHN G ABIZAID

This book has, in my opinion, good proofs to show that the earth is flat and always remains stationary, while the sun, moon and stars are always in motion. It also contains some testimonials on the subject. JOHN G. ABIZAID.

Boston, Massachusetts. May 13, 1910.

48 Shawmut Ave., Boston, Mass.,
June 6, 1912.

John G. Abizaid,
   121 Tyler Street.

Dear Sir:

I have read your book on the "Flat Earth," and am quite pleased with it. It would be a good book for school children and for men and women of all ages who are ready and willing to think for themselves. I do not see for a moment how any man or woman can be intelligent and at the same time honest and believe that the earth is round and not flat—or as we see it. Any and all expanse of water cannot be other than flat upon its surface, and all land has but one surface, and difference only according to its various heights from the sea level.

That the sun is very high, small in comparison to the earth is certainly true.

The earth is too large and dense to move in the least, while the sun is always moving. For to teach otherwise is most absurd.

I shall be pleased to aid you in any way I can to unfound the falsity of a round and moving earth.

                                  Yours sincerely,
                                       WILLIAM PEIL.

THE ENLIGHTENMENT OF THE WORLD.

This book contains proof that the earth is flat and stationary, while the sun, moon and stars are in constant motion.

Also letters showing the testimony of Scripture on the subject.

By JOHN G. ABIZAID.

Boston, Mass., U. S. A.

PREFACE TO THE SECOND EDITION.

This book has common words and phrases, so that everyone may understand its meaning.

I wrote it first in Arabic, then translated it into English, because nearly all Americans can read and write their own language.

My reason for printing this book is not to make money or fame, but because I wish to show people who differ from my views where their mistakes lie.

Everything in this book is true. What I think on the subject is not guess-work. You will find in it a good foundation for belief, good ideas, and absolute proofs. You do not need to accept it merely on faith, but you may see with your own eyes and feel with your own bodies. Thus you may know it is the truth; and as long as you have eyes to see and a body to feel and brains to think with, use them and you will know the facts.

Christians believe the Bible. In this Book of Holy Writ are proofs given by the prophets that the world is flat and stationary, and that the sun, moon and stars are always in motion.

There are many, however, who do not accept the truths of the Bible, and for this reason I give other proofs.

There are some things that need witnesses, but other truths are their own witnesses; they bear the proof in themselves.

CHAPTER I.
THE WATER.

You know that there is more water on the earth than dry land. Here is the first proof:

The water proves that the earth is flat, level and stationary. Water is liquid. It runs always down and seeks its level, and never runs up unless by power. And the water of the ocean is every way pretty near joined together on the earth, and they call it liquid. It will not stay on a round earth. Take a glass of water and pour it on a round ball and if the water stays on, then the earth must be round, but if the water falls off then the earth must be flat.

This is the proof: Take a pan and fill it with water and see if it is higher in the center than on the sides. If this is true, then the water of the ocean might be round, but if the water in the pan is flat and level, then the water of the ocean must also be flat and level.

The water of the ocean is level and can be nothing else, because it is liquid, which proves that the land is flat and not round. The land is not exactly flat and level, for there are mountains and valleys. But it makes no difference, for if the mountains were in the valleys it would leave the land flat and level and higher than the water. I have been thinking of the question for a long time.

At last I have found out for myself from the water and the sunbeams, the sun's rising, the sunlight, etc., that the earth cannot be round and in motion, as books and teachers have taught. The teachers offer proofs to show that the earth is round, but the proofs they offer amount to nothing. When the children go to school, the teacher tells them the world is round, and, of course, they believe it, and they do not ask how it is round. They are young and they know very little. When they grow up they still believe the earth is round and in motion, and so in turn they teach this to others.

The water is flat, which proves that the earth is flat and stationary.

From its fruit you will know what kind of a tree it is.

You can tell by looking at a building whether it has a good foundation or not.

A judge will not take witnesses who are far away, but those that are near him, so that he can speak to them and examine the case and find out the truth.

The proofs about the earth are found the same way. The nearer the proofs are to the mind the better we can understand them. The proofs in this book are very good and can be understood by anyone. Don't pay attention to a far away proof. It may not be the truth.

The sun cannot throw the earth's shadow on the moon, because sometimes they are very near to each other. This you can tell in the evening and in the morning by seeing the moon and the sun at the same time.

You ought not to believe everything the professors say, because they do not know everything.

The man who said that the earth was proved round by the sight of a row of ships upon the ocean made a mistake, and what he says is not true.

Just because he measured and looked at the subject in his own way, he has not proved it but guessed at it. From his guess he thought the land and water are round; but the water will not stay round, as he thinks.

They claim in the geographies that when the last ship is out of sight, it is a proof that the water is round, but it is a proof that the water is flat. You will see if you look and measure in a good many different ways, as is done in this book.

If he was wise enough he would measure the steamers from different points and see that it always comes to the same thing, like the Fifteen Puzzle which will count fifteen when added in any direction.

Try and insert the figures 1 to 9, inclusive, so they will total 15, adding across, up and down, and diagonally. The key to the puzzle is to have the right figure in the center.

If you take the five out of the center you cannot add them fifteen every way; try and see.

In this way, measuring the steamers from different places proves that the earth is flat, and if you measure from many places you can prove nothing else.

It is the same way with the steamers. If you don't leave one in the center you will think the water of the ocean is round. But if you try this proof of the steamers you will know the water of the ocean is level and flat.

A person standing on the shore sees a steamer full size when it is near him, but when it goes out farther it keeps growing smaller until it is out of sight. They think this is good proof that the water of the ocean is round, but it is not a good proof, because water is liquid and will not stay round. The distant view shows you that. It is the same as with a book; you cannot read it five yards away from your face, but near your face you can read it plainly.

A person standing on the seashore will see a large steamer, but as it goes out farther it grows smaller, and the mast of the steamer will be shorter, too; but if a person stood on the opposite shore he would see the steamer large when near him and as it goes out farther it grows smaller, which gives proof that the water is flat and level, and not round: the far distance will show you that.

There is no chance for anybody to say that the earth is round, for as long as the water is flat and level and straight, it proves that the land can be nothing but level and stationary, and the sun in motion.

I have read books on geography, and I find they claim that the earth is round and in motion.

I have been thinking over this matter, and have read considerable in regard to the subject.

I do not blame the children or teachers, but the first man who said the earth is round. He told the people the earth was round, but did not know for sure; he only guessed at it. We do not have to believe such guess-work as long as we have eyes to see and bodies to feel and brains to think.

They say the earth is round, *both* land and water, *and that they are always* moving around the sun. But if you ask them how the water keeps in its bed while the earth is turning, they will tell you to take a pail of water and swing it fast and see, for the water will not come out until the pail is slowed down.

Who has seen the earth turn around in the way that a person turns a pail? No one. And as long as no one has ever seen the earth turn, you need not believe that it does so.

Who is turning the earth? Is it turning on its axis and what kind of an axis is it? No one knows. Is it iron, steel, wood or tin? You do not want to believe such statements that the earth is standing on an axis, for no one has seen it. It is all guess-work and nothing seen by human beings. Some people say the world is round and that God is turning the world, but very few believe it. If they do they should pay attention to his prophecies, and ought to believe them as they say about the movement of the sun and standing of the earth, and so forth. If you wish to know what the earth stands on, I can say, that, and I believe what I say, the earth rests on water just like a ship floats on the water. This is a good proof without doubt. I know the earth is large, and being so tremendous large, must necessarily be of tremendous weight. And a heavy body, it has no power to move in the sky around the sun anywhere. Also it cannot lay or turn on nothing as they think. Try something and let us see if you can make anything stand or turn daily on nothing itself. I am sure you cannot do that. Also they have no right proof to it as they believe. The swing of the pail is not a correct proof either, because the man swings the pail. But the pail cannot swing itself because it has no power. Also the earth cannot move on nothing because it has no power to move.

Try and put some water on the outside of a pail, and I am sure you will see the water on the ground before you begin to swing the pail. The water of the ocean is the same thing, for if the earth is round, as they say, there would be no water on the earth, as all of it would fall off.

If the earth is round and revolving we should feel it; also the wind would be coming from one direction only, and not from north, south, east and west.

I have read a story in the geography about the earth being round, as they claim. It says when the ship is near the shore a person on shore will see the mast, high and big, but as the ship goes further out, he will see it low and smaller as if it were going down hill. If a man were standing on the steamer deck, however, he would see the land low, while he would be high.

If there were any high mountains near the shore, and you were on a steamer some distance from the coast, when you looked toward the land, the mountains would seem to you to be low and small, but they are not. It is the distance that makes them seem so.

The water is flat and level always and cannot be anything else because it is liquid. You know, of course, what liquid is, and what the word means.

Here is another proof to show that the water of the ocean is flat and level: —

Suppose five steamers are in a circle, each just within eyesight of the others, with another steamer in the middle. If you are in the middle steamer looking out at the others, they will look as if they were very low and in a hollow, while you seem to be the highest. But if you were to leave your own steamer and go in turn to all of the others in the circle it would then look to you as if the steamer you were in first were very low.

Whichever steamer you are on seems to you at the time to be the highest. It is not true that one steamer is higher than the other, or that the water of the ocean is round. It is flat, and nearly always straight. It is the distance that makes the other seem lower.

Here is still another proof: If these steamers are in a straight line, on the ocean, as far apart as eye can reach, and you are in the middle one looking at the others on either side, — then they will look as if they were in a hollow, while you seem to be the highest; but it is true that to the people on the other steamers it looks as if your ship were low in the water and their own the highest.

Do you think there are hills and valleys in the ocean? No! No! This cannot be true, because the water is liquid, and therefore must be level.

These are proofs to show that the water is not round as the geography says, but that it is really flat and level.

I think I have given enough proofs to change your minds about the earth being flat, not round.

The water proves that the land is stationary, flat and level.

CHAPTER II.

THE LAND.

The land is solid and all in one piece, joined together under the ocean all over the earth.

The earth cannot be round and in motion as they claim in the geographies, but flat and stationary, because the water is liquid and level and it proves that the land is flat and not moving.

Another proof which shows that the earth is standing still and not moving is to be found in the wind, which they call the atmosphere.

If the earth is turning around daily it will give wind from one direction only.

If you see a train running you will find out that it is giving wind in one direction. But when it stops the wind stops with it.

Nobody ever felt the wind blowing at the same rate daily, but they will feel the wind blowing at times very hard and at other times very gently, and often from different directions, and sometimes there is a very little wind. This is good proof to show that the earth is stationary. If the earth was in motion in one direction, it would make the wind pressure in the same direction always.

You will find my proofs right if you will only pause and think them over.

This is another proof to show you that the earth is not moving:

If, when you go up in a balloon or airship, starting from a certain place and going as high as you can and when you come down find that place gone from under you, then you may believe that the earth has moved; but if you find yourself in the same place, or not so very far away from your starting point, then it is plain that the earth is stationary and not moving at all, and this is what really happens.

If, moreover, the earth is turning around and the birds or airships are flying the same way, they will find themselves over the same place they were before. If they fly in the opposite direction of the turning of the earth, they will find themselves flying very fast and always over different places. If you think this over you will find these proofs correct.

After all, the earth cannot be round, because most of it is water, and water cannot be anything but level, as I have given proofs in Chapter I.

Take a pan of water and turn it upside down, and I am sure the water will fall out. Also, if the earth turns around, the water of the ocean would destroy a great many cities, towns and countries. But the earth cannot be turning or in motion. If it is turning, we should know it immediately, because we have eyes to see, bodies to feel with, and brains to think with.

Suppose you were riding on a train, car or boat, you would know that you were on something that is shaking. When the train or car goes up hill or down hill, you can easily tell, even with your eyes closed whether it is going up or down.

It is the same way with the earth. If the earth is turning around or is in motion, we would know it from many different things, as I said before.

They say that there is gravity in the earth which holds everything on it when it turns around. They give these proofs to the people and children in school, and most of them believe such foolish things as that. Their proofs do not amount to anything, but it is all guess-work, because nobody ever felt it or saw it.

I was surprised to know that some people believe in things that were never seen or felt. But the things they can see and feel they pay no attention to. They say if you drop anything it will fall down to the ground, they only think the law of gravitation takes it down.

Oh, I am sorry for them! They ought to know better than that. They ought to know that heavy things, and all things that have no power like dead bodies, will go down themselves. But some things have power, like the bird, or light things like smoke, and so forth; they will go up, or any way they like. If there was a law of gravity it would bring them down also. Most of the people know up from down, and ought not to believe this foolish thing. I feel sorry that the children are taught about gravity, and that the earth is round and travelling around the sun, and also that the sun is larger than the earth.

There is no gravity at all, for if there was, a person could not move a step, and if a bird was on the ground it could not fly in the air again, because the gravity would hold it back. If there is gravity, it is not everywhere, and cannot hold everything, such as gas, balloons, airships, etc.

Here is another proof to show that the earth is flat and stationary: —

If the earth is round and in motion, as they claim, we should know, for when it turns around, at times we should find our heads and feet and the world above us, as though a person were standing on a ceiling with his head downwards and his feet up.

If we were held fast to the earth by gravity, or tied by ropes, we should feel it, and know which way the earth is turning with us.

But there is no such thing as gravity, because no one ever felt it. You must not think that you are an intelligent being because you know a great many things. I am sure that no person knows everything in the world. You ought not to believe everything that people say — that the earth is round and in motion, and that the sun is stationary and larger than the earth, etc., until you have examined the facts and found out the truth and follow it.

I think I have given you enough proofs to change your minds about the earth's being round and in motion.

You will find in Chapter III proofs about the sun going around in a circle above the earth.

CHAPTER III.

THE SUN, MOON AND STARS.

In this chapter you will find proofs to show that the sun, moon and stars are always in motion.

I have proved in Chapter I and II that the earth is flat and stationary.

As long as the earth is flat and stationary, it will prove that the sun is in motion over the earth, and traveling in a circle, lighting up all the countries that it shines upon. It cannot light up all the world at once, because the earth is larger than the sun.

People think the sun is large because they see it through a spy-glass. The spy-glass, you know, makes everything seen through it look large, even when it is not large, and you can see the sun round and move, circling above the earth; you will know that without any proof.

Do not believe that the sun is stationary and larger than the earth. For it cannot be larger than the earth. (See illustration of sun rising and sunbeams.)

If the sun is larger than the earth, it would light up all the world at once.

Think this over and you will find that what I say is correct and the proofs are correct also.

If you are in a room with windows towards the rising sun, you will find that the sun always throws its rays down near the floor and not on the ceiling.

I am sure that the rays of the sun will never touch the ceiling either morning or evening. It makes no difference if you have your room on a high mountain, the sun will be higher. Its own light proves that for itself.

This will prove that the sun is high, but if you see the sunlight on the ceiling, that will mean that the sun is rising from a low place, or that the earth is round and in motion. But if you cannot see the sunlight on the level ceiling, that will prove that the sun is always high, and the earth flat and stationary.

The sun is always high and turning in a circle above the earth.

In winter the sun goes far from us, which makes the days short and cold, but in summer when it comes back near us, it makes the days long and warm.

You will know when the sun is near you and when it is far from you, from its light; also you can tell which way it is turning in a circle above the earth from its rising and from its light, and also you know that by observation. And if you examine the sunlight you will find the sun is traveling in a circle.

Some people say that if they start from a certain place and go around the earth, that they would pass under the earth and come back at the same place they started from. They think that the earth is round like an orange; also that they have passed under the earth. The earth cannot be round, for I have given a great many proofs to show that it is flat and level.

They follow the compass, and the compass always points to the middle of the earth which we call north. They travel in a circle on a level place near the equator and they think they are going around the earth in a circle, but they are really traveling on flat and level ways. It cannot be any other way. If they started on a steamer and went south without a compass, they would get lost and find nothing but water, ice and darkness.

Here is another good sign that shows that the sun is very high and smaller than the earth: —

When the sun is rising in the United States of America, it will look as though it is rising from a low place, but the people in Europe will see the sun very high over their heads about noon time, also the people in Asia Minor will see the sun very low in the afternoon.

Do not think that there is a hollow between Asia Minor and the United States because the sun looks low to the people in Asia and to us (U. S. A.), but high to the people in Europe? No! No! There is no hollow, but it is all level, for the great distance from here to the sun shows you that.

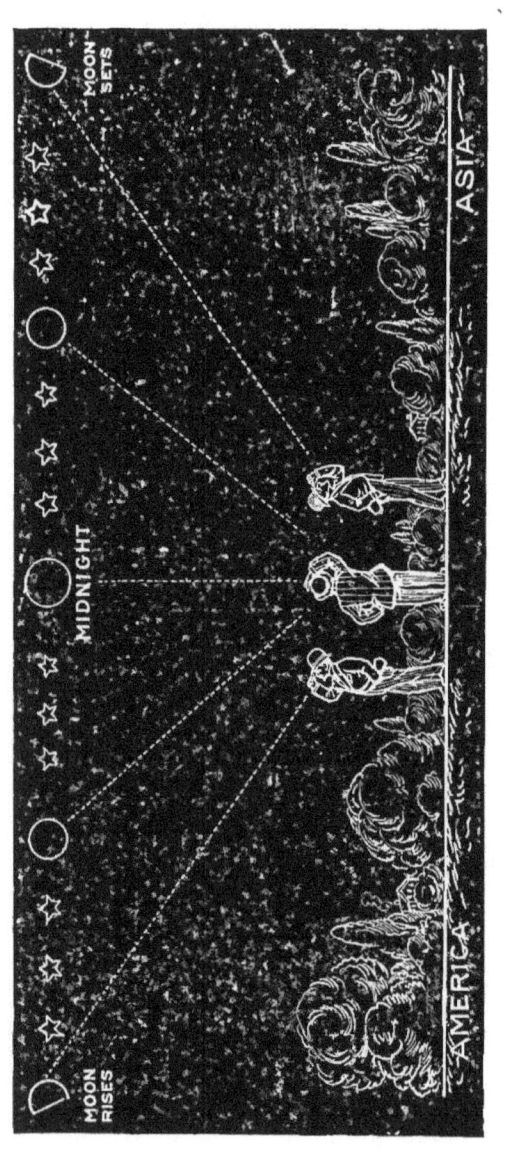

You will get good proofs if you look with me closely for the moon rising. The moon, of course, encircles the earth When the sun comes to here it will be day, and when the sun goes from here it will be night

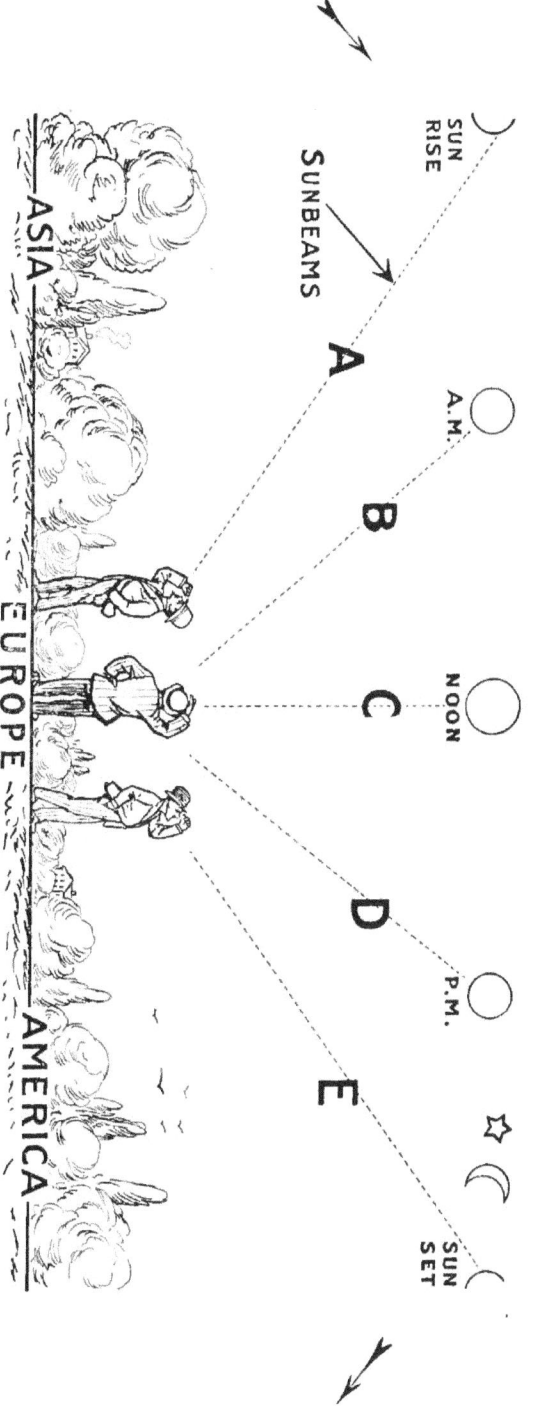

You will get good proofs if you look with me closely for the sun rising and sunbeams. The sun, of course, encircles the earth when the sun goes from here it will be night, and the place where it goes will be day

EXPLANATION OF THE CHART

The meaning of A, B, C, D and E on the lines of the sunbeams in the ilustration of the position of the sun.

The first line of the sunbeam, A, means when you first see the sun in the morning. It seems to you the sun is rising from a low place. If you examine any shadow or your own shadow, you will see the shadows long when the sun is coming from a great distance; but when the sun comes near you, your shadow will be short. Examine the sunbeams and your own shadow; you will soon know that the sun is round, high and in motion, smaller than the earth and moving in a circle above it.

B means in the A. M., when the sun has come nearer to you, and is shining more directly over you and you feel it warmer. Your shadow is shorter and is passing by you. You will see that this is correct if you will take time and watch it.

C means midday. At that time the sun is nearest to you. You will see it above your head, and you will feel warmer and your shadow will be shorter than at any other time.

D means in the P. M. You will see that your shadow is growing longer because the sun is moving from over you.

E means that the sun is going to a great distance from you. You feel cooler, and your shadow is growing longer and turning in a circle. If you measure the sunbeams and your shadow before sunset, you will see that the sun is always high and in motion, passing in a circle above the earth from one country to another, and also you will see the moon and stars before sunset and after.

In the illustration of the sun rising and the sunbeams, you will see that the sun is round and smaller than the earth. Also that it is moving in a circle, above every place reached by its light. When the sun leaves a place it changes from day to night, and in the place where it was night it will be day. The illustration shows this.

Pay attention to these proofs: the first shows that the sun is very high; the second shows that the sun is smaller than the earth, for if it is larger than the earth the people in Europe would see the sun above their heads, while the people in America would see it above their heads, and both at the same time; the third shows that the earth is flat and level.

This is another proof to show you that the sun is going around in circles above the earth: —

When the sun is rising you will never see it coming straight to you, but going in a circle always to the right above your head. You will find that from the shadow of your house, or the sunlight when it shines into your room, etc.

I know that the sun is smaller than the earth and in motion high in a circle above the earth. You will know that by looking at the sunbeams.

Here is a good proof about the sun that everybody can try at home: Place your lamp on the table and place your hand at the side of the lamp, and you will see the shadow of your hand on the wall, and if you place your hand over the lamp you will see the shadow of your hand on the ceiling, and if you place your hand below the lamp you will see the shadow of your hand on the floor. if you watch the sun beams you will know that the sun is smaller than the earth and always in motion in a circle high above the earth.

And after all, if you want to know which moves, — the earth or the sun, — you must take time to stop a while and watch them both.

Here is another proof to show you that the earth is flat and stationary, and the sun in motion: —

If you can see or feel the earth turning down on one side and up on the other, then you are right, and the earth is in motion and the sun stationary; but if you cannot see or feel this, then the earth is stationary and the sun is in motion; and if the earth is always level on every side of you, it means that the earth is flat.

If you will take time to think you will find out the truth for yourself, — when I have shown you the way. I do not want you to believe it because I say so, but I want to show you how things are and if you will pay attention you will know what is right as well as I do.

If the sun is stationary, we would feel the earth turning, from many different things.

First the houses and posts would lean to one side with the land, and we would see one direction low and the other direction high. If you see the land low on one side and high on the other side, then the sun is stationary, but if it is not, then the sun is in motion. If you cannot see or feel the earth turning, then it is stationary, and the sun is turning. You will know from the water of the ocean that the sun is moving above it, for if the earth is turning upside down, there would be no water in the ocean.

They say that the sun is bigger than the earth. But I found out from the sunbeams that the sun is smaller than the earth. If the sun was bigger than the earth, the people that live in Asia, the people that live in Europe, the people that live in Africa and in America would see the sun above their heads at the same time. There will be noontime all over, it will make no difference. If the earth were flat or round we would never see the sun over our heads when the people in other countries had noontime. Suppose there were ten people under one umbrella, the same size as one we use now, do you suppose they would all see the umbrella over their heads? No; they cannot see it above their heads because the umbrella is too small to cover them all; but if they have a big umbrella it will cover all ten people and all would see it above their heads. If there were an umbrella as big as a town it would cover all the people that live in a town, and all the people who live in the town would see it over their heads. And that is the way with the sun. If the sun were bigger than the earth it would shine on the whole earth, and all the people in the whole earth would see it over their heads at the same time. This is a good proof from the sunbeams and the umbrella, and you will find it a good proof that the sun is smaller than the earth and in motion, and the moon and stars are also smaller and in motion. The earth is flat and stationary, as I have proven it before.

The sun we can compare with a parasol — if the parasol was large enough it could cover the whole world — just so the sun. It is not large enough to throw light over the whole world at one time.

The Indian is looking at the proof of the parasol. He compares the sun with the parasol by saying that if the parasol were large enough, "I could see it here above my head, too." Also the same with the sun — if it were large enough it could cover the earth; and the people, no matter where they were located, could see it at the same time above their heads.

They used to say about 500 years ago that the earth was flat and stationary and the sun was in motion. But they had no proofs like those I have in my book. This is the reason they changed their minds. But now you must wake up and judge the truth for yourself.

Read everything I say in this book and you will find it just right. There is no guess-work in my book as in the geographies. You do not have to believe what I say in this book, because you have eyes to see with, bodies to feel with and brains to think with.

The geography says that the earth is turning, but no one has ever seen it turn, or felt it, but they can see the earth flat and stationary; the geography also says that the sun is stationary, but no one has ever seen it standing on anything at all.

There is no use believing such foolish talk as we read in the geographies. The geography has no good proofs, but only guess-work, which amounts to nothing at all. There is no sense believing professors, because they made a big mistake by telling the people that the earth is round and in motion and the sun stationary.

We do not blame them for this mistake, because every one makes mistakes.

This book is called "The Enlightenment of the World," because it will wake every one up, and correct this mistake.

The sun does not RISE from a low place, as it appears to do.

Examine your own shadow, or any shadow, in the morning, and again in the afternoon, and you will find the sun always is higher than you, just as the sky is above you. The earth and sky do not meet, as they seem to do. It is only the distance that makes it appear so. Anything at a distance from you seems low, anything near seems high. This is a proof that the earth and sky do not meet. If you stand in a field on a foggy day, you will see that near at hand the fog is higher than you are, but a little further off it looks low and touches the ground.

No matter where you stand the fog appears in the same way, and the earth and sky do not seem to meet, and that is just like the fog around you. Examine it and you will find that what I say is correct.

I started from Boston in a steamer going to New York City.

As I neared the city, I happened to see two bridges ahead of me. Just as I was passing under the first bridge, I found that it was very high, then I looked at the other bridge and I found that it looked very low, almost touching the water.

But when we reached the second bridge I found it as high as the other bridge, so I looked back at the first bridge we passed under and I found that it looked very low, almost touching the water.

Then I knew that both of the bridges were high, but that it was the distance that made them seem low.

It is the same if you go on land. For instance, suppose there is a long, clear road, with telegraph poles the same distance from each other. You look at the one nearest to you, and you will find it high, but the next one looks lower and next one still lower, and so they seem to be growing shorter and shorter until you cannot see them at all.

But it is not true that the poles are lower, as they look; it is only the distance that makes them seem so just like masts of the ship on the water. This may also apply to the sky.

The sun is the same way.

When the sun is rising and setting it looks as though it were very low. It is not low, but high always, and it is the distance that makes it seem low. When the sun is over our heads it looks very high because it is near us.

The sun, being of a limited capacity, can throw its rays only on a certain portion of the earth at one time. It is similar to an umbrella with a limited capacity to cover a certain number of persons. Of course, you can make an umbrella to cover hundreds of people, if needed, but the sun remains a certain size always, and can neither be made large or small, for it is the work of God, whereas the umbrella is the work of human beings.

The earth has been, and will always be, flat, and there are several proofs and those proofs I claim.

You will find in this chapter that I have given some proofs about the sun.

Pay attention to every word you read in this book. You must not believe anyone who says the earth is round, for you are not crazy, nor a child. Do not believe without a correct proof. Children may believe stories, but now that you are old enough, wake up! You cannot make water nor anything else stay on a round ball. It is the same way with the earth. You cannot make the ocean stay on a round earth without falling off, no matter what people may tell you. There is more water than dry land on the earth, and the water being liquid, cannot be anything else, as I proved in Chapter I.

Do not let anybody fool you by telling you that the earth is round. As long as there is no proof that the earth is round, you must believe that it is flat and stationary. If you have a good mind why don't you try to get the best knowledge and follow it? If you are smart, why don't you wake up and have the best for yourself? If I were you I would not allow anybody to fool me. Don't be afraid to change your mind when you find that you have been mistaken by following the wrong way. If you wish to know whether the earth is flat or round, read this work carefully, and you will find proofs that the earth is flat and nothing else.

If you wish to buy books and maps of the flat earth, send me word and I can supply them.

In this book we have given you facts that show that the water, the wind, the land, and the sun itself give you their own proofs. Use your own body to feel with, your own eyes and brains to see and understand these proofs.

If anyone should wish to have a further explanation of my ideas, I will answer his questions, provided I am paid for my trouble.

JOHN G. ABIZAID.

BOSTON, MASS., U. S. A.

THE END.

Dear Friends, —

I should like to say this to you.

If you should like to read my Book, kindly read it two or three times carefully, and think over every sentence and proofs, and you will find that I am right.

Don't agree without your being positive.

I don't want you to believe the earth is flat just because I am saying so in my book; but I want you to think over the proofs and try them and find out the truth for yourself.

Now, I would like to ask for some testimony on the subject. I wish the names and addresses of those who agree with me so that I could help in stopping them teaching the small children that the earth is round, and such foolish things as the sun being stationary, and etc.

I like the truth. I also wish the people to like the truth, and for that reason I wrote my book to explain the flat earth subject and so forth. I have proofs and I have some testimonials from educated people, and I desire some more testimonials or some good poetry on the subject from any one who agrees with me. I should be glad to have their names to print in the third edition of my book; also if anyone, having other good proofs on the flat earth subject, et cetera, will send same to me, I will print it and their name also. If I die my children will build upon the same foundation.

That's the reason I am interested to get your opinion.

Yours truly,

JOHN G. ABIZAID.

No. 121 Tyler St., Boston, Mass.

TESTIMONIALS

Cardinal's Residence, 452 Madison St., New York,
Mr. John G. Abizaid, April 8, 1912.
Dear Sir:

His Eminence, Cardinal Farley, desires me to acknowledge the receipt of your brochure, "The Enlightenment of the World" (2d edition), and to say in reply that he congratulates you on the evidence of thought, painstaking and research shown in the work. Yours truly,

JAMES LEWIS, Secretary.

64 Wavehee Rd., Liverpool, England,
Mr. John Abizaid, January 26, 1911.
121 Tyler St., Boston, Mass., U. S. A.
Dear Sir:

I have received your wonderful book from a friend in Boston and I read it with great interest. You have excellent foundations. It is indeed the "Enlightenment of the World," and I hope you have every success in your excellent work.

You may use this testimonial to the best of your advantage.

I am, dear sir,
Yours sincerely,
PROF. H. B. NEWTON,
(Master of Science.)

AMERICAN PRESS WRITERS' ASSOCIATION.
A. F. Hill, General Secretary and Treasurer.
13 Isabella St., Boston, Mass.,
John G. Abizaid. June 24, 1910.
Greeting:

Your book, "The Enlightenment of the World," is before me. Using the flag of the United States of America is a good plan.

Proper words in proper places may lead us right. Water of the ocean is straight at its surface in long distances. Water flows from high to lower grades.

May you go ahead and win great success in helping to overthrow one of the great delusions made by men of education.

The sun is of less size than the earth; were it not so the rays of the sun would shine vertical upon all parts of the earth on the side of the earth toward the sun.

During a number of years I have known of the globe-earth delusion being a failure. Prove it over and over again to increase the great army in favor of the flat earth facts.

Honest doubters need education in Flat Earthism until they are wise and know that they know. Prove it in the most easy ways, to reach as many as you can of the thinking and knowing wise people.

Respectfully,
AURIN F. HILL,
(Architect.)

John G. Abizaid.   September 24, 1910.
Dear Sir:
Again I have read your important book, "The Enlightenment of the World," by John G. Abizaid.
I am glad of your good work to remove the foolish globe-earth delusion from minds of the people.   Yours truly,
Boston, Mass.   AURIN F. HILL.

                81 West St., New York City, U. S. A., May 20, 1911.
Dear Sir:
I have received your wonderful letter and I thank you very much for your explanation. It interested me very much. You deserve to have more success for your useful book. Accept my love and best wishes.
                     Respectfully yours,
                             N. A. MOKROZEL,
                        Publisher Al-Hoda, " The Guidance."

Mr. N. A. Mokarzel of 81 West Street, New York City, owns and issues "Al-Hoda," or "The Guidance," the best Arabic newspaper in this country at this time.

He is a man of good education.

After he had read the "Enlightenment of the World" he was so impressed with the truths it presented that he published a long article in his paper about it, stating many of its arguments.

His article was published June 25, 1910.

                              Salem, Oregon, March 23, 1911.
FEAR GOD .... A. G. Gy to Him
KEEP THE COMMANDMENTS OF GOD AND THE FAITH OF JESUS
          Copyright June 1, 1909, by Lewis Hain
             Full of the Everlasting GOSPEL
                New Jerusalem, U. S. A.
Mr. John G. Abizaid,
Dear Sir:
I have read your book last night, "The Enlightenment of the World." The proofs given in it are true and they are the very same as mine have been for years. If you please so you may forward one of your World's Maps and one of the Routine Maps of the SUN.

If they will prove true too, then we will agree together to reform the world.   Respectfully yours,
Route 8, Box 115, Salem, Ore.   LEWIS HAHN.

                                  Plymouth, Sept. 13, 1910.
My Dear Brother:
I read your pamphlet with very great interest, and while I cannot accept its conclusions, yet I very much admire its independence of thought and its all-round general ability.
Thanking you for sending it to me, I am,
                     Very sincerely yours,
Rev. John P. Bland, Cambridge, Mass.   J. P. BLAND.

Denison House, 93 Tyler St., Boston,
April 5, 1911.

My Dear Mr. Abizaid:

We have been much interested in the writing and publication of your book and want to congratulate you on the interest it has aroused in the public. It presents a theory of the universe that receives little attention now. With good wishes,

Yours truly,
HELENA S. DUDLEY.

Academy for Science, New York, N. Y.,
March 21st, 1911.

Mr. John G. Abizaid,
121 Tyler St., Boston, Mass.
Dear Sir:

Having heard so much about your book, I decided to procure a copy and found it full of interesting reading.

The subject is a good one, and with the aid of your book, it makes it wonderfully plain.

It is true that everything above the earth moves—the clouds, the sun, etc., and that the earth is motionless.

I am distributing your copy among my friends, and am hoping for good results. Wishing you every success, I remain,

Yours truly,
WILLIAM J. HUTCHINSON, M. S.
"A. F. C.," New York City, N. Y.

Daniel J. Faour & Bros., 63 Washington St.,
New York, N. Y., June 16, 1910.

John G. Abizaid,
121 Tyler St., Boston, Mass.
Dear Sir:

We received your book, "The Enlightenment of the World." We did not get a chance to read it, but from the title it indicates that it is a wonderful book.

You ought to have thankfulness upon your application.

Yours sincerely,
DANIEL J. FAOUR & BROS.

SECOND LETTER.

November 3, 1910.

Dear Sir:

The book of your autobiography we have read, and from our point we found it built on an excellent foundation.

Yours sincerely,
DANIEL J. FAOUR & BROS.

Boston, Jan. 2, 1912.

Mr. John G. Abizaid,
Dear Sir:
It gave me great pleasure to read your valuable book upon "The Enlightenment of the World." It is indeed very interesting. Hoping that you will meet a tremendous success, I remain,
Yours truly,
PHILIP K. NAOUFAL, Secretary.

60 Hudson St.
Secretary "Hudson Club" of Boston.

LEWIS PENNINI, STEAMSHIP AGENT.
Notary Public and Justice of the Peace.
27 Broadway Extension, Boston, Mass.,
August 6, 1910.

Mr. John G. Abizaid,
Boston, Mass.
Dear Sir:
I have read with the utmost attention your "Enlightenment of the World" and find it is a book of great importance, and which deserves the most attention of the great men of scientific work.

It shows plainly what you state and I firmly believe the whole of it to be true and interesting.

I admire greatly your "Enlightenment of the World," and I shall speak to all my friends to get a copy of the same.
Respectfully yours,
LEWIS PENNINI.

July 29, 1911.

Dear Sir:
I think your efforts to prove that the earth is flat are commendable.
W. W. RICH, Printer.

434 Main St., Charlestown, Mass.

Dear Sir:
That the earth is flat and stationary was believed for centuries and has not as yet been absolutely proved to be otherwise. I do not believe the earth is round and the fallacious theory should be exploited.
Yours truly,
HABEEB CURY THRUBUY.

Tonnurine, Mt. Lebanon, Syria.

Boston, Mass., July 14, 1910.

Mr. John G. Abizaid.

Dear Sir:

Have read your book on position of the earth and see nothing but simple facts. The Bible is the oldest and best book in the world. It is as explicit on the question of the world as on any other subject: Gen. 1:9. God said, let the waters under the heaven be gathered together unto one place, and let the dry land appear, and it was so. Shall we want any other proof than that? Here it is; 14th Verse. And God said, let there be lights in the firmament of the heaven to divide the day from the night: And let them be for signs and for seasons and for days and years, Gen. 1:14, 15. The 15th Verse you will please read, for it is important. Read the 16th and 17th and 18th. Do you need any more proof? Here it is; when Moses led the children of Israel out of Egypt two thousand five hundred and thirteen years from creation at Mount Sinai Jesus gave the Ten Commandments. The second Commandment places it just where Gen. 1:9 when He spake and it was done. You may go where you will on this broad earth and you will find it as the word says. We must not forget that it is a destroyed earth for it is not what God made in the first place. It was mostly land—now it is two-thirds water to one-third land. I will refer you to the 24th Psalm, 1st and 2d Verses, also Ps. 104:1-22, Ps. 136:6, Ps. 102:25, 26, Isaiah 42:5. There are over a hundred passages that prove that this earth is not flying in the air. Let us believe in the old book and be governed by it. May God bless you in your work is my prayer.

Yours very truly,

J. B. THOMPSON.

Dr. J. B. Thompson, Burroughs Pl., Boston.

Boston, Mass., U. S. A., May 29, 1911.

John G. Abizaid,
  Boston, Mass.

Dear Sir:

I have heard about your book, "The Enlightenment of the World." I know you have a good idea on the flat earth subject and good, correct proofs. I agree with you because I found out your work is right, and I wish the people to pay attention to your book and agree with you, too.

Very truly yours,

SEGEAN JOSEPH GEAGEAH.

9 Hudson St., Boston.

Denison House, 93 Tyler St., Boston,
April 19, 1912.

My dear Mr. Abizaid.

I wish to thank you for the copy of your book, "The Enlightenment of the World." It is a very interesting statement of your theory. Please find enclosed payment for the same. I should like one more copy.

Yours truly,

HELENA S. DUDLEY.

I received a letter from Prof. Davis of the Harvard University, Cambridge, Mass., dated April 27, 1911, saying that to print my book would cost a considerable sum of money; also that the earth turns round as some think, or stands still, as you think; still it is the same hard-working place.

I think Prof. Davis wants me to stop printing my book, by saying it will cost me a sum of money. And he told what the other people think, but he did not say what he thinks.

Also he says the earth is the same hard-working place. He did not say which way is right. I think he knows I am right on the flat earth subject, but he doesn't want to agree with me because it will be hard for him; he does not want to say anything on the subject for the reason that it would destroy his teaching.

HARVARD UNIVERSITY.
Geological Museum, Cambridge, Mass., U. S. A.,
Seismographic Station.
J. B. Woodworth, in charge.
February 23, 1911.

John G. Abizaid.
Dear Sir:

The press of many studies has prevented my acknowledging before now the receipt of your booklet and letter.

I am afraid I am too old now to learn that the earth is not a sphere-like body.

To be frank I do not see that you have proved that the world is not just as I regard it—and there we are; you don't see it as I see it.

It does not matter much, does it? Whether it is round or flat since we see so little of it.

I hope you prosper in your new place of business.
Very truly yours,
J. B. WOODWORTH.

Boston, Mass., April 3, 1911.

Professor Woodworth,
Harvard University, Cambridge, Mass.
Dear Sir:

Answering yours of February 22nd, 1911, would like to say that at that time I was very busy so I could not answer you until the present time.

As you stated in your letter that you are too old to learn, I am old and I learned how to read and write the English language. As a fact, the older a person gets the more he knows, and I think that you ought to know whether the earth is ROUND or FLAT.

You also said that I don't see the earth as you see it, but I am sure that I do, but I don't see or feel that this earth is round or in motion and going around the SUN.

You did not see or feel this earth of ours going around the SUN. You also said that my booklet did not have any proofs in it. There are a great many proofs in it to make you change your mind, but you did not read it carefully.

Now you want to read it carefully and pay attention to every proof that it contains and you will find that I am right.

As to the difference it makes whether the earth is round or flat, why it makes a lot of difference between the true and the false.

Take my advice and read my book carefully and you will find the truth and you will let other people know the truth also.

Kindly don't forget and write me and let me know whether you have changed your mind or not, and oblige,

      Respectfully yours,

         JOHN G. ABIZAID.

121 Tyler St., Boston, Mass.

         August 27, 1910.

Mr. John G. Abizaid,
 Boston, Mass.
Dear Sir:

I am interested in your book and would like to get a copy of it. How may I do so?

         E. A. SCANLON.

Care of Night Desk, Boston Globe.

       Boston, January 30, 1911.

Mr. John G. Abizaid.
Dear Sir:

I agree with you in every respect save one which is that there is a force of gravitation which you fail to explain; that there is such a force neither you nor I can doubt. This, I think, disproves your theory.
The Boston Globe.

         E. A. SCANLON.

Boston, Mass., April 20, 1911.

Mr. E. A. Scanlon,
  The Boston Globe,
Dear Sir:

Answering your letter of Jan. 30, 1911. You say you agree with me in every respect, save one which is that there is a force of gravitation. I am sure that there is no gravity on the earth as they believe. You ought to know that if you throw a dead body, it will go the way you give it power; as much as you give it power it will go, and when the power goes out from it it will come down itself. But if you throw a live body like a bird or an airship, and so forth, it will go where it pleases any way it likes. This is a good proof to show you the earth is a dead body; it has no power and no gravitation at all. If no water comes out of the earth or from the sky nothing will grow on the earth.

Yours truly,
JOHN G. ABIZAID.

Brookline, Mass., May 7, 1910.

John G. Abizaid,
  121 Tyler St., Boston, Mass.
Dear Sir:

I have read your manuscript through, but I could not correct it, very well, for you being a foreigner and I an American, and the expressions used by us are so different that I would have to rewrite a lot of it, to meet with my hearty approval. However, you have some good ideas, but if you expect to sell your book to the people all over the earth, you will have to take out some and add some to it.

I would be glad to exchange with you when your book is ready for the market. Respectfully yours,

CHARLES W. MORSE

Charles W. Morse, author of "Is the Earth a Level Stationary Plain or a Whirling Globe?" "Is the Earth in Motion or at Rest?"

Salem, Oregon, April 18, 1912.

Mr. John G. Abizaid.
Dear Sir:

I have received your book (2d edition), and have mailed it to the Governor of this State. All the proofs given in the same by you in regard to a flat and stationary earth and of rotating planets are correct and true, and I fully agree with you on the subject, because the earth never was formed in a rotating ball or globe whatever.

I am yours truly,
LEWIS HAHN,
High Priest and President of the Divine Reformation—
the Divine Embassy.

The Midnight Sun

A SPLENDID PROOF THAT

THE EARTH is NOT A GLOBE,

BUT A VAST OUTSTRETCHED

AND CIRCULAR PLANE.

PRICE 2D.   BY "ZETETES."   POST FREE.

"THE LAND OF THE MIDNIGHT SUN."

The above is the title of an interesting book by Paul B. Du Chaillu, in which he describes his journeys through Norway and Sweden, Lapland and Northern Finland. In this book the writer unconsciously gives us proof that the earth is not a revolving globe such as the Astronomers teach, although of course he tries to explain the phenomenon of the midnight sun in harmony with the astronomical theories he was taught at school. While we have no space here for these theories we shall try to find room for the *facts* brought before us; then we shall proceed to shew how these facts conflict with the globe-earth doctrine, and how they harmonise with the truth that the earth is a motionless plane, with sun revolving daily above and around the North Centre, commonly but erroneously called the north "pole."

In his preface M. Du Chaillu says; "The title of the book is derived from one of the most striking phenomena in the north of the country, and one which I witnessed with wonder and admiration on many occasions." In chapter v. he states how, between the 13th and the 18th of June, he sailed "towards the midnight sun" in a steamer leaving Stockholm for Haparanda, "the most northerly town in Sweden," on or "near the right bank of the picturesque Torne river." The passage lasting about three days; while, he says, "The Bothnia was not yet free from ice." He proceeds to describe

The Journey.

"As the voyage drew to a close, and we approached the upper end of the Gulf of Bothnia the twilight had disappeared, and between the setting and rising of the sun hardly one hour elapsed."

Haparanda "is in 65° 51' N lat., and forty-one miles south of the arctic circle. It is 1° 18' farther north than Archangel, and in the same latitude as the most northern part of Iceland. The sun rises on the 21st of June at 12.01 a.m., and sets at 11.37 p.m. From the 22nd to the 25th of June the traveller may enjoy the sight of the midnight sun from Avasaxa, a hill six hundred and eighty feet high, and about forty-five miles distant, on the other side of the stream ; and should he be a few days later, by driving north on the high road he may still have the opportunity of seeing it."

This intrepid explorer then describes his journey overland from Haparanda to the Arctic sea, "the distance as the crow flies being over 5° of latitude to the most northern extremity of the land," but by the route about 500 miles. The country is inhabited by Finns, who are cultivators of the soil. The Laplanders roam over the land with their herds of reindeer. The summer climate is delightful, and during the period of *continuous daylight* one can travel all night if he pleases."

Strange Nights.

Speaking of a station called Pajala, M. Chaillu says ; "From the high hills on the other side of the stream at this place one may enjoy the sight of the midnight sun a few days later. How strange are those evening and morning twilights which merge insensibly into each other ! to travel in a country where there is no night, and no stars to be seen ; where the moon gives no light, and, going further north, where the sun shines continuously day after day ! The stranger at first does not know when to go to bed and when to rise ; but the people know the hours of rest by their clocks and watches, and by looking at the sun."

We may mention that at Ranea, which skirts the Baltic, M. Du Chaillu was told they had snow on the ground so late as the 2nd of June, after a winter during which the thermometer had fallen to 40° and 45° below zero ; yet at the time of his visit he saw garden peas "about two inches above the ground which would be fit for the table at the end of August or the beginning of September." Referring again to Pajala he says ; "In these latitudes the snow has hardly melted when the mosquitoes appear in countless multitudes, and the people have no rest night or day." "The traveller is surprised to meet so many comfortable farms, with large dwelling houses, which with the barn and cow-house are the three prominent dwellings."

"Between the stations of Kunsijärvi and Ruokojarvi (*Järvi* means lake in Finnish) we crossed the Arctic circle at 66° 32' N, or 1,408 geographical miles (?) south from the pole, where the sun shines for an entire day on the 22nd of June, and the observer will see it above the horizon at midnight, *and due north*. After that date, by journeying north on an average of about ten miles a day he would continue to see the midnight sun till he reached the pole. On the 22nd of September the sun descends to the horizon, where it will rest, so to speak, all day long; on the following day it disappears till the 22nd of March."

"When returning southwards at the same rate the traveller will continue to see the midnight sun in his horizon till he reaches the Arctic Circle, where for one day only, as we have seen, the sun is visible."

The Sun's Motion.

Further quoting from these interesting travels we read;—" The sun at midnight is *always north of the observer*, on account of the position of the earth (?) IT SEEMS TO TRAVEL AROUND IN A CIRCLE, requiring twenty-four hours for its completion, it being noon when it reaches the greatest elevation, and midnight at the lowest. Its ascent and descent are so imperceptible at the pole, and the variations so slight, that it sinks south very slowly, and its disappearance below the horizon is almost immediately followed by its reappearance."

After giving the modern astronomical " explanation " of these northern phenomena, an explanation founded on half-a-dozen unproved and unprovable assumptions, the writer naively and unconsciously owns that *appearances* are against these assumptions. He proceeds ; " The nearer any point is to the pole the longer during this time " (from the vernal to the autumnal equinox) " is its day. The number of days, therefore, of constant sunshine depends on the latitude of the observer; and the farther north he finds himself the greater will be this number. Thus at the pole " (the north centre ?) " the sun is seen for six months ; at the arctic circle for one (whole) day ; and at the base of the North Cape from the 15th of May to the 1st of August. At the pole *the observer seems to be in the centre of a* GRAND SPIRAL MOVEMENT OF THE SUN, which, further south, takes place north of him." (*Italics ours*).

Thus we see, that in spite of educational bias and Newtonian belief, the truth will unconsciously and innocently crop up in any description which is true to the facts of Nature. But before we criticise these phenomena further we prefer first to give all the facts which the interesting writer of *The Land of the Midnight Sun* has so carefully gleaned for us. He goes on to describe

How the Sun is seen.

"We have here spoken as if the observer were on a level with the horizon; but should he climb a mountain, *the sun of course will appear higher;* and should he, instead of travelling fifteen miles north, climb about 220 feet above the *sea level* (!) each day, he would see it the same as if he had gone north; consequently if he stood at the arctic circle at that elevation, and had an unobstructed view of the horizon, he would see the sun one day sooner. Hence tourists from Haparanda prefer going to Avasaxa, a hill 680 feet above the sea, from which, though eight or ten miles south of the arctic circle, they can see the midnight sun for three days."

"There are days when the sun has a pale whitish appearance, and when even it can be looked at for six or seven hours before midnight. As this hour approaches the sun becomes less glaring, gradually changing into more brilliant shades as it dips towards the lowest point of *its course.* ITS MOTION is very slow, and for quite awhile it apparently follows *the line of the horizon,* during which there seems to be a pause, as when the sun reaches noon. This is midnight. For a few minutes the glow of sunset mingles with that of sunrise, and one cannot tell which prevails; but soon the light becomes slowly and gradually more brilliant, announcing the birth of another day—and often before an hour has elapsed the sun becomes so dazzling that one cannot look at it with the naked eye."

Nature Asleep in Sunshine.

Again, ascending the river Muonio, on the last day of June, M. Du Chaillu says; "I came to Kicksisvaara, the first boat station situated on a hill commanding a fine view of the country, and overlooking the river Muonio. The people were all asleep as it was midnight; the sun had become paler and paler, its golden glow shedding a drowsy quiet light over all the landscape, and a heavy dew was falling; the house-swallows had gone to their nests, the cuckoo was silent, and the sparrows could not be heard." "How beautiful was the hour of midnight! How red and gorgeous was the sun! How drowsy was the landscape; Nature seemed asleep in the midst of sunshine. Crystal dew-drops glittered like precious stones as they hung from the blades of grass, the petals of wild flowers, and the leaves of the birch trees. "Before two o'clock the swallows were out of their nests, which they had constructed on the different buildings of the farm. How far they had come to enjoy the spring of this remote region! I did not wonder that they loved that beautiful but short summer, or that they came year after year to the Land of the Midnight Sun."

Civilization North.

At a short distance from latitude 70°, near a place called Wind, on the banks of the Alten, a few miles from the sea, our traveller and writer says; "I could hardly believe I was so far north, the birds were so numerous." Near this place at Bosekop he found a village of "scattered farms, with a church, a school, several stores, and a comfortable inn." Bosekop is the seat of a fair, and "in winter is a place of great resort for the Laplanders; court is also held here." Here too he met with a "small society of educated people," with whom he spent a pleasant evening, and had a game of Tag. He says; "I liked the game amazingly; at 11 p.m., the sun shining brightly, they bade me good night, and went to their homes, leaving me full of admiration at their simplicity, innocence, and gentle manners." There also, "in 70° of north latitude, in the quiet parlour of the hotel at Bosekop," he delivered a lecture, by request, on his travels in the Equatorial regions of Africa!

Of the Alten Fjord he says; "There is no part of our *globe (!)* where vegetation is so thriving *at so high a latitude* as on the Alten Fjord." He might have said that there is nothing at all like it in equal latitudes south!—*How is this pray ?*—" Near Bosekop, rhubarb, barley, oats, rye, turnips, and potatoes grow well, also carrots, strawberries, currants and peas. "The thermometer sometimes rises to 85° the warmest temperature during my stay being 63° in the shade, the coolest 55°." Looking over a dreary waste, he says; "from the top of the hills the midnight sun can be seen as late in the season as on North Cape, but the scenery is not so impressive."

A Farewell View.

But we must conclude, for the present, with a brief description of the final view, from the island of Mageröe, the most northern land in Europe. The north Cape is its northern extremity. On the 20th of July, M. Du Chaillu hired a boat and landed on the island. He proceeds;—"After a walk of several miles I stood upon the extreme point of North Cape, in latitude 71° 10', nine hundred and eighty feet above the *sea-level*." Sea "level." (Hear, hear!). "Before me, as far as the eye could reach, was the deep blue Arctic Sea, disappearing in the northern horizon. Wherever I gazed, I beheld Nature bleak, dreary, desolate; grand indeed, but sad. A sad repose rested upon the desolate landscape, which has left an indelible impress upon my memory."

"Lower and lower the sun sank, and as the hour of midnight approached, it seemed for awhile to follow slowly *the line of the horizon;* and at that hour it shone beautifully *over* that lovely sea and dreary land.

As it disappeared, behind the clouds, I exclaimed from the very brink of the precipice, Farewell to the Midnight Sun."

"I had now seen the midnight sun from mountain tops and weird plateaus, shining over a barren, desolate, and snow-clad country; I had watched it when ascending or descending picturesque rivers, or crossing lonely lakes; I had beheld many a landscape, luxuriant fields, verdant meadows, grand old forests, dyed by its drowsy light; I had followed it from the Gulf of Bothnia to the Polar sea as a boy would chase a will-o'-the-wisp, and I could go no further."

"I now retraced my steps to where we had left our little boat. The men were watching for us; it had begun to rain, and when we got back to Gjœsver I was wet and chilly, and my feet were like ice. I was exhausted, for I had passed two-and-twenty hours without sleep, but to this day I have before me those dark rugged cliffs, that dreary silent landscape, that restless Arctic Sea, and that serene midnight sun shining OVER ALL; and I still hear the sad murmur of the waves beating upon the lovely North Cape."

Proof that the Earth is not a Globe.

Having given the facts connected with this interesting enquiry, we now proceed to show how those facts utterly conflict with the globular theory, and how beautifully they harmonise with the plane-earth truth. To do this effectually we must have recourse to diagrams. As these increase the cost of printing we hope our friends will make it up by doing what they can to increase the circulation of our paper. We willingly give our services, but we cannot expect the printers to do so. We will now refer to diagram 1, which represents the sea-earth world according to the globular theory.

Let A C B D represent the "globe," rotating upon its "axis" A B. (see next page). The line C D will represent the circle of the equator midway between the "poles" A and B.

The line F G will shew the position of the tropic of Cancer said to be $23\frac{1}{2}°$ north of the equator, which is the highest north declination the sun attains on or about midsummer day, June 24th. Let P represent the position of the sun directly vertical over this tropical line at this period. In this position it would be mid-day on the side of the earth next the sun along the meridian L F N; and it would be midnight on the opposite side along the meridian M D O.

Let L.M. represent the Arctic Circle said to be $23\frac{1}{2}°$ from the North "Pole" A, or about $66\frac{1}{2}°$ of north latitude; which latitude, or circle, runs across the northern parts of Norway and Sweden or Scandinavia.

THE EARTH AS A GLOBE.

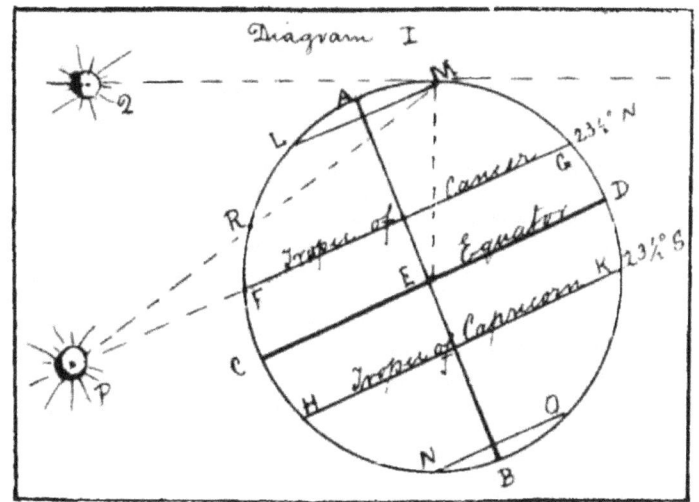

Now we are credibly informed by travellers that in this latitude, and at or about the above mentioned date, a spectator at M can see the sun at midnight, above the horizon, looking directly over the north "pole" in the direction M Q. The horizon is a straight line tangential to the surface of the sphere at the point of observation, and it must therefore be placed at right angles to the dotted line E M running from the centre of the sphere to the latitude and position of the observer.

But we have already alluded to the fact that the sun is never seen directly over any part of the earth north of the tropic of Cancer; that is, the sun is never more than $23\frac{1}{2}°$ north of the equator. Persons living further north than this have always to look in a *southerly* direction for the sun at noon; and it ought therefore never to be seen to the *north* of them at any time, so we must place the sun in the diagram somewhere on the line P F G. Let it be placed at any point P. Now it is manifest that for an observer at M, near the latitude of Haparanda, to see the sun at midnight at P, over the tropic at Cancer, he would have to look *downwards* and be able to see right THROUGH THE "GLOBE" for about five or six thousand miles along the dotted line M R !! I am not aware of any traveller who claims this ability; nor yet that the "globe" to oblige the astronomers, becomes transparent at this period! I am not aware that any spectator of the phenomenon of the midnight-sun has to look *down* at all upon this gorgeous spectacle. The traveller sees it *above* his horizon, and the higher he ascends the higher the sun is seen. Therefore *the earth cannot be a globe*; and thus the midnight sun is a splendid and periodic witness to the fallacy of this absurd unscientific and infidel hypothesis.

Further Assumptions needed.

We are well aware of the further assumptions the astronomers make to get over these difficulties; and we are quite prepared to meet them when occasion requires. They have first to remove the sun millions of miles from where we know and can see that he is; and then they have to assume that he is millions of times larger than he is. In fact assumptions vitiate their whole system. For the midnight sun to be seen, as it is, by a spectator at the point M looking directly over the north "pole," it would have to be placed somewhere on *or above*, the line M Q, say at Q. The further off the sun is placed from the "globe" and the greater divergence there would be between its proper place at Q, above the northern horizon, and its hypothetical position at P. If the spectator could look right through the earth and sea the sun ought to be found on the line G F P to satisfy the conditions of the globular theory; but as a matter of fact it is found many thousands (and according to astronomical ideas many millions) of miles north and away from where it *ought* to be. I fear that the sun has not yet been converted to the Newtonian way of thinking or of acting. Its course of conduct is rather inconsistent with modern scientific " belief "—and there are philosophical creeds as well as religious " beliefs "—and it is very well known that the behaviour of the moon is even more outrageous, considered from an astronomical point of view. There may be some little excuse for the moon in her wayward wanderings, considering her changeable character and the sex generally applied to her; but surely the sun ought to keep his place better with respect to the " globe " than to go out at nights staring at travellers nearly at the "north pole." But perhaps, if they could only see it, he is staring with astonishment at some of their unphilosophical ideas; and if their "scientific" consciences be not utterly seared he must stare them out of all countenance with such ideas.

There must be something sadly wrong *somewhere*, for both luminaries regularly to shew their smiling faces in positions both when and where they ought never to be seen. How is it? Perhaps " gravitation " gets a bit slack at times, and kindly allows them these little excursions! However, we planists have no need to complain, although it rather frets the Astronomers. Why should the sun not visit the north pole, and make a considerable stay there too, for the benefit of Arctic explorers? But here is the strange part of the question, Why is he, *and why are they*, so partial to the *north* " pole " ? Why not try the *south* sometimes in the same way? It seems rather strange : does it not ? Very ! How is it that vegetation, flowers, fruits, birds, animals, men, civilization, &c. cannot be found so far south as they can north? The Plane truth explains it. However we will now proceed to show how simply the phenomenon of the Midnight Sun can be explained in harmony with the truth that the earth is a vast outstretched and motionless plane with the sun circling above it in a spiral orbit around the North Centre.

THE PLANE TRUTH.

The earth and sea together form a vast circular plane. The surface of standing water has been abundantly proved to be *level*. We cannot repeat the evidence here; but those who want it may find the evidence given in an excellent book by "Parallax" (Dr. Birley) which has never yet been answered. This book though out of print at present may be reprinted before long, or as soon as the necessary means are available. Oh! Is there no one with sufficient means, *and sufficient love of the truth*, to do himself so great an honour and the truth so great a service? (But see a letter in the April number of the "Earth Review.") However, to our subject. As water *is* level, the earth *must* be a plane.

THE EARTH AS A PLANE.

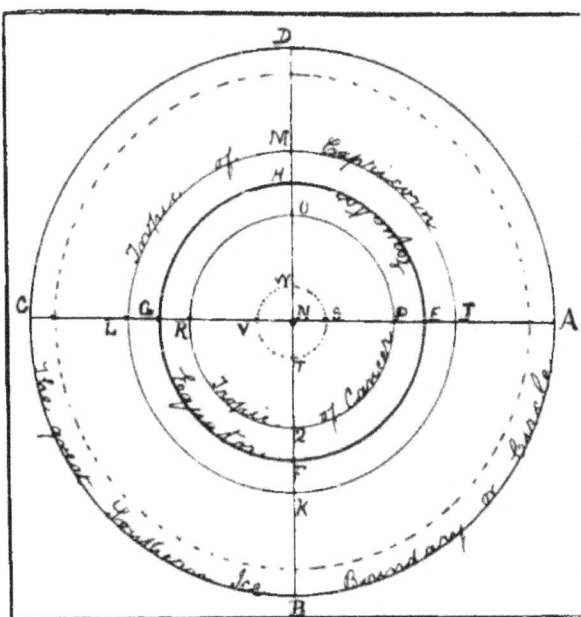

Let A B C D represent the great circular plane, with N for the north centre. The thicker circle E F G H will represent the equator or sun's daily path at the equinoxes in March and September, half way between the North Centre N. and the outer Southern ice circle A B C D. All countries inside the equatorial circle have North latitude; and all outside it South latitude. Let the outer and thinner circle J K L M represent the tropic of Capricorn, or the sun's expanded and daily path in our mid-winter, and the New Zealand mid-summer; and the inner and thinner circle P Q R O the tropic of Cancer, or the sun's contracted and more northerly path or circle at the time of our mid-summer and the southern midwinter. The small dotted circle S T V W will show the position of the Arctic circle, and the larger dotted circle near the outer circumference, the Antarctic circle.

"Degrees."

Now a glance at this diagram will reveal another very popular fallacy in connection with this subject. There cannot be ninety "degrees" of the ordinary geographical extent, between any point on the equator and the north centre. The number and the length of "degrees" of latitude north and south of the equator have been "calculated" on the *assumption* that the earth is a globe. But as the "level" of the surface of the sea proves the earth to be a plane these "degrees" are so far misleading. If we allow 360 degrees for the equatorial circle E F G H, there would only be about $114\frac{1}{2}$ of *such* "degrees" in its diameter say from E to G, or F to H; and only about $57\frac{1}{4}$ of *such* "degrees" in its radius, or from the equator to the so-called "pole," or North Centre. So that if we take all the "degrees" as equal, the distance from any point G, on the equator, to the North Centre, N, instead of being 6,250 miles, or one quarter of a meridional *circle* of 25,000 miles in circumference, as the astronomers assume, it would really be only about 3980, or a little under 4,000 miles. We should have to substract about one-third. But more about this "degree" delusion another time. See *Earth Review* for April, 1893.

THE SUN'S SPIRAL PATH.

Now when the sun is on or over the equator, say at the point G, it is acknowledged that its light extends to the North Centre, at the point N. Therefore the distance G N represents the distance which the sun's rays can pierce through our atmostphere, in a *northerly or southerly* direction, so as to show the full body of the sun to an observer north or south. Hence when the sun is on the tropic of Capricorn in our mid-winter, say at the point L, its direct rays cannot be seen beyond the point V in the Arctic Circle V W S T. Hence all who live within the Arctic Circle at this season of the year are in darkness as far as the sun's direct rays are concerned, the distance L V being the same as the distance G N. But when the sun's daily circular path has contracted towards the north so as to bring that luminary to the point R in the tropic of Cancer at our midsummer, then it is evident his rays must shine right across the whole Arctic Circle from R to S, the distance again being the same as that from G to N.

A Plain Proof.

So that if the earth be a plane with the sun moving over it as already described, a spectator on or near the Arctic Circle at the point S ought to see the sun at midnight at the point R as he looks over and across the North Centre. But this is just what the spectator in such a position *does* see according to the abundant evidence already adduced. Therefore the earth is again clearly and abundantly PROVED TO BE A PLANE.

In such a position on a plane the spectator although in a high northern latitude, must necessarily look still further *north* to see the sun at midnight as he circles round the North Centre; but on a globe, as we have already seen, where the body of the sun never attains more than 23½° north declination, a spectator in such a position, 66½° north latitude, would, (if he could see the sun at all) be compelled to look *downwards* through the "globe" and in a *southerly* direction. This cannot be done, and if it could the sun would not be found there; therefore again the earth is not a globe.

A Faithful Witness.

Thus the sun in his movements becomes a grand and solemn witness to the truth of God and a stationary and outstretched earth. As M. Chaillu, in spite of his astronomical education and bias, is constrained honestly to confess that it seems to be the sun and not the earth which revolves. He says "It," the sun, "*seems to travel around in a circle*, requiring twenty-four hours for its completion." Hear, hear! And since by plane triangulation the sun can be proved to be a comparatively small small body and not more than three thousand miles away, we need not wonder at this. It is surprising how near the truth our Arctic explorer comes when, forgetting his astronomy, he simply and honestly describes the phenomena he witnessed. He further says;—"At the pole the observer seems to be in A GRAND SPIRAL MOVEMENT OF THE SUN, which further south takes place *north* of him." Well done M. Chaillu! We thank you for your honest and noble testimony. It agrees with that of the inspired Psalmist when he said; "The heavens declare the glory of God; and the firmament sheweth his handywork · · · In them hath He set a tabenacle for the SUN which is as a bridgegroom coming out of his chamber, and rejoiceth as a strong man TO RUN A RACE. *His* going forth is from the end of heaven and HIS CIRCUIT unto the ends of it, and there is nothing hid from the heat thereof." Psa. 19 : 1—6.

Let us then, in conclusion, again unite with the Psalmist, in his song of Praise; "To Him that by his wisdom made the heavens; for His Mercy endureth for ever. To Him that STRETCHED OUT the earth *above the waters*; for His mercy endureth for ever. To Him who made great *lights*; for his Mercy endureth for ever. The sun to rule by day; for His Mercy endureth for ever. The moon and the stars (all "lights" only) to rule by night; for His Mercy endureth for ever." Psa. 136 : 5—9.

N.B.—For further information and leaflets on this subject enclose 2½d. to the writer, Albert Smith, Plutus House, St. Saviour's Road, Leicester. See Advertisements in *The Earth Review*.

The following TABLES (copied from *The Land of the Midnight Sun*) give the dates of the appearance and disappearance of the Midnight Sun within the Arctic Circle.

THE CONTINUOUS NIGHT.

WHERE THE SUN IS LAST SEEN.			WHERE THE SUN IS FIRST SEEN.		
Bodö	...	December 15	Bodö	...	December 28
Karasjok	...	November 26	Karasjok	...	January 16
Tromso	...	,, 25	Tromso	...	,, 17
Vardo	...	,, 22	Vardo	...	,, 20
Hammerfest	...	,, 21	Hammerfest	...	,, 21
North Cape	...	,, 18	North Cape	...	,, 24

THE CONTINUOUS DAY.

Where the Midnight Sun is **first** seen.	Upper Rim.	Half Sun.	Whole Sun.	Where the Midnight Sun is **last** seen.	Whole Sun.	Half Sun.	Upper Rim.
Bodö	May 31	June 2	June 4	Bodö	July 8	July 10	July 12
Karasjok	19	21	May 22	Karasjok	21	22	23
Tromso	18	19	20	Tromso	22	24	25
Vardo	15	16	17	Vardo	26	27	28
Hammerfest	13	15	16	Hammerfest	27	28	29
North Cape	11	12	13	North Cape	30	31	Aug. 1

The So-called " Mistakes of Moses," a Satire on " Science."   7d. post free.
Carpenter's 100 Proofs the Earth is not a Globe.   ...   1s. 1d.   do.
The " Earth Review." (to be had from address on previous page) 2½d. do.

ONE HUNDRED PROOFS

FROM THE SCRIPTURES

Against the Teachings of Modern Science.

1.—The Hebrew word, *Bara*, to create, which is making something out of nothing. In the beginning of time, before which all was eternity, God created the heavens and earth, no evolution, no geological progression.

Creation was the work of the Divine Jehovah—God the Father being the architect and planner—God the Son executed, and carried it out, as the Father's Master Workman. God the Spirit breathed Life and Light *upon all.*

2.—When He set a circle upon the face of the deep, when He *marked out* the *foundations* of the earth; then was I by Him as a Master Workman.—*Prov.* viii. 29-30. (R V.)

3.—It is He that sitteth upon the *circle of the earth* and the inhabitants thereof are as grasshoppers.—*Isa.* iv. 22. And woe to the grasshopper that presists in contradicting the Arch-angel.

4.—Thou hast established the earth, and it abideth, nay, standeth—stands fast—*Psalm* cxix. 90.

5.—For He hath founded it upon the seas, and established it upon the floods.—*Psalm* xxiv. 2.

6.—Like the earth which He hath *established* for ever.—*Psalm* lxxviii. 69.

7.—The Lord reigneth: the earth is also established that it cannot be moved.—*Psalm* xciii. 1. Not even by the modern astronomers.

8.—To Him that stretched (or laid out) the earth above the waters; for his mercy endureth for ever.—*Psalm* cxxxvi. 6. Not doubled it up into a ball, to roll three ways at one time.

9.—He hath also *established* them for ever and ever; He hath made a decree which shall not pass.—*Psalm* cxlviii. 6.

10.—The Jehovah by wisdom hath *founded* the earth.—*Prov.* iii. 19. Not cast it as a shot from a melting furnace.

11.—Who hath established all the *ends* of the earth.—*Prov.* xxx. 44.

12.—God Himself, He hath established the earth, He created it not in vain. He formed it to be inhabited —*Isa.* xliv. 18. When the earth was first created it was passable, habitable, and navigable in *every inch of it*—no curse then.

13.—He hath established the world by His wisdom; and hath stretched out the heavens by His discretion: parallel with the earth.—*Jer.* x. 12.

14.—The world also shall be established that it shall not be moved.— *Psalm* xcvi. 10. Dr. Gratton Guinness sends it rolling on after it is a new creation. What folly.

15.—For the *pillars of the earth* are the Lord's, and He hath set the world *upon* them.—*I. Sam.* ii. 8. The astronomers are learned enough to tell us that there is only an imaginary axis.

16 —Which shaketh the earth out of *her place*, and the pillars thereof tremble.—*Job* ix. 6. The earth has *her place* then, not *her orbit*, which would require no pillars.

17.—The earth and all the inhabitants thereof are dissolved; I bear up the pillars of it.—*Psalm* lxxv. 3. And will re-establish it.

18—And after these things, I saw four angels standing on the fou corners of the earth, holding the four winds of the earth.—*Rev.* vii, 1. What corners are there to a round ball?

19.—And the earth was *without form.*—*Gen.* i. 2. Not even globular. What a mistake!

20.—And God called the dry land earth; and the gathering together of the waters, He called the seas. What, not a globe in either case? Oh dear!

21.—While the earth *remaineth*, seed time and harvest, cold and heat, and summer and winter, and day and night shall not cease.—*Gen.* viii, 22. Not while the earth revolveth, or travelleth, or is flying through space to please imagination.

22.—Then the earth shook and trembled.—*II. Sam.* xxii, 8. But that she *always should be flying* through space 120 times swifter than a cannon ball, the prophet did not understand.

23.—The measure thereof is larger than the earth, and broader than the sea.—*Job* xix. 9. The earth then is long not oblong, oval or globular.

24.—Hast thou perceived the breadth of the earth? declare if thou knowest it all.—*Job* xxxviii. 18. Breadth, not circumference, mind.

25.—The Lord reigneth; let the people tremble: he sitteth between the cherubims; let the earth be moved.—*Psalm* xcix, 1. That she would always be doing according to the astronomers. No need for a special order.

26.—The earth shall *reel to and fro* like a drunkard, and shall be removed like a cottage; and the transgression thereof shall be heavy upon it; and it shall fall and not rise again.—*Isa.* xxiv, 20: That is, at the second advent, but not till then will it reel and shake.

27.—Thus saith the Jehovah, the heaven is My throne, and the earth is My footstool.—*Isa.* lxvi, 1. Who ever heard of a footstool always rolling away. The earth will be Christ's everlasting footstool. Praise His name.

28.—At the noise of the taking of Babylon the earth is moved, and the cry is heard among the nations.—*Jer.* l. 46, and xlix. 21. Moved at the end; but not till then. *Through Rome shall come the end of the times.*

22.—They are the eyes of the Lord, which run to and fro *through* the *whole* earth.—*Zech.* iv. 10. Not round and underneath it, but through it at one time.

30.—And they answered the angel of the Lord that stood among the myrtle trees, and said, "We have walked to and fro through the earth, and behold *all the* earth sitteth still, and is *at rest.*—*Zech.* i. 11. A wonder they never saw it revolve on its axis.

31.—The earth and the works that are therein shall be *discovered*; as the most ancient manuscripts read.—*II Peter*, iii. 10. The foundations among them.

32.—And I saw a great white throne, and Him that sat on it, from whose face the earth and the heaven fled away, and there was found no place for them.—*Rev* xx. 11. But according to modern astronomy it is always fleeing away, no one knows where.

33.—And the channels of the sea appeared, the foundations of the world were discovered, at the rebuking of the Lord, at the blast of the breath of His nostrils; He sent from above, He took me; He drew me to Himself out of many waters.—*II Sam.*, xxii. 16, 17. To be fulfilled at the advent, see also *Psalm* xviii. 15, 16.

34.—Thou wentest forth for the salvation of Thy people; for the salvation of Thine annointed. Thou woundest the head out of the wicked, by discovering the foundation unto the neck.—*Hab.* iii. 13. Driving out satan as the god of this world, and preparing everything for Himself, in the inhabited earth to come.

35.—Fear before Him, all the earth; the world also shall be *stable*, that it be *not moved.*—*I Chron.* xvi. 30. A stable earth, not moved, is very different to the wobbling merry-go-round of the astronomers.

36.—The world also is established that it *cannot* be *moved*. Thy throne is established of old, Thou art from everlasting.—*Psalm* xciii. 1, 2. *Cannot be moved* is a strong statement. Who is to cancel it? See also *Psalm* xcv. 10, which refers to the future.

37.—The devil showed him all the kingdoms of the world, and the glory of them.—*Matt.* iv. 8.

38.—Of old hast Thou laid the foundations of the earth; and the heavens are the work of Thy hands.—*Psalm* cii. 25. This is Daniel's psalm, who was doubtless a wise astronomer, and had his wisdom from above.

39.—Mine hand also hath laid the *foundations* of the earth, and my right hand hath spanned the heavens: when I call unto them they *stand up* together.—*Isa.* xlviii. 13. Not *revolve* together.

40.—The burden of the word of the Lord which layeth the foundation of the earth, and formed the spirit of man within him.—*Zech.* xii. 1.

41.—And, Thou, Lord in the beginning hast laid the foundations of the earth, and the heavens are the works of Thy hands.—*Heb.* i. 10. St. Paul believed in foundations then?

42 to 52 —I will utter things which have been left secret from the foundation of the world.—*Matt.* xiii. 35. See also *Matt.* xxv. 34, *Luke* xi. 50, *John* xvii. 24, *Eph.* i. 4, *Heb.* iv. 3, 9, 26, *I Peter* i. 20, *Rev.* xiii. 8, xvii. 8.

53.—Then the Lord answered Job out of the whirlwind, and said, Where wast thou when I laid the foundations of the earth, declare if thou hast understanding.—*Job* xxxviii. 1, 4. Our modern astronomers would here contend there are none, only an imaginary axis. What folly to be wiser than God—and think they could teach God something.

54.—Whereupon are the foundations thereof fastened? or who laid the corner stone thereof? Whereupon where the sockets made to sink.—*Job* xxxviii. 6. Job did not know, and God did not then reveal it.

55.—They know not. neither will they understand; they walk on in darkness: all the foundations of the earth are out of course.—*Psalms* lxxxii. 5; see also *Isa.* xxvii. 11; *Psalm* lxxxii. 5.

56.—He laid the foundations of the earth; that it never should move at any time.—*Psalm* civ. 5. Prayer book version and margin *Heb*. He hath founded the earth upon *her bases*, not axis. General Drayson quoted this text in argument with the writer, E.B·, and promptly admitted that according to Psalms, it was a *fixed earth*; and "if you put scripture above geometry, then I cannot argue with you, for it is a fixed earth." To which we replied, "that *we certainly should*, as we did not believe the Creator would give the wrong impression of *His own works*."

57.—And it shall come to pass (at the second advent) .... the foundations of the earth do shake.—*Isa.* xxiv. 18.

58.—Hast thou not understood from the foundations of the earth. *Isa.* xl. 21. Every solid building rests on a solid foundation.

59.—And forgetest the Jehovah, thy Maker, that hath stretched forth the heavens, and laid the foundations of the earth.—*Isa.* li. 13.

60.—That I may plant the heavens, and lay the foundations of the earth ; and say unto Zion, Thou art My people.—*Isa.* li. 16. That is, in the new creation.

61.—Thus saith the Lord ; If heaven above can be measured, and the foundations of the earth searched out beneath, I will also cast off all the seed of Israel for all that they have done, saith the Jehovah. —*Jer* xxxi. 37. Here the Lord stakes the salvation of Israel upon the challenge, *if heaven above can be measured*. He knew it to be impossible, or would not have made the challenge.

62.—Hear ye, O mountains. the Lord's controversy, and ye *strong foundations of the earth*; or, "ye enduring foundations of the earth."—*Micah* vi. 2. (R.V.)

63.—The Sun is *always declared* the perpetual traveller—*never the Earth*—by the mighty Creator Himself, and the wisest man that ever lived, Solomon. The sun also ariseth and the sun goeth down and hasteth to his place where he arose —*Eccl.* i. 5. (R.V.) Is there a wiser man than Solomon among modern astronomers?

64.—And God said let there be lights in the firmament of the heaven *to give light upon the earth*, to divide b.tween day and night, so *as to shine upon the earth.*—*Gen.* i. 14 The heavenly bodies are then entirely subservient to the earth, and the perpetual servants of the earth. Who would think of making the servant 1230 times larger than the master ; or the sailor boy 1230 times more important than the admiral ? as the astronomers make Jupiter 1230 times bigger than the earth. What insanity !

65.—And let *them* be for signs, and for *seasons*, and for days and years —*Gen.* i. 14. Not let it. the earth. be for seasons, by a useless revolving round the sun ; but let *them*. The earth is entirely passive as regards the *seasons*, as passive as a drum ; it has no more to do with their regulation than the Queen has to do with regulating the affairs of the Court of Russia. Everything in the economy of nature is based upon the plan of philosophical necessity, and there is not the slightest necessity for the earth's pathway to regulate seasons, since the sun's path through the twelve signs of the Zodiac does it *all*. This is acknowledged by God Himself to Job.

66.—Therefore we read of the precious fruits brought forth by *the sun* —*not by the earth*, which is entirely helpless in the matter—no sun, no fruit —*Deut.* xxxiii. 14.

67.—Joshua knew that the sun *was the traveller* as well as Moon therefore on a specially important occasion, for the service of Jehovah, he commands the sun to stand still: "Sun, stand still upon Gibeon ; and thou, Moon, in the valley of Ajalon." And the sun stood still, and the moon stayed, until the people had avenged themselves upon their enemies.—*Joshua* x. 12. The perpetual controversies there has been in Christendom on this one event is truly appalling. Simply because the fixed laws of nature have been misrepresented by modern astronomy, some going so far as to say, in the Portsmouth Town Hall, that the light was continued without the sun. Who ever heard of the light of a whole day being supplied without the presence of the sun? What subterfuges do the difficulties of unbelief leave people to adopt. The difficulties of unbelief are always greater than those of faith. Dr. Adam Clark regretted that he had commenced his commentary, and puzzled his brain for a whole fortnight over it; but Dr. Gill, a great predecessor of the Rev. Charles Spurgeon, took a wiser course for his inimitable commentary, and wrote: "It was a most wonderful and surprising phenomena to see both luminaries standing still in the midst of heaven. How this is to be reconciled with the Copernicum system, or that with this, I shall not enquire." Quite right, Dr. Gill. *Lux* himself, would stand on firmer ground if he followed Dr. Gill, and believed Joshua before the astronomers.

68.—David was a bright astronomer, with the mind, inclination, and opportunity for studying the heavenly bodies, while watching o'er his flocks by night, and his 19th Psalm gives results—" The heavens declare the glory of God ; and the firmament sheweth his handywork. There is no speech, nor language ; *their voice cannot be heard*. In them hath he set a tabernacle, a tent for the sun. Which is as a bridegroom coming out of his chamber, and rejoiceth as a strong man to run his course. His going forth is from the end of the heaven, and his circuit unto the ends of it : and there is nothing hid from the heat thereof." Then David compares it to the infallible Word of the Lord.—*Psalm* xix. (R.V.)

> I sing the wisdom that ordained
> The sun to rule the day ;
> The moon shines full at His command,
> And all the stars obey.

69.— Deborah, the prophetess, prayed with the same understanding: "So let all thine enemies perish, O Jehovah: but let them that love him be as the sun when he goeth forth in his might.—*Judges* v. 31.

70.—" To him that made the sun to rule the day ; for his mercy endureth for ever."—*Psalm* cxxxvi. 7-8. Not the earth to rule the day, which makes no interference with the sun in any way whatever !

71.—David held his views of astronomy to the end of his life, and on his dying bed declared, "The God of Israel said, the Rock of Israel spake to me: one that ruleth o'er men righteously, that ruleth in the fear of God, he shall be as the light of the morning when the sun riseth. A morning without clouds, *i.e.*, in the times of the restoration of all things, which God hath spoken by the mouth of all His holy prophets since the world began."—*Acts* iii. 21. 2 *Samuel* xxiii. 3-4.

72.—So Isaiah prophesies that when the nations of the earth which are saved, are in their triumphant state in the new creation, when the new Jerusalem shall be a praise *on* the earth, the old Jerusalem being now done with for ever; Thy sun shall no more go down; neither shall thy moon withdraw itself: for the Jehovah shall be thine everlasting light, and the days of thy mourning shall be ended. Thy people also shall be all righteous; they shall inherit the earth for ever. The Jehovah will hasten it in Christ's time." *Isaiah* lx. 20.

73.—God declared by Amos that He would cause the sun to go down at noon, and so darken the earth in the clear day.—*Amos* viii. 9.

74.—Habakuk in prophesying concerning the end of all things, says: "The sun and moon *stood still* in their habitation."—*Hab.* iii. 11. or shall stand still.

75.—Christ Himself declared the sun to be the traveller, in His sermon on the mount—who dares to contradict Him, at their cost— "That ye may be the children of your Father which is in heaven: for he maketh his *sun* to *rise* on the evil and on the good, and sendeth rain on the just and on the unjust."—*Matt.* v. 45.

76 to 90.—Fifteen texts speak of the sun as the traveller, viz.: *Joshua* i. 4; *Judges* viii. 13; ix. 33; xiv. 18; xix. 14; 2 *Samuel* iii. 35; *Job* ix. 7; *Psalm* civ. 22; *Jeremiah* xv. 9; *Jonah* iv 8; *Micah* iii. 6; *Nahum* iii. 17; *Malachi* iv. 2; *Ephesians* iv. 26; *James* i. 11.

91.—Thou hast set all the borders of the earth; thou hast made summer and winter."—*Psalm* lxxiv. 17. What borders are there to a globular earth, or to an artificial globe, which only revolves when it is pushed?

92.—Daniel prophecies that subsequent to the destruction of the fourth beast, or the Romish Antichrist, the kingdom, and dominion, and greatness of the kingdom *under the whole heaven*, shall be given to the people of the saints of the Most High, whose kingdom is an everlasting kingdom, and all rulers shall serve and obey him.— *Daniel* vii. 27. So the greatness of creation is *under* the heaven, not *above* it, nor above the firmament of stars, making the earth but a point, a speck, or mere pea, compared to them, as the ridiculous modern astronomy teaches. A point may have existence, but no magnitude. The earth may have existence, but no magnitude. Away with such rubbish!

93.—How would it read, "In the beginning God created the heaven and the point; and the speck was without form and void; and darkness was upon the face of the pea."— *Genesis* i. 1-2. A rare astronomical paraphrase.

94.—The stars at the second advent are to fall to the earth as figs from the fig tree. Whoever heard of a fig being ten million times bigger than the fig tree?

95.—He made known his ways unto Moses (*Psalm* cii. 7), but we never read that God has made known his ways in Creation to the modern astronomers. Therefore we prefer the Pope well informed, to the Popes uninformed.

96.—"If ye believe not his writings, how will ye believe my words." So Christ confirmed the infallibility of Moses writings, for ever.

97.—So Abraham declared to the rich man : "They have Moses and the prophets, let them hear them." *Luke* xvi. 29.

98.—All Scripture is given by inspiration of God, *i.e.*, God spake in the prophets as well as by them, and they wrote as they were borne along by the Holy Ghost ; therefore, whatsoever their writings teach, is divine, and not human ; and their right interpretation cannot err.

99.—"And if any man shall add or take from the words of the book of such prophecies, God shall take away his part out of the book of life, and out of the holy city, and from the things which are written in this book."—*Rev.* xxii. 18-19. One minister wrote in *The Christian World* the other day, that, "while acknowledging that the doctrines and grace, or faith, was inspired, they were not obliged to acknowledge that the chronology and astronomy were inspired." Supposing we were to reverse it, where could the influence of the Bible be? We believe it is *all inspired*, from Genesis to Revelation.

100.—"We ought to obey God rather than man," (*Acts* v. 29), and endeavour, as an highly civilised and educated nation, to banish such erroneous teachings and deceptions from our shores, and thus show other nations that —

        Britons never will be slaves!
        Rule Britannia! Britannia! Rule the waves.

101.—Behold, I will bring again the shadow of the degrees which is gone down in the sun-dial of Ahaz, ten degrees backwards. So the sun *returned* ten degrees, by which degrees it was gone down. *Isa.* xxxviii. 8. There is scarcely a more demonstratable proof of a fixed earth than the sun-dial, no *certain* time could be foreshadowed on a revolving dial.

102.—" And all the host of heaven shall be dissolved, and the heavens shall be rolled together as a scroll: and all their host shall fall down, as the leaf falleth off from the vine, and as a falling fig from the fig tree."—*Isa.* xxxiv. 4, *Rev.* vi. 13. Are leaves millions of times bigger than the vine or tree?

103.—" Canst thou bind the sweet influences of Pleiades, or loose the bands of Orion?" Pleiades, the seven stars which point to the north star, being central in the heavens are especially mentioned as having a special influence and centre. Orion a constellation just before the sign of Taurus; it consists of about 80 stars.

104.—" Canst thou lead forth the Mazzaroth in their season? or canst thou guide the bear with her train?"—*Job* xxxviii. 32. Here the Creator Himself acknowledges the primary cause of the seasons to the signs of the Zodiac, as " Mazzaroth " is the Chaldee name for Zodiac, and the great assertion which follows confirms it.

105.—" Knowest thou the ordinances of heaven? canst thou establish the dominion thereof in the earth?"—*Job* xxxviii. 33. Not the earth to have dominion over them for seasons, or light, or anything else; but they over the earth for all this business. So let's hear no more about the earth regulating the seasons by its supposed revolutions. Let God be true, though every philosopher a liar. God is not a man that He should lie, or waver, nor the son of man that He should change.

Thus we have found from analogy, nature, scripture and experience that there is not an atom of truth in the measurements, distances, quantities, nor theories of modern astronomy, therefore God will at last turn wise men backward and make diviners, who are mostly mere materialists, mad. " What I know not, teach Thou me." Indeed, there never were any arguments urged in favour of materialism, infidelity, or rationalism, but better was always produced in support of truth. The stability of Christ's times will be the abundance of salvation, wisdom, and knowledge, without deception.

P.S.—Incidentally this pamphlet is compiled through the spasmodic *melee* after the third lecture at the Albert Hall. A gentleman friend called one evening to see how the lecturer was; and asked for scripture proofs on the subject. There were none at hand; but when he left, the thought entered our mind to see what texts there were on the subject, personally. So finding over 100, conferred with my friend who thought it advisable to publish them. So if any good is effected by the same, we must thank the disturbers, whose conduct was greatly exaggerated by the local Press, the 40 programmes being multiplied by thousands; and thank the supporter, who so kindly enquired after the welfare of the lecturer.

THE POPULARITY OF ERROR,

AND

THE UNPOPULARITY OF TRUTH;

HAVING SPECIAL REFERENCE TO THE OLD COPERNICAN AND LATER NEWTONIAN THEORY OF THE ROTUNDITY AND REVOLUTION OF THE EARTH; SHOWING IT TO BE AS DEVOID OF TRUTH AS IT IS UNSUPPORTED BY ONE TITTLE OF SCRIPTURE EVIDENCE OR AUTHORITY.

COLLATED AND ABRIDGED,

BY PERMISSION, FROM

"ZETETIC ASTRONOMY,"

SECTION 14,

(BY "PARALLAX.")

BY

JOHN HAMPDEN, ESQ.

LONDON:
SIMPKIN, MARSHALL, AND CO., STATIONERS' HALL COURT;
AND
J. NISBET AND CO., BERNERS STREET, OXFORD STREET.

Swindon:
ALFRED BULL, PRINTER, VICTORIA STREET.

Price 6d., Cloth Boards, 1s.

1869.

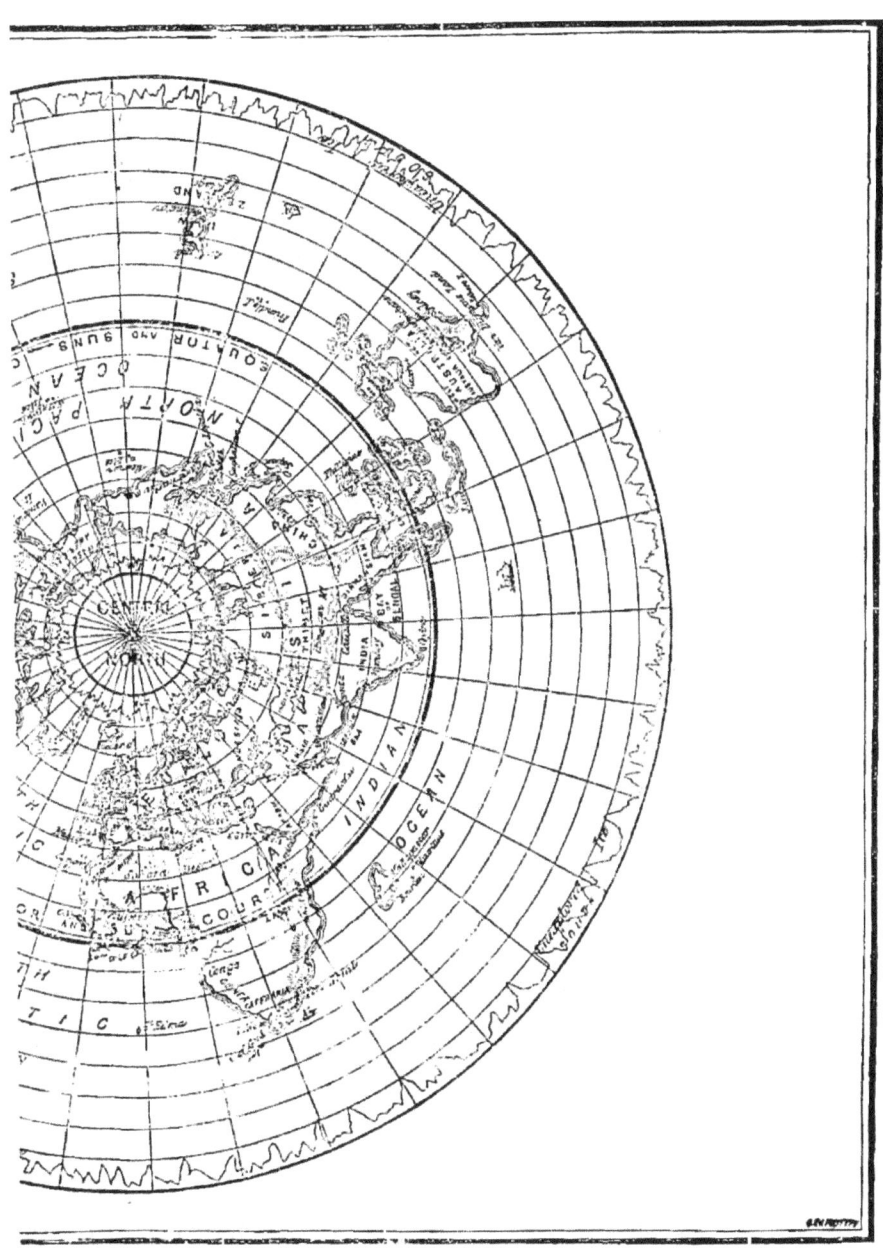

THE POPULARITY OF ERROR,

AND

THE UNPOPULARITY OF TRUTH.

Whether this applies to social, mechanical, or divine science, the result is for ever the same. Proud man rebels against whatever would dispel or expose his own ignorance and folly; and, almost without a single exception, those who have braved the bigotry and prejudices of the world have met with nothing but reproach and resistance instead of the aid and approval they deserved. How many, who have been duped or flattered into exposing their prejudices as objectors and opponents to the appointment of Dr. Temple to the see of Exeter—on the assumed ground of his unsound views—have the slightest knowledge of his writings, or the remotest idea wherein his heresy consists? How surprised would almost every individual of them be to be told that he himself was daily rejecting the testimony of the Mosaic records, and only escaped an equal amount of censure and obloquy, such as has been poured forth on the head master of Rugby, by the fact that almost all the so-called "Christian world" has shared in the disbelief to which I am about to refer!

What are called the Newtonian and Copernican theories respecting the rotundity and revolution of the world, are quite as much at variance with the inspired records as is any statement ever made by a Colenso or any of the Essayists and Reviewers. However trivial or unimportant the subject may appear in itself, yet the fact of its being unsupported by and directly contrary to the Word of God ought to render it of unspeakable interest to all who wisely consider that the minutest departure from the spirit of what Moses and the prophets have written, to be as prejudicial to the whole scheme of revelation as if it referred to an article of faith. If Moses wrote doubtingly or uncertainly about one single point in the history of the creation, no one can justly blame an avowed sceptic for throwing discredit on the whole. I would, therefore, call attention to the following facts:—First and foremost, the word "world" is used rather over 260 times in the Bible; the word "round" is *never once* applied to it. Not a single expression is used, from Genesis to Revelations, suggestive of the idea that the earth is other than a stationary plane; and no hint is to be gathered by the most prejudiced advocate of the Newtonian theory as to its rotundity and revolution. Expressions are constantly used which would be downright nonsense if the earth were a revolving globe.

These facts alone ought to render any further arguments superfluous. But, such is the tendency to throw discredit on the language of inspiration, that an appeal to scientific research and to the unanswerable logic of facts seems to be imperative. Now, what do we see? First, that not a single experiment has ever been made in support of the Newtonian theory but what would equally, if not more forcibly, apply to its opponents. The Newtonian argues that, looking across the ocean, the water appears convex; when asked to look to the right and left, he is obliged to confess that it is horizontal, though the distance surveyed be in both cases the same. That the doctrine of the earth's rotundity cannot be mixed up with the practical operations of the civil engineer and surveyor has been peremptorily decided by a Parliamentary enactment, that "to prevent waste of time and money, which has frequently attended the operations of those who made their calculations according to the prevailing theory of the convexity of the earth's surface, every survey in this or any other country should be carried out according to the horizontal *datum*, as no other method has proved satisfactory, or can be adopted without involving an unnecessary destruction of property, and more or less complete failure of the work in progress." [No. 44, Standing Orders of the House of Commons.] Can anything be more conclusive?

The next experiment is even more decisive still. Take an artificial globe or wheel of any dimensions possible; there will be only one single spot where a level can be obtained, and that under the condition of absolute and complete repose. **Any** inclination, backwards or forwards, would instantly disturb the level which had been obtained at its extreme apex or highest point. But just reduce this experiment to practice, and take the theodolite to any part of the habitable "globe" as it is called, and ten thousand levels can be made wherever a yard of still water can be found, at any point of the compass, by day or by night. *The absolute and undeniable fact that all waters upon the face of the earth are horizontal to each other, is a positive proof that the earth cannot be a sphere, and cannot revolve on an axis.*

Let me proceed to ask a few practical questions :—

Has any navigator ever asserted that he has sailed round or seen anything he could call the "South Pole?"

Has anyone ever crossed the North Pole?

Why is the smallest earthquake so perceptible, while we cannot feel the violent revolutions of the earth, going at the rate of over 1100 miles per hour at the Equator, and more than 700 feet per second in England?

Have any of the navigators who have declared that they have "sailed round the globe" ever been bottom upwards, the sky where the water ought to be; or had they any other *proofs* that when they were midway, their decks were not as level as when they left the English harbours?

If these wonderful navigators had never seen a globe with a map of the earth and sea on it, would they ever have ventured to declare that they had "sailed round the world" perpendicularly?

When a little child runs "round" the loo table in the drawing-room, is anyone insane enough to believe that he went across the top and down underneath the legs or pedestal and up again to complete the circle?

Would not the easiest and least expensive method of getting to any distant place be to ascend in a balloon on a very still day, and remain suspended till the revolution of the earth brought round the distant land to which they were bound, when they could descend, and save at least 95 per cent. of their passage money and all the risks of a sea voyage?

But I have not the patience to "answer fools according to their folly," or I might proceed to expose the absurdity of every theory which has been devised to bolster up this preposterous system of Sir Isaac Newton and his predecessor, Copernicus, endorsed and accepted by men wise in their own conceits, but sheer infidels when brought to the test of Scripture. The Word of the living God, the Creator of Heaven and Earth, does not give the slightest shadow of authority in support of such a notion. Not a verse, or a line, or a syllable can be produced calculated to convey such an impression, but uniformly the reverse; for, however wicked mankind were said to be, it was never contemplated that their folly and ignorance would require instruction on a subject which could hardly admit of misconstruction or mistake.

Copernicus himself, the author and originator of this fanciful and purely fictitious theory, had the honesty thus to speak of his own so-called discovery; (and for this and for the greater part of what follows, I am indebted to a most interesting and instructive little book, entitled "Zetetic Astronomy," by Parallax; Simpkin and Marshall, London). "Copernicus admitted," the author remarks, "It is not necessary that hypotheses should be true, or even probable; neither let anyone, as far as hypotheses are concerned, expect anything certain from astronomy; since that science can afford nothing of the kind; lest, in case he should adopt for truth things feigned for another purpose, he should leave this study more foolish than he came to it. . . . The theory of the terrestrial motion was nothing but theory, valuable only so far as it explained phenomena, and not considered with reference to absolute truth or falsehood."

Happy would it have been for the followers of this great man had they exercised the candour and integrity which always accompanies real genius. Instead of which, they have defended and maintained what the inventor himself declared to be but mere hypothesis, with a bigotry and positiveness which he so emphatically repudiated and discouraged. A plausible theory never loses by age, and if it is fortunate enough to secure the advocacy of some noted authorities, it is forthwith received and maintained with far greater zeal and pertinacity than if it had proceeded from the pen of inspiration itself. The author of the book above mentioned (page 76) challenges its advocates to show a single instance wherein a phenomenon is explained, a calculation made, or a conclusion arrived at without the aid of an avowed or implied assumption!

The very construction of a theory at all—and especially such as the Copernican—is a complete violation of that sound and legitimate mode of investigation which is the result of a careful and experimental enquiry and unbiassed observation. The doctrine of gravitation, which is said to extend through all space, and to influence all celestial as well as terrestrial objects, is but a specimen of " that pride and ambition which has led philosophers to think it beneath them to offer anything less to the world than a complete and finished system of nature " so-called. It was said, in effect, by Newton, and has ever since been reiterated by his disciples—Allow us, without proof, the existence of two universal forces—centrifugal and centripetal, or attraction and repulsion—and we will construct a system which shall explain all the mysteries of Nature which inspiration has failed to demonstrate or left imperfectly detailed. . . The earth we inhabit was called a *planet;* and because it was *thought* to be reasonable that the luminous objects in the firmament which were called *planets* were spherical and had motion, so it was only reasonable and plausible that, as the earth was a planet, it too must be spherical and revolve. And to the sun was given properties which the mind of the Almighty Creator had failed to conceive, or were too scientific for Omnipotence to comprehend!

And, further, the earth being a globe and inhabited, it would follow almost as a matter of course that the planets were worlds like the earth, and inhabited by sentient beings! What reasoning! Assumption upon assumption; and the conclusions derived from such fictitious premises employed again to substantiate the first assumptions! Such a medley of fancies and falsehoods, extended and intensified as it is in theoretical astronomy, is calculated to make the unprejudiced enquirer revolt in horror from the impious fabrication which has been palmed upon him, and to resolve to resist its further progress as far as his influence and energies can be made to extend. For their patience, perseverance, and ingenuity, let the inventors have all the praise which is their due. But their false reasoning, and the advantage which they have taken of the general ignorance of mankind, and the universal desire to be " wise above what is written," should be resisted and denounced with a determined and avowed antagonism.

By the most simple and direct experiments it may be shown with certainty that the earth has no progressive motion whatever. And the advocates of this interminable and perplexing arrangement are challenged to produce a single instance of so-called proofs of these motions which does not involve an assumption—often a glaring falsehood—but always an hypothesis which is not or cannot be demonstrated.

The sizes, the distances, the velocities, and periodic times which these theorists attach to the various bodies, are all glaringly fictitious, because they are only such as a false assumption creates a necessity for. It is geometrically demonstrable that all the visible luminaries in the firmament are within a distance of a few thousand miles—not more than the space which stretches between the North Pole and the Cape of Good Hope; and the principle of measurement—that of plane triangulation—which demonstrates this important fact is one which no mathematician, demanding to be considered a master in the science, dare deny or impugn for a moment.

All these luminaries, then, and the sun itself, being so near to us, cannot be other than very small compared with the earth we inhabit. They are *all* in motion over our heads, and giving days and times and seasons to the inhabitants of the world, which is alone immoveable, and—as plainly as the language of inspiration can describe it—" standing in the waters," " founded on the seas," and " stablished upon the floods." This is a plain, simple, scriptural, and in every respect demonstrable philosophy, agreeing with the evidence of our senses, borne out by every fairly-instituted experiment, and never requiring a violation of those principles of investigation which the human mind has ever recognised and depended upon in its every-day life. The modern, or Newtonian astronomy has none of these characteristics. The whole system, taken together, constitutes a most monstrous absurdity. It is false in its foundation; irregular, unfair, and illogical in its details; and in its conclusions inconsistent and contradictory. Worse than all, so wholly and entirely devoid of scriptural authority as to make it a prolific source of irreligion and of atheism, of which its advocates are unwittingly, but practically, supporters. By defending or endorsing a system which is directly opposite to that which is taught in connection with all the religious and divinely inspired intelligences of the prophets and preachers, both of the Old and New Testament dispensations, they lead the more critical and daring intellects to question the authenticity of Sacred History throughout, to ignore the wisdom, and deny the very existence of a God!

The doctrine of the Earth's rotundity and motion is now shown to be unconditionally false; and therefore the scriptures which assert the contrary, are, in their philosophical teachings at least, *literally true*. In practical science, therefore, Atheism and denial of scriptural authority have no foundation. If human theories are cast aside, and the facts of nature, and legitimate reasoning alone depended upon, it will be seen that religion and true philosophy are not antagonistic, and that the hopes which both encourage may be fully relied upon. To the religious mind this matter is most important, it is indeed no less than a sacred question, for it renders complete the evidence that the Jewish and Christian scriptures are true, and must have been communicated to mankind by an anterior and supernal Being. For if after so many ages of mental struggling, of speculation and trial, and change and counter-change, we have at length discovered that all astronomical theories are false, that the Earth is a plane, and motionless, and that the various luminaries above it are lights only and not worlds; and that these very doctrines have been taught and recorded in a work which has been handed down to us from the earliest times; from a time, in fact, when mankind could not have had sufficient experience to enable them to criticise and doubt, much less to invent, it follows that whoever dictated and caused such doctrines to be recorded and preserved to all future generations, must have been superhuman, omniscient, and, to the Earth and its inhabitants, pre-existent.

To the dogged Atheist, whose "mind is made up" not to enter into any further investigation, and not to admit of possible error in his past conclusions, this question is of no more account than it is to an Ox. He who cares not to re-examine from time to time his state of mind, and the result of his accumulated experience, is in no single respect better than the lowest animal in creation. He may see nothing higher, more noble, more intelligent and beautiful than himself; and in this his pride, conceit, and vanity find an incarnation. To such a creature there is no God, for he himself is an equal with the highest being he has ever recognised! Such Atheism exists to an alarming extent among the philosophers of Europe and America; and it has been mainly fostered by the astronomical and geological theories of the day. Besides which, in consequence of the differences between the languages of Scripture and the teachings of modern Astronomy, there is to be found in the very hearts of Christian and Jewish congregations a sort of "smouldering scepticism;" a kind of faint suspicion which causes great numbers to manifest a cold and visible indifference to religious requirements. It is this which has led thousands to desert the cause of earnest, active Christianity, and which has forced the majority of those who still remain in the ranks of religion to declare "that the Scriptures were not intended to teach correctly other than moral and religious doctrines; that the references so often made to the physical world, and to natural phenomena generally, are given in language to suit the prevailing notions and the ignorance of the people." A Christian philosopher who wrote almost a century ago in reference to remarks similar to the above, says, "Why should we suspect that Moses, Joshua, David, Solomon, and the later prophets and inspired writers have counterfeited their sentiments concerning the order of the universe, from pure com-

APPENDIX.

BY THE EDITOR.

Now it may justly be asked, how much longer are our Geological and Geographical Professors, our Schools and Colleges, to instil into the minds of our sons and daughters these monstrous absurdities, these unscriptural notions, upon the mere authority of semi-infidel philosophers and enthusiasts, whose only delight seems to have been to ignore, if not to bring dishonour and discredit on divine revelation? I say again, that it is no wonder that the doctrinal portions of the sacred Word are held in such disesteem—regarded with such mistrust, and openly scouted by the more profane, when ordained ministers of the Gospel can persist in deliberately rejecting the matter-of-fact details contained in the very first pages of our Bibles; and seem to consider it as the evidence of an enlightened mind that the mystical philosophy of their fellow-worms should be held by them in greater reverence than the plain and simple statements of the Egyptian scribe. Who can read those well-known manuals, "Curiosities of Science," without being shocked at the temerity of both ancient and modern professors, as they gravely narrate the result of their wonderful experiments in attempting to pry into what God has not revealed and what man has never yet been able thoroughly to unravel or explore?

The thinking men of England are slowly being awakened to the fact that the Church's divinity consists chiefly in a medley of Popish and Pagan mummeries and ceremonials—which, under the specious disguise of "our incomparable liturgy," have been palmed upon the nation till we are almost carried back to the darkness, ignorance, and superstition of pre-Reformation times. Shall we, then, any longer submit to be fooled by an infidel science, which has for centuries forced us to acquiesce in the impious hallucinations of a few crazy enthusiasts, whose proper asylum would have been a madhouse had not their dupes been as insane as themselves? Let this groundless fraud be at length resisted, and let our children no longer be taught that we are spun through the air like cockchafers, at the rate of thousands of miles per hour. Let those who have lately occupied two-thirds of our public journals by their reckless and intemperate abuse of Dr. Temple's opinions, be forced to acknowledge the daring impiety of their own enunciations of sacred science, and publicly admit to the world at large how greatly they have been deluded and blinded to the simple teachings of the Word of God. It is no vain or unimportant question—Is human philosophy to supersede divine revelation? Is the prescriptive applause of centuries to render us insensible to the inquiry whether God or man is to be trusted? If the Earth be indeed a globe, then the whole history of the flood is palpably imperfect and untrue. Unless the Earth were a Plane, Moses invented all the particulars connected with that event, from the beginning to the end. "Forty days' and forty nights'" rain could not have half flooded as many acres; and when, somehow or other, the waters did "cover the earth" and "the tops of the highest hills," they could not possibly have dispersed again by any means which the most scientific skill could account for or devise. "Oh, the sun and the wind dried it." Why, then, has not "the sun and the wind" dried up the seas during the space of nearly 6000 years? No; these waters have not diminished by one single hogshead since the day the flood was at its height! The rain of the "forty days and forty nights" merely helped by its weight to submerge the plane of the Earth below the level of "the great deep;" and the flood therefore consisted of the irruption of the salt water from the ocean, and not, as is supposed, of the fresh water from the sky. The sea-sand, the sea-shells, and shingle which are found hundreds of miles inland prove this. Then, at the end of so many days, "God made a strong wind to pass over the earth, . . . and the waters returned" —mark the expression "returned"—"from off the Earth continually." They *could* not have "returned" from off a globe, unless we could see the waters stacked up into mountains, with the dry earth lying at their base. When the weight of the water was displaced by the high wind, the earth rose again like a submerged vessel, and its surface was drained of its moisture like the decks of a ship heaved up from beneath.

Others dwell on the plea that the shadow on an eclipsed Moon shows the spherical form of the Earth But they omit one very essential fact, which is, to prove that the shadow must be that of the earth at all. And, unfortunately for these sages, the eclipse of the Moon has been known to occur while the sun was yet visible above the horizon; thus showing that the Earth's shadow had nothing to do with the phenomenon on those occasions. That there is a shadow, no one denies; but it is sufficient for our case that it cannot be that of the Earth.

Then, again, of the disappearing of the decks of an outward-bound ship. A long line of gas lamps at night on a perfectly level road or esplanade, proves that the apparent sinking of the more distant ones is a mere optical delusion, and clearly explains the apparent sinking of the ship's hull to the least intelligent observer.

But what I would more earnestly enforce than any scientific reasoning I could employ, is the consideration of those scripture passages which the author of "Zetetic Astronomy" has referred to in support of his theory. In the Bible, the word "world" occurs over 260 times, and the word "earth" over 350 times. On no single occasion is the remotest idea of its being a globe, and having motion, ever expressed! Is anyone insane enough to believe that such an extraordinary arrangement could have been designed without a shadow of a reference being made to it? Instead of which, the diurnal motion of the *sun* is spoken of scores of times. Its "rising," its "going down," its "standing still," its "returning," and many other expressions implying motion, are familiar to every reader of the scriptures. But the impious philosophy of the day has the audacity to declare God to be a liar, and man alone trustworthy. God says He made "two great lights;" man says no, He only made one —the second is but a reflected light! God says He "founded," He "established," He "formed the Earth upon the waters," "upon the great deep." Man says, He did nothing of the kind, but the waters rest upon the Earth. And in many other instances, which the readers of the *Essays and Reviews* will recollect, giving the Almighty the lie in every statement He ever made.

Both Isaiah, Job, Solomon, and David, in all their references to the Sun and to the Earth, speak of the motion of the one and the immobility of the other. So does every writer, from Moses to John of Patmos. Dare we, then, venture to accuse these inspired historians of ignorance, or rather of making statements directly contrary to the evidence of their senses? No! May our united answer be, "Let God be true, and every man a liar" who speaks not according to His word. The science of the day is for the most part a rivalry between men who can invent the most incredible theories. For instance, speaking of "the velocity of light," which they say is about 200,000 miles a second, or eight times round the world in the twinkling of a tomtit's eyebrow! Again, Mons. Arago asserts that "several million rays of light can pass simultaneously through the eye of a needle without interfering with each other." He should have added, that during their passage through the archway they tied themselves into knots, and never "jostled!" The "velocity of light" is simply, to the eye, what touch is to the body. But they tell us that a star or planet can, from its immense distance, continue shining thousands of years after it has been smashed to atoms, because its quicker-than-lightning speed has not yet spent itself in its passage through the sky! These are some of the mildest specimens of lying fiction which our philosophers gravely propose for our instruction. But even these might be swallowed with greater ease than the theory of the Earth's revolution round its axis, and at the rate, too, of 700 miles per second! when everyone knows that the wind from a railway train, going at not more than 40 miles per *hour*, will knock a strong man down, if he stood within reach of its action; yet we see an unfledged linnet or tiny moth may repose, without a ruffle on its down, on a floating tendril within a few yards of what I may justly compare to a flash of lightning! Is it possible that mankind has listened to such astounding statements, and endorsed these out-Heroding.Herod attacks on our credulity, without a single protest or demur, for a period of 2000 years, if we are to believe the writings of Aristotle? If the world had but given a fiftieth part of the credence to the simple story of a Saviour's love, or even to the A B C of the history of the Earth's creation as has been vouchsafed to those vain and idle theories of the schools, what a different state of things should we have witnessed this day! Anything new, anything improbable, anything that tends to throw a doubt on the Word of the living God, is greedily seized and used by the sophist and infidel as an argument against the truth of that Word, before which, shortly, every stubborn knee shall bend, and in support of which every tongue shall confess that the Word of God is perfect, and that man alone is vile.

Note.—Those who are unacquainted with the general tenor of Sir Isaac Newton's writings, may take exception to some of the epithets used with reference to the philosophy with which he was identified. I will, therefore, make only one extract, which affords a fair specimen of the *animus* which seems to have influenced all his principles. He argues as if he considered the account of the Creation in Genesis as purposely adapted to the comprehension of the semi-barbarians by whom it was likely to be read. "Had Moses," he says, "described the processes of Creation as distinctly as they were in themselves, he would have made the narrative tedious and confused amongst the vulgar, and become a philosopher instead of a prophet." This is the genius whom all the divines of the present day delight to honour!—a man who dared to imply that Moses was merely a representative of one of the stump orators of the period, whose chief desire it was to make his history as popular as possible, and carefully to avoid all attempts to adhere to simple truth if, by doing so, he might be charged with being "righteous over-much." A worthy sire of the Colenso breed was the far-famed Sir Isaac Newton, and well have his admirers sustained his reputation. God grant, however, many may be found to aid us in dissipating this sham hero-worship —" to clutch the monster Error by the throat, to lead opinion to a loftier sphere, and blot the era of delusion out!"

In the course of any discussion which this subject may lead to, it is insisted upon that our opponents adhere to their own arguments; if the Earth be a globe, let it be treated as such, with all its consequences. When a fly walks upon the ceiling, he does not pretend to be walking on the floor or on a table; if he did, he would inevitably fall; but he knows he is upside down, and he therefore brings a very different set of muscles into play, which he has no use for when standing *on* his legs. Now, our friends at the so-called Antipodes must prove to us that nature has endowed them with the same advantages as the fly. If they cannot do this, they must not be allowed to argue that they can do as the fly does. A plane is a plane, and a globe is a globe. The meanest animals know how to distinguish the two, and man must not and shall not confound them.

Again, our mechanical instrument makers tell us that a mariner's compass can only work when kept perfectly horizontal. But no sooner does a navigator go to sea, than he finds, according to the globe in his cabin, that he is sailing one day at right angles to the spot from which he set out, and in the course of a few more, finds himself upside down, in relation to the docks from which he sailed. Yet he comes home, and gravely tells us his compass has been face upwards all the time; and no one has the courage to tell him that either he or his optician must have lied! No; rather than have their pet theory overthrown, bigots will resist their own convictions, and betray greater folly than the gnats and spiders of our barns and cellars, who never pretend to do more than their Almighty Creator intended they should.

The Editor has already been taunted with advocating a theory which is countenanced by some of the Popes and Romish Cardinals. If Satan himself asserted it, he would be obliged to confess that, for once, his majesty spoke the truth.

We are not insensible to the fact that matters of the gravest import must engage the minds of our legislators during the coming year; but we nevertheless urge the deepest consideration of this subject. How many of our difficulties may be attributed to the dishonour done to God's Word by us as a Church and a nation, we need not here discuss. If there is one thing more than another of which an honest and honourable mind is jealous, it is the amount of credit attached to his word. Yet here we have been, for hundreds of years, telling the Almighty that He does not know the shape of His own world, or that if He knew it, He has deceived us on the subject.

Our readers need not be reminded that no attempt has been here made to refer even to a tenth of the scientific points relating to this question. The velocity of the Earth's course in its orbit round the Sun, estimated by the Newtonian philosophers at 1000 miles per minute, has not been taken into any account. Agreeably to this theory, every creature on the face of the "globe" is doomed to a fate, compared with which Mazeppa's flight must be simply play. Tied to a cannon ball, lashed to a driving wheel, chained to a thunder-bolt, would hardly illustrate the fearful situation of those fated to an existence on this "terrestrial globe." Talk of the calm sleep of the grave—why, were it not for the massive tombstones which considerate survivors place over the remains of the departed, coffins would be flying through the air like rockets, and nothing in or on the face of this Earth would know the meaning of repose. If anyone has ever watched the smoke leaving its long track in a continuous line from the funnel of a steam engine, either on the Earth or water, when, instead of 1000 miles per hour, the speed of the locomotive has been under 15 or 30, how will they account for the lazy wreathings from the cottar's hearth, on which poets dwell so fondly, hanging as it does around the roofs, or else curling in a straight stream to the sky? Would any but the maddened brain of the drunkard venture to assert that that cottage was being shot through the air like lightning, and that the inmates never breathed the same atmosphere for two consecutive moments, or till a whole twelvemonth had passed by? What do our medical professors mean when they impose on the credulity of their confiding patients and send them here and there for "change of air?" If a bar of iron had life, would not the "change of air" to which the Almighty had doomed it be sufficient to waste its form to a very thread? The orbital and axial motion of this so-called "globe" must be seen to be the most monstrous lie that the brains of man ever invented! We say it with all due fear and reverence, that such a statement, if contained in the Word of God itself, would lead us to reject the whole of its revelation. But every Christian Protestant points with pride to the fact that nothing is there proposed for our belief which is contrary to the evidence of our senses. It would be just as easy to believe that this Earth was made by fairies, as that it is a globe, round which mankind were made to crawl like ants, or hold on like monkeys to a palm tree in a gale of wind!

We are proud and happy to say that the subject is finding favour with some of the most learned professors at home and abroad, who can no longer resist the palpable claims it has on their assent. Scores and hundreds have acknowledged our arguments to be unanswerable, and their only or chief reason for not openly avowing their convictions is that the science of ages would be dashed to the ground, and the geography of the world need a radical reform. But surely the additional confirmation it would afford to the truth of God's Word must silence all objection; and the improved knowledge of what has been hitherto and truly termed a "boundless ocean" cannot fail to remove many difficulties which have more or less baffled our most experienced navigators. When it is known that the great deep has a beginning and has an end, formed by that mighty hand which "placed the sand for the bound of the sea, by a perpetual decree, that it cannot pass it; and though the waves toss themselves, yet can they not prevail; though they roar, yet cannot they pass over it"—impious and daring must those minds be that resist such testimony, and charge the prophet with deceit and falsehood rather than forego the applause of their fellow-men, and say that they have gone where the Creator of Heaven and Earth has distinctly declared they could not and should not go.

JOHN HAMPDEN.

Swindon, Wilts, December, 1869.

DESCRIPTION OF THE MAP.

The least intelligent of our readers will hardly need much explanation to understand the chart an amateur friend has kindly sketched out for us. Take it to any part of the world, the mariner's compass uniformly points to the central North. Navigators have announced the strange fact that inside the frozen belt of Northern icebergs, varying from 80 or 90 to 100 miles in breadth, is an unfrozen sea, upon whose bosom no craft of man in any shape has ever rested, the distance over the ice rendering the transport of any vessel physically impossible. The cause of its being unfrozen has never been ascertained, and no surmise can be offered beyond the supposition that a submarine volcano or hot springs must cause a higher temperature of the waters at that point. The well-known length of the day alternating between the Northern centre and the Southern circumference is caused by the contraction and expansion of the Sun's orbit—nearer to the North in summer, and more distant in the winter months. This will be understood more clearly on reference to the larger work, where phenomena of day and night, summer and winter are fully explained.

Passing over the Equator, we come to the frozen extremities of the world—South, South, South all the way round. Facing the Northern centre from any point of the Southern circumference, of course, to the right is East, and to the left is West. By sailing due East or due West, the ship returns again from an opposite direction to that in which it set out on its voyage. But it can neither pass the Northern, much less the Southern barriers of icebergs. What lies beyond the outer circle, no man has yet dared to explore; and but a few miles into the airy regions of space may the bold aeronaut ascend; so into the icy barriers below, the most daring adventurer is told—"Thus far shalt thou go and no further." Further knowledge would be too much for the finite mind of man to bear. His all-wise Creator, in pity to his frame, has given him limits, which he cannot, dare not over-pass. "Such knowledge is too wonderful and excellent for him; he cannot attain to it." Let him be content with the wonders his Lord has seen fit to reveal. But the space above him, beneath him, and all around, is, in this stage of his existence, hidden from his view. J. H.

THE SHAPE OF THE WORLD,

BY

A. E. SKELLAM.

May be had of the WRITER, 20, Elmsleigh Road, Wandsworth, S.W., or Mr. J. WILLIAMS, *Secretary*, Zetetic Society, 32, Bankside, London, S.E.

Single copies free by enclosing stamp to cover postage. 1/- *per* 100.

HARRISON & SONS, Printers in Ordinary to Her Majesty, St. Martin's Lane, London, W.C.

THE SHAPE OF THE WORLD.

On this subject there are two great schools of thought; one which teaches that the world is a globe of 25,000 miles circumference, rotating on its axis once in twenty-four hours at the rate of 1,000 miles per hour, and revolving in an orbit round the sun once a year at the rate of 68,000 miles per hour; the other teaches that the earth is a vast irregular plane, stretched out from the Central North and standing in and out of the waters, the surface of which is horizontal. One of these two views must be wrong, and the other right.

Now that the world cannot be a globe will be evident from the following FACTS :—

1. The horizon ALWAYS appears on a level with the eye, no matter to what altitude we ascend. Aëronauts assert that the horizon H H, Fig. 1, always appears on a level with the car,* and that it (the horizon) seems to rise as they rise, the deepest part appearing to be immediately under the car, and that instead of the surface of the earth appearing convex, as it should if it were a globe, *see* Fig. 2, it appears concave as in Fig. 1.

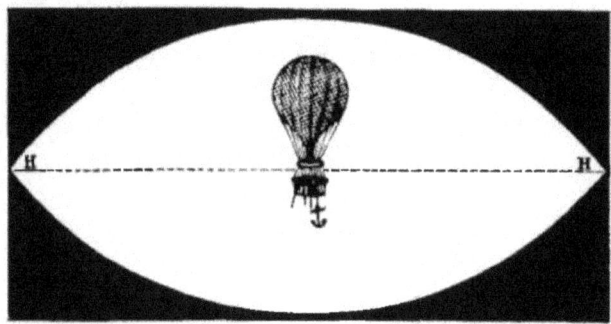

FIG. 1 represents what is seen by Aëronauts.

FIG. 2. What should be seen if the earth were a globe.

---

* See Mr. Glaisher's (of Royal Observatory, Greenwich) Report in *Leisure Hour*, October 11, 1862.

Perspective shows that parallel lines appear to meet in the distance, as, for instance, railway lines, but diverging lines never can appear to meet, no matter how slight the divergence. If A B, FIG. 3, represent the skyline, and C E D a part of the earth's surface, if curved the earth would appear to be curved from C to E to an observer at A, and the horizon instead of being "on a level with the eye" would always be below, and sink lower the higher the observer ascended.

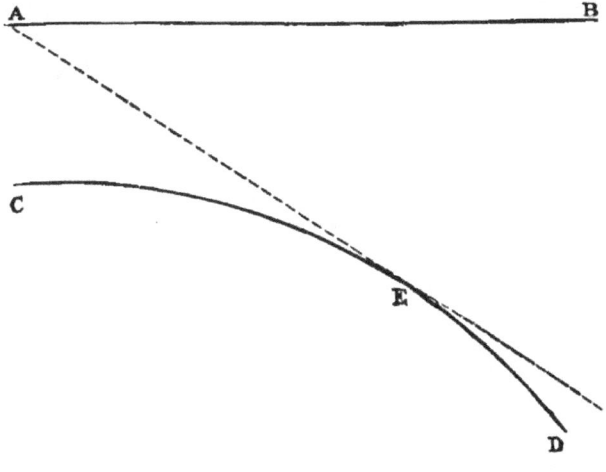

FIG. 3.

2. The horizon, at sea, to the right and left of an observer, always appears as a straight line, but if the world were a globe it (the horizon) would dip to the right hand and to the left, and show the arc of a circle. If a plank or straight edge, about 12 feet long, be set level above the water, and the observer get at the back of the plank or straight edge, so that his eye be on a level with it, he will see the horizon as a perfectly straight or horizontal line.

3. The lights from many lighthouses are visible at a much greater distance than they could possibly be if the world were a globe 25,000 miles in circumference.

*The flame of the Clare Island light can be seen, in clear weather, a distance of 31 statute miles. The dip in 31 miles is 640 feet ($31^2 \times \frac{2}{3} = 961 \times \frac{2}{3} = 640\frac{2}{3}$ feet). The altitude of the light above high water is 341 feet, and is seen from the deck of a vessel 15 feet above sea-level ($640-341-15 = 284$ feet). After deducting the altitudes the light should be 284 feet below the horizon if the world be a globe.

*Beachy Head light can be seen 26 statute miles away. The dip in 25 miles is 450 feet ($26^2 \times 8 \div 12 = 450$ feet 8 inches), altitude of light above high water 284 feet, vessel 15 feet ($450-284-15 = 151$ feet). The light should be 151 feet below the line of sight if the globular theory of Copernicus be true.

---

* See "Admiralty List of Lights," Part I., 1893. Published by Hydrographic Department, Admiralty, and sold by J. D. Potter, 31, Poultry, E.C.

*Cromer light is 274 feet above high water, and can be seen a distance of 26 statute miles; vessel 15 feet above high water. The dip in 26 miles is 450 feet; less altitudes (450−274−15 = 161 feet) leaves 161 feet to be accounted for if the world be a globe. There are many examples similar to the foregoing. The following note in the Admiralty List makes the case all the stronger against the globular theory :—" Under certain atmospheric conditions, and especially with the more powerful lights, the GLARE of the light is visible considerably beyond the radius given, which is calculated for the ACTUAL FLAME of the light."

4. No allowance is made by engineers for curvature in making canals and railway cuttings, and is, in fact, forbidden by Act of Parliament, which states that all plans, &c., shall be made from " a datum horizontal line, which shall be the same throughout the whole length of the work."† If the earth were a globe this allowance would be indispensable.

5. The surface of standing water has been proved, beyond all question of doubt, to be horizontal and not convex as it should be if the world were a globe. Water always finds its level, the surface of which, at rest, is always found to be horizontal. A convex surface is not a level or horizontal surface, although some people would have us believe it is.

6. The Midnight Sun being seen ONLY in the northern regions is evidence against the globular theory. It the world were a globe, the midnight sun would be seen in the southern regions in December as it is seen in the northern in June. The author of a book entitled " The Land of the Midnight Sun " says :—" At the pole the observer seems to be in the centre of a grand spiral movement of the Sun." Why does this NOT occur in the southern regions? These few facts alone are sufficient to prove that the world is not a globe, but a vast irregular plane stretching out from a north centre to a south circumference in every direction ; surrounded by water and ultimately in the extreme south by impassable barriers of ice. Vasco-de-Gama in his " Voyage to the South " says :—" The waves rise like mountains in height. The winds are piercing cold and so boisterous that the pilot's voice can seldom be heard, whilst a dismal and almost continual darkness adds greatly to the danger." How far this gloom and darkness and storms extends is not known. All we know is that the most daring have been stopped at the entrance to this gloomy, and what seems to be, forbidden region of the world.

If these lines cause any to think and search after Truth, the writer's object will have been attained. Truth is not injured by enquiry and test, but is like a cube always right side up.

A. E. SKELLAM.

Wandsworth,
August, 1893.

---

* See "Admiralty List of Lights," Part I., 1893. Published by Hydrographic Department, Admiralty, and sold by J. D. Potter, 31, Poultry, E.C.
† See No. 14 Standing Order of Houses of Lords and Commons on Railway Operations, for Session 1862.

Reprinted from "THE EARTH."

THE SOUTHERN MIDNIGHT SUN;

By Zetetes.

In a late number of the *Windsor Magazine* we had an account by Dr. Cook, Surgeon of the Belgian Antarctic Expedition, of the experiences of the crew of the *Belgica* in the south polar ice, about latitude 71º, and averaging about the same number of degrees west longitude. These experiences are interesting as showing the great perils these hardy voyagers endured in their daring expedition. But the chief point of interest to Zetetics is found in what professes to be a photographic picture of "The Midnight Sun: Christmas, 1898;" showing also the *Belgica* frozen in the great ice pack, and never moving, except as she moved with the whole ice-field, from March 4th, 1898, to February 14th, 1899."

A few of our planist friends have been unnecessarily disturbed by this picture, and the account to which it refers; and some of our globularist opponents have been prematurely elated by it. One of the latter, thinking the discovery of a Southern Midnight Sun was a clear proof of the sphericity of the earth wrote a letter and triumphantly demanded to know of the planists

"WHAT WILL YOU DO WITH THIS?"

In fact our friend did not put the question quite so politely as this. But if he will excuse us polishing it a little for him, to make up for his want of courtesy, we will, as far as our health and the editor's space permit, proceed to reply. In the meantime we can make some allowance for our opponent, as his head might be a little giddy through the globe, like a monster fly-wheel, turning him under; living as he does at the "antipodes," and, of course, hanging head downwards at the time of his writing! And this antipodean predicament was the position of the explorers, according to our astronomical friends.

But first let me tell all the globularists what true Zetetics will *not* do with this fact, if it be a fact, which we are under

THE SOUTHERN MIDNIGHT SUN.

no anxiety to deny. First, then, we shall not lose confidence in another fact, which our opponents conveniently, persistently, and quietly ignore, namely, the fact that the surface of still

WATER is LEVEL, and the EARTH therefore a PLANE!

This great fact has so frequently been attested in Zetetic literature (and the editor of *The Earth* may again give testimony to it) that I shall not dwell upon it, especially as no astronomer has ever seriously attempted to disprove it. I only retort in the words of the question asked above, What will our opponents do with this fact? And echo answers, What? I venture to predict that they will not even attempt to deal with it.

But we will attempt to deal with their fact; and in the first place we will proceed to show that the Southern Midnight Sun, from the position of the *Belgica*, as reported,

COULD NOT BE SEEN ON A GLOBE?

To make our argument clear we refer to the following diagram.

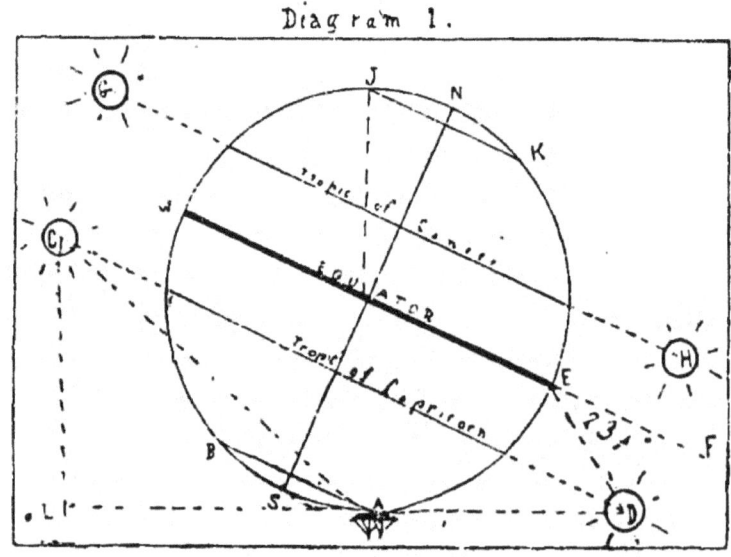

Diagram 1.

THE SOUTHERN MIDNIGHT SUN.

Let N E S W represent the sea-earth globe of science; N S the "imaginary axis;" and E W the equator. At $23\frac{1}{2}°$ north and south draw lines representing the tropics of Cancer and Capricorn respectively. Produce these lines indefinitely, say to C and D, and G and H. Produce the equatorial line W E to F. And at the point E, on the equator, at the surface of the earth, draw the line E D making an angle of $23\frac{1}{2}°$ south of the vertical line E F. Where the line E D intersects the produced tropic of Capricorn place the sun at D. This shows the sun's position about Christmas time, when in the southern solstice, as seen from the equator at $23\frac{1}{2}°$ south of the vertical. The sun of course is seen from the *surface* of the earth, and not from the centre, where our astronomical friends cunningly place the aforesaid angle. But we are not ignorant of their devices. The angle ought to be made where it is seen, on the equatorial surface, not down below in Hades! We have therefore placed it there. This proves the sun to be comparatively near the earth, as we have already proved by mathematical demonstration in a previous article on the *Sun's Distance*. Thus one fact corroborates another.

We have now to point out the relative position of the observers in the *Belgica*. According to Dr. Cook's report the vessel had reached about $71\frac{1}{2}°$ south latitude. In the diagram the line A B represents the southern, or Antarctic, circle, at $66\frac{1}{2}°$ S. Therefore the explorers had passed this line by about 5° nearer the south "Pole." The vessel was therefore a little to the south of point A, where we have placed it in the diagram. From this point draw the line A D a tangent to the sphere at the place of observation. Strictly this line would fall a little below the sun's place at D, but we give our opponents this amount to make up, and more than make up, for whatever refraction they might consistently claim. The point D then marks the position of the sun to the crew of the *Belgica* at mid-day, that is barely on their horizon! The point C, on the opposite side of the globe, marks the position of the sun at mid-night, twelve hours later. To see the sun at mid-night from the position of the explorers the observer would either have to look round the great curve of the earth down a "dip" of some four thousand miles, or he would have

THE SOUTHERN MIDNIGHT SUN.

TO LOOK DOWN THROUGH THE SOLID GLOBE,

as represented by the line A C, a distance of some four or five thousand miles! If our friends claim the ability to see through either land or water for four or five thousand miles, or to see the sun when below their horizon some thousands of miles, as represented by the "dip" from L to C, then I will yield, and confess that the southern midnight sun could be seen from the position assigned upon a globe. But if not —and notwithstanding the temerity of the astronomers in making some of their preposterous claims and hypotheses, we hardly think they will claim this ability—then, if not, I claim the fact that the midnight sun has been seen in extreme south latitudes as another proof that

THE EARTH IS NOT A GLOBE!

Thus we have so far answered our friend's defiant query, and shown the globularist what we can "do with this fact." And in the same way it can be proved that it would be impossible to see the northern midnight sun at G, from the point K, if the earth was a globe. But as the writer gave this proof some years ago in a pamphlet entitled *The Midnight Sun* (north), reprinted from the *Earth not-a-globe Review*, the reader is referred to it for the elaboration of this splendid proof that the earth is an extended plane.

If our friend at the antipodes, or if any of our astronomical friends, or foes, will deal with the demonstrated fact that the surface of still water is level, absolutely level, we will, God willing, deal further with this last reported fact of the Southern Midnight Sun, and offer an explanation of the same on purely Zetetic lines.

THE
SUN STANDING STILL

AT THE COMMAND OF JOSHUA,

OR

Ancient *versus* Modern Science.

BEING **A REPLY** TO A

" Lecture by the Rev. W. W. Howard, Liverpool."

BY

ALBERT SMITH, LEICESTER.

From THE EARTH (*Not-a-Globe*) REVIEW,
January, 1894.

"If ye believe not his Writings, how shall ye believe my Words?"—
The Christ.

PRICE **2D.**   POST FREE.

A Reduction of 20 p.c. on 5s. worth and upwards.

To be had from "Zetetes," Plutus House, St. Saviour's Road, Leicester, England.

H. BANBURY & Co., PRINTERS, LEICESTER.

THE
Sun Standing Still.

IF any proof were needed that the Bible teaches the docrine of a stationary earth and a moving sun and moon, it is given in the tenth chapter of the book of Joshua. Here it is recounted how Joshua, the leader of the Israelites after the death of Moses, and the armies of Israel fought against the five kings of the Amorites and their armies, the LORD also casting great hailstones down from heaven upon the enemies of His chosen people. "Then spake Joshua to the LORD in the day when the LORD delivered up the Amorites before the children of Israel, and said in the sight of Israel ;—

Sun, stand thou still upon Gibeon ; and thou Moon, in the valley of Aijalon.

And the sun stood still, and the moon stayed, until the nation had avenged themselves of their enemies." *Rev. Ver.*

Now although this account is evidently quite as historical as the account of the rest of the Israel's doings and battles, yet because the teaching conflicts with the views of men and the theories of modern astronomers it is tortured and twisted by laboured " explanations " to mean anything and everything but what the words naturally mean on the face of them. And, as though to prove that all these fanciful " explanations " are off the track, no two expositors are perfectly agreed, or give exactly the same explanation of the passage. They are only alike in one laudable but misguided intent, and this is, to save the Scriptures from reproach and to "harmonize" the account with the theories of modern astronomy and the views of so-called "scientists." It never seems to enter the minds of these well-meaning expositors to question the truth of this modern " science," but only how most plausibly to "reconcile" with it ancient and Bibical Cosmogony. This is not as it ought to be. We shall make no such futile attempt, neither shall we pause to vindicate the character of Israel's God, who will, we believe, do this Himself perfectly when the day of final judgment arrives ; but we shall proceed to shew the unsatisfactory nature of all attempts at reconciling the Bible with modern astronomical theories, and boldly challenge any man, either scientist or sceptic, to give us one reasonable and practical proof that the earth has any of the awful motions attributed to it by them. If they cannot do this, and we have hitherto asked for the proof in vain, then we have both right and reason to believe that Joshua was correct in believing, with other Bible worthies, that the motion of the sun, and not of the earth, was and is, the cause of day and night.

The latest effort we have seen at impossible reconciliation calls forth these remarks. We give the writer credit for sincerity and devotion. As he has sent us a copy of his pamphlet we thank him for it, but he must excuse us pointing out clearly and conscientiously where his effort, like that of others has failed. His pamphet is entitled "Joshua commanding the Sun to stand still. The miracle explained and defended. A lecture by the Rev. W. W. Howard, price 3d., to be obtained from the author, 47, Heman's street, Liverpool,"

We cordially agree with the opening paragraph ;—

"The subject we have to discuss to-night has engaged great attention for ages. Believers in revealation have explained and defended the wonderful occurrence with great learning, zeal, and ingenuity, and infidels have made it the favourite object of their scorn and raillery. Many theories have been advanced with a view to give satisfaction to faith and remove doubt ; and the way in which the event is still regarded to-day, both among believers and unbelievers shews that not any of them have met with much success."

This is quite true, especially the closing sentence ; and we think the present effort is doomed to like failure with former efforts. And for the same reason, viz ; lack of faith, on the part of "believers in revelation" in not receiving the account as it stands, and ignorance of true science on the part of infidels, and others, who unreasonably revile what they do not understand, and who credulously believe any absurd theory if propounded in learned jargon and uttered in the name of "Science." Thus the "Christian" has generally much too little faith in the All-wise God and His Revelation to believe it, so he explains it away ; and the infidel has a great deal too much faith in ever erring mortals and their philosophy, so he proudly scorns and rejects it. But, of the two, the infidel is the more consistent ; for the Christain expositor, like himself, unquestionably accepts those astronomical theories which makes the Word of God of none effect, while the sceptic does not believe in a Divine Revelation. But Zetetics can boldly challenge the truth of those theories, yea, more, they can shew that even as theories they are false to Nature, as well as to the Scriptures ; and so the infidel's raillery is checked—and in all reason it ought to be—until he becomes sufficiently instructed to offer some decent proof in support of his position.

Let him try, for instance, to give proof of the earth's supposed motion; as we have allowed some to try in public meetings, and the laugh is soon turned to the other side, See our So-called "Mistakes of Moses," under heading, *The Book Wrong*, which gives an instance which really occurred, in Blackburn, once when the writer was lecturing there. But we do not wish to satirize honest doubt, but rather to suggest reasons for thorough enquiry and christian belief.

FOUR LEADING THEORIES.

Referring to the printed lecture before us we find that Mr. Howard selects *four* as the leading theories by which this miracle has been explained, and which even he himself cannot accept. The first is called

"THE POETICAL THEORY."

Those who accept this theory, he says, suppose that the hours of sunlight did really appear to them to be lengthened? Someone afterwards expressed his feelings in poetry, "with the usual poetical license," whatever that is, and incorporated his poem in a book of military songs called "The Book of Jasher." We reject this exposition for the same reasons as the writer; because, "firstly, there is possibly a more reasonable view; and, secondly, the genius of Hebrew poetry lends no confirmation to its position." And we further cordially agree with him when he adds;

> "I have sought all through the Bible and have not discovered one instance of a natural event being exalted into a miracle by any of its bards." . . . .
> "This enquiry into the veracity of Hebrew poetry has amazed me—made me feel how, contrary to the general view, in all their highest inspirations, the Bible bards kept a clear eye upon the sober truth."

This, we think, is well and truthfully spoken. The second theory, he says, is called

"THE SPIRITUAL THEORY."

There are those who hold that God, at the command of Joshua, allowed the sun and moon to go on their journey as usual, but in their places "two other bodies of a spiritual kind were slipped in so stealthily that the Israelites were unaware of what was done." This theory, commonly held by Swedenborgians, the writer very properly rejects as charging God with deception, and assuming an impossibility. He gives his reasons, which those who are interested to know can find by obtaining the pamphlet. Our space compels us to be brief. The next exegesis reviewed is, thirdly,

"THE OPTICAL THEORY."

Under this heading Mr. Howard says ;—

> "It is true that light is refrangible, and also that we see, not as we think,

always straight and direct, but on lines of light. When light, in its flight, strikes a medium denser than that it has been travelling through, it is turned aside somewhat, and we are led to think that objects are not where they really are. If you thrust a stick into water it appears to bend at the surface of the water . . . We may also say that the stars are never where we seem to see them in the heavens, but where they were when the light we see them by left them."

So far we have been happy to agree with Mr. H., but from this he begins to flounder unconsciously in the meshes of absurd and extravagant philosophical theories. He re-affirms the popular fallacy that the sun is seen in the morning " eight minutes before he is above the horizon," that the light from some stars " would require thousands of years to cover the distance between us," and that

A "star or nebulæ might be completely annihilated, and yet it would not seem to disappear from its position in the universe till its last beam of light had reached us, and that might be 20,000 years or even longer " !

He further affirms that " the axis of the earth is inclined to her orbit," that the " pole " dips so that "anyone living at the north pole would see the sun 12 or 13 days time before *he* actually rose above the horizon " (!) and moreover that "this would follow from the atmosphere bending the light beams, and the *north pole* rising by gentle graduation into the zone of day " ! *Italics ours.* The writer innocently calls this contradiction "a fact," and says ; " From this fact some have argued that the light rays of the sun and moon were bent, at Joshua's petition, to give him an extra 12 hours light to exterminate the enemy." And he quotes James Austin Bastow who supports this view in his Bible Dictionary. However, this theory, though " plausible " is rejected as "delusive," there being a vast difference "between the refraction of a few degress on the one hand and that of half a circle on the other." We are then informed that

"The fourth theory is the Astronomical one."

Here of course, the tangle becomes greater than ever. We are told that

" The rotary motion of the *earth* was arrested, the arrested motion was prevented becoming heat, the water in the oceans, seas, lakes, and rivers was kept from obeying its natural laws, and the solar system was guarded against injury."

The writer, while agreeing, of course, with the " science " of the above paragraph, sympathises with men like Huxley and Tyndale, in their refusal to accept such an explanation, adding that Professor Tyndale, in *Fragments of Science*, remarks ;

"There is a scientific imagination as well as an historic imagination ; and when, by the exercise of the former, the stoppage of the *earth's rotation* is clearly realised, the event assumes proportions so vast in comparison with the result to be obtained by it, that belief reels under the reflection. The energy here involved (in the " scientific imagination "?) is equal to that of six trillions of horses working for the whole of the time employed by Joshua in the destruction of his foes. The amount of power thus expended would be sufficient to supply every individual of an army a thousand times the strength of that of Joshua, with a thousand times the power of each of Joshua's soldiers, not for the few hours necessary to the extinction of a handful of Amorites, but for millions of years."

These calculations are all very pretty, but they are worse than useless as the Bible does not speak of "arresting the *earth's* motion," but of the *sun* standing still. Hence they are utterly beside the mark; but the above quotation serves to shew how men of " science " are led away from the Scriptures by unfaithful expositors and a false philosophy until, as Tyndale confesses, "Belief reels under the reflection." While christian men and so-called " Reverend Divines," who are paid to defend the Holy Writings, play into their hands by ignorantly, or cowardly, yielding the claims of unfounded astronomical theories so utterly subversive of Bible teaching and true Natural Science. However, it is only fair to the writer of the pamphlet under consideration to say that he rejects this " explanation " also; although, at the same time, he holds those astronomical theories by which it is supported. He also makes the same mistake of talking about the *earth's* motion being arrested instead of that of the sun, for he says;

"Why did not the ocean overflow the land? Run with a pail of water until you come in contact with a wall, and observe the effect upon the liquid, how it will dash over the side: and the sudden stoppage of the rotary motion of *the earth* (!) would naturally send the sea almost all over the dry land . . . You know the shaking you get with the violent stoppage of an express train going at sixty miles an hour, and we ask you, please, to *fancy* the result to us, and to all cattle, dwelling houses, monuments, and even trees, if the *earth*, which at the equator *moves nearly* 1,100 *miles an hour*, was brought quickly to a stand still."

Now that is altogether and utterly irrelevant. When will professed defenders of the Bible let it speak in its own terms? What infidel could wrest the Scriptures more from their plain literal and grammatical sense? The American infidel Ingersol writes just in the same strain respecting this miracle in his so-called " Mistakes of Moses." But is it not rather a mistake, and a grave mistake, of Ingersol, Tyndale, Howard & Co., to speak of the Bible arresting the *earth's* motion, when the account says nothing whatever of the kind; but distinctly tells us that it was the *sun* and moon which stood still? They may charge the Bible, if they like, with being contrary to modern science; but we should retort that it is both illogical and unscientific to condemn the Bible on such a charge until the " science " in question has first been shewn and proved to be true. Let them first prove the earth has any motion, before talking about the "arresting" of it. And we want something better than *Foucault's* pendulum experiment for this—especially as different pendulums will sometimes oscillate in opposite directions !—and more especially as practical experiments have already proved that the earth has no such motions as those attributed to it. The account of these experiments may be found in Parallax's great work, "Earth not a Globe." We have no space now to quote these experiments, as we are at present only engaged in shewing up the inconsistency of those who wrest the plain statements of the Holy Scriptures to suit the fanciful and absurd theories of modern "Science," falsely so-called. They may yet appear in the *Earth Review* in due course, if our friends will only come forward and sustain our hands in this unequal conflict. Some of them have already appeared.

THE LATEST EXPOSITION.

But our readers will naturally be anxious to know what is the final "explanation" given by the writer in question, who acknowledges that he had previously been "utterly bewilded with every attempt either to explain the miracle, or to explain it away." We shall let him speak for himself. He says ;—

"I have now a FIFTH VIEW to lay before you, which appears to be both rational and simple." . . . "My *belief* is this: Joshua and his men having walked all night, as the 9th |verse tells us, would be tired next morning, but God caused a great trembling to spread itself amongst the foe, and there was an easy victory. When the war had pursued the Amorites some distance, hailstones fell upon them and did much damage. At the approach to Bethhoron the hailstorm increased in fury; and Joshua, seeing the devastation produced, and being cognisant of the fatigue of his men, *prayed Heaven to let the hurricane go on* till total and irreparable disaster was inflicted."

We refrain from saying all we think about this so-called "explanation," as the writer is evidently both sincere and devout ; and he says that "it flashed across my mind many years ago, when I was on my knees." But we think it doomed to the same failure as the rest, and and for similar reasons ; it is not true to the sacred narrative. It reminds us of what the editor of the *Daily Chronicle* said of Dr. Geikie's book, *The Bible by Modern Light.* "He makes assertions which have the charm of novelty, but also the vice of inaccuracy." (See fuller remarks from the *D.C.* in another page). This is the case with the present attempt. We have no record that Joshua "prayed Heaven to let the hurricane go on." This is an assertion, not of the narrator, but of the "expositor." Joshua prayed for the *sun* to "stand still." not for the *hailstorm* to proceed, and we are told that "there was no *day* like that, before it or after it, that the LORD hearkened unto the voice of a man for the LORD fought for Israel." But to get rid of this fact our expositor says ;

"The chapter (10th of Joshua) is made up of two accoonts, the one historical the other poetical. The poetical extends from the 12th to the 15th verse. The rest is historical."

This is oracular and authoritative! Mr. Howard comes back after all to a "*Poetical Theory*," although such a theory was the first one he so conclusively rejected. This only proves the impossibility of explaining the account in harmony with modern science *on any theory*. In short *the narrative needs no explanation in itself;* IT ONLY NEEDS BELIEVING! And, as " all men have not faith," let anyone of those without try to prove, if he can, that the account is not in harmony with the facts of Nature. This would be straight forward and reasonable; but to wrest the Scriptures, to twist and torture their language until it is made to mean anything the writer wishes, is neither strictly honest nor truly scientific. The very attempt to do so only serves to shew the unconscious influence and injurious effect modern astronomy has had on the minds of otherwise good and honest searchers after truth. Only let the incubus of this superstition (and we use the word "superstition" advisedly as of something standing above, or outside, natural facts) only let this incubus be removed from their minds, and the skill such writers manifest might do credit to the expository science they affect; but while their minds are, consciously or unconsciously, enchained by the trammels of a false philosophy, imposed upon them while they were too young to question it, they will not only " wrest the Scriptures," as they do, but writhe as it were in the meshes of a critical snare evidently laid for us by the Arch Deceiver of mankind. We have need to pray that our minds, and that the minds of our " Ministers," may be delivered from this "snare of the fowler." The miracle under consideration shews that God hears prayer, and answers it; but when He does *He* never flashes ideas or interpretations across the mind which are out of harmony with the general statements of that Divine Cosmogony revealed in his Holy Word.

"To the Law and to the Testimony; if they speak not according to this Word, it is because there is no light in them." Isa. 8 : 20.

JOSHUA CORRECTED.

Before concluding our paper let us briefly consider the validity of some of the reasons given for this novel interpretation. Firstly, the employment of a hailstorm was a "means already in operation, and in every way capable of securing the end in contemplation." This is so utterly beside the question that we dismiss it at once. We might deny the hailstorm itself on such flimsey grounds. Secondly, we are told that " the language of the inspired penman suits this theory, *and no other* !" We will content ourselves with putting a note of exclamation after that!

Then "It is poetical, and all poets are allowed some latitude in their descriptions." Our expositor ought to be a poet of no mean standing for he evidently claims a poet's privilege ! He says the account is extracted from the Book of Jasher, which seems to have been made up of martial odes," intended to "develop patriotism and faith in God." If Mr. Howard had not prefixed the title "Rev." to his name, a title which his Master has practically forbidden (Matt. 23 : 8 vs.) we might have thought this the suggestion of a sceptic, that "faith in God" could be developed by the poetical recounting of a false miracle ! But supposing that Mr. H's bare assertion that "the poetical portion extends from the 12th to the 15th verse" were true, what has he already told us respecting the genius of Hebrew poetry ?

"I have sought all through the Bible and have not discovered one instance of a natural event being exalted into a miracle by any of its bards. Great occurrences which are wonderful in themselves are greatly adorned, but left free from all miraculous elements. . . . . This enquiry into the veracity of Hebrew poetry has amazed me—made me feel how, contrary to the general view, in all their highest inspirations, the Bible bards kept a clear eye on sober truth—a remark, I think, which applies to the poets of no other nation."

Thus his own words are sufficient to answer the supposition that the account in question is a "poetical" figment. But we do not admit that three verses are poetical. They seem to us just as historical as the rest of the chapter, and ancient Israel believed them to be so. We believe that Mr. H. would never have objected to them as equally historical with the rest of the chapter were it not for the absurd idea that we are living on a vast globe, turning us all head over heels once every twenty-four hours, and so alternately bringing day and night. This appears from his further remarks. He says ;

"The first remark I have to make upon these words, as here rendered, is that if the prayer had been answered the day would not have been lengthened. To lengthen the day the *earth* must either slow in her rotatory motion or stop it altogether ; and Joshua, had he wanted more hours of light, should have said, 'EARTH pause in thy revolution upon thy axis, or go slower.' Thus you see our Versions take all the meaning out of Joshua's prayer. *Our View* shows its point and beauty."

This would really be amusing to Zetetics if the matter were not otherwise so serious, and the writer evidently so earnest. He calls poetry, Hebrew, and astronomy all to his aid. He says that the Hebrew word *dom* never means to "stand still." It may not be again so translated, not exactly, and yet it may have this meaning. We think it has. The root word is *damam*. The writer admits it is once translated "tarry" 1 Sam. 14 : 9. Athough the word sometimes may be rendered *be silent*, this passage clearly shews it also means to *stand still*. It reads, "If they say unto us, *Tarry* (*damam*) until we come to you, then we will *stand still (amad)* in our place." This latter word *amad* is the very Hebrew term used in Hab. 3:11, which again speaks of the sun standing still! Is this wrong also? We have faith in the translators to believe that they understood Hebrew as well, if not better, than the writer; and they, while giving various shades of meaning in the margin, give unmistakably the right meaning in the text, "Sun *stand* thou *still*," for we read "the sun *stood still (amad)* in the midst of heaven." v. 13. Mr. H. says the latter term means to *rise up*. But it can not mean this only, for Parkhurst gives the primary meanings, "*To stand, stand still, stay, remain.*" This Hebrew Lexicographer also says that "The Seventy generally render the verb by *istemi* to stand, and its compounds." As it may interest the reader we will give the translation from the Septuagint, shewing, how ancient Greek translators, untrammelled by modern astronomical theories, understood this passage;

"Then Joshua spoke to the Lord, in the day in which the Lord delivered the Amorite into the power of Israel, when He destroyed them in Gabaon, and they were destroyed from before the children of Israel. And Joshua said, Let the sun stand over against Gabaon, and the moon over against the valley of of Aelon. And the sun and the moon *stood still*, until God executed vengeance on their enemies."

Italics of course are ours. Those who wish to pursue this point further will find the same Hebrew word (*amad*) translated " stand still," or its equivalent, in the following passages;—Josh. 3:8, 17; 10:13; and 11; 13; 1 Sam. 14:9; and 2 Sam. 2:23 and 28; &c.; as also in the remarkable passage referred to in Hab. 3:11. It plainly appears, therefore, unless the translators did not understand Hebrew, that "stood still" is a correct and frequent translation of *amad*; and doubtless it never would have been called into question as applied to the sun were it not for the baseless theories of modern astronomy. These are at the bottom of the whole contention. The passage had to be harmonized with a philosophical, or rather an *un*philosophical, theory; so the translation must be altered to suit! As Mr. H. remarks;

"When once a theory takes holds it grows apace and wields a power over future ages that is seen in expositions, annotations, and translations . . . till the original modicum of truth is distorted or lost in the process."

And again, we quote with approval;—

"The Bible itself will have to be studied anew in its own light; and when this is done, and we get back to its grand and simple truths unmixed with false views from extraneous sources, we shall be delighted with what it is and what it has to tell us."

This is good advice, if followed. And amongst the grand and simple truths of the Bible will be found that the sun has motion (Psa. 19:4); that the earth (or *land*) rests on "foundations" (1 Sam. 2:8);

and that it is so established "that it should not be removed for ever." Psa. 104:5., &c., &c. Yet in spite of this good advice, and the fact that the Scriptures do teach the Plane system, the writer speaking about his new theory or explanation says;—

"Our theory disposes of an old infidel objection to revelation. Sceptics sneer at the Scriptures because as they say, they inculcate the Geo-centric system of astronomy. instead of the true (!)—the Helio-centric; and this miracle has ever been the prop of their charge. 'See,' they have said, 'when Joshua wanted the day lengthening, he commanded the sun and moon to stand still, thinking falsely (?) that they circled round the earth every 24 hours; whereas it is the *earth* (oh!) revolving round on her own axis, that makes day and night' But *our theory* will put an end to this, and prove that Joshua knew what he was doing."

Vain hope! No mere "theory" will put an end to the infidel's sneer. Our plan is not to oppose *theories* or quibbles to the sneer of the sceptic, but *facts*; and then let him sneer if he can for shame. If the infidel can prove that water is convex, or that the earth really tumbles at all, land and water, topsy-turvey once every twenty-four hours, then he has a right to sneer at Joshua's ignorance; but if he cannot, and the pages of the *Earth Review* are open for any respectable effort, then *we* shall sneer at *his* ignorance, his lack of reasoning power, and his consummate folly for allowing himself to be duped out of Eternal Life over the simple and plain facts of Nature! We have a word also for the Christian. Why should you allow infidel theories respecting the universe, its form and its origin, to blind your eyes to the facts you see, or may see, around you, and to the harmonious teachings of that Divine system of Cosmogony revealed in Holy Writ? You need not attempt to make truth "reasonable"; it is reasonable, to the unfettered and really free thinking mind. Neither need you attempt to "explain" a miracle; it is above you. While the attempt to "defend" a miracle is puerile and absurd. A miracle is its own defence. All you have to do is to *believe* it, when attested. Defending a miracle is like a child defending a giant, or a fox defending a lion! But if you cannot believe your Bible, and if you are too indifferent or too ignorant to go into the proofs offered around you, then honestly join the infidel party, and prove the Bible is wrong in its Creation and its Cosmology, that is *if you can*.

We shall conclude our paper with a quotation from Josephus, a Jewish writer and historian who lived in the first century of the Christian era, and who was doubtless well acquainted both with the language of the Jews and the remarkable and miraculous history of Israel. Respecting the miracle in question he writes;—

"Joshua made haste with his whole army to assist them (the Gibeonites), and marching day and night, in the morning he fell upon the enemies as they were going up to the seige; and when he had discomfited them he followed them, and pursued them down to the descent of the hills. The place is called Bethhoron; where he also understood that God assisted them, which He declared by thunder and thunder-bolts, as also by the falling of hail larger than usual. Moreover it happened that *the day was lengthened* that the night might not come on too soon, and be an obstruction to the zeal of the Hebrews in pursuing their enemies" . . . . Now that the day was lengthened at this time, and was longer than ordinary, is expressed in the books laid up in the Temple."

ANTIQ. B. V. C. I. S. 17.

In a note under this paragraph Mr. Whiston, the learned compiler of Josephus' works, while hesitating what explanation to give the miracle says;

" The fact itself was mentioned in the Book of Jasher, now lost, Josh. 10 : 13, and is confirmed by Isaiah (28 : 21), Habakkuk (3 : 11), and by the son of Sirach (Eccles. 46 : 4). In the 18th Psalm of Solomon, ver. *ult.* it is also said of the luminaries, with relation no doubt to this and the other miraculous standing still and going back, in the days of Joshua and Hezekiah. 'They have not wandered from the day He created them, they have not forsaken their way, from ancient generations, unless it were when God enjoined them (so to do) by the command of his servants.' See Authent. Rec. part I, p. 154."

> " Hear the just law, the judgment of the skies,
> He that hates truth shall be the dupe of lies ;
> And he that *will* be cheated, to the last
> Delusions strong as Hell shall bind him fast."

"CRANKS,"

OR

THE FALSE THEORIES OF "SCIENCE"
versus
THE TRUTH OF NATURE AND THE BIBLE.

By "Zetetes."

ANYONE but moderately acquainted with Theoretical Science knows that much which passes current in these days as "Science" is opposed to the plain teachings of the Bible. But this fact gives neither the "Scientist nor his disciple, any concern whatever; because, while he has been taught to doubt, or to discredit the teachings of the Bible, he has never been sufficiently sceptical, or perhaps sufficiently instructed, to doubt the teachings of what is presumptively called "Science." And in fact the ordinary mortal who dares to question anything advanced in the name of "Science" is considered to be either an ignoramus, or what is called a "crank;" and especially so if he dare to stand forth in defence of the Holy Scriptures against so called scientific teaching. But why should "Science" be exempt from criticism and enquiry any more than the Bible? The way some professed Christians treat the Scriptures, where they are opposed to the theories of Science, is discreditable in the extreme; not only to the authority and inspiration of the Scriptures, but discreditable to their own profession and understanding. They profess to believe that the Bible is inspired of God, yet they apologize for its language, as though the writers were uninspired and ignorant of the fundamental facts of Creation. But may not this ignorance possibly be ours, not theirs? Such conduct is highly reprehensible on the part of those who are in positions where they are paid to defend and to advance Bible teachings and doctrines; and especially where they have never paused to inquire whether the discrepancy found between "Science" and the Bible is due to the ignorance of Bible writers or to the fallacies and unfounded theories of a "Science" which, as Paul says, is "falsely so called."

The word "Science" means *knowledge,* knowledge of the facts of Nature, &c.; but no doubt much that now passes for "Science" is not *knowledge* at all, but mere theory, or scientific guess-work. We have nothing whatever to say against the *facts* of science or Nature, but no *fact* in Science, and no real fact in Nature, can be found opposed to, or inconsistent with Bible teaching—but we have a right to question mere scientific *theories,* especially when those theories are opposed, as many of them undoubtedly are, to plain Bible doctrines. Take for instance the theories of Modern Astronomy, which is supposed to be one of the "exact sciences." Recognized Astronomers have differed as much as a hundred millions of miles respecting the distance of the sun alone; yet the sun's distance is the very elastic "measuring rod" of all other astronomical distances. Honesty must confess that the theories of modern Astronomy are directly opposed to Bible teachings; but it never enters the mind of the Scientist, and seldom even that of the professed Christian, to inquire which side is right. No! The claims of "Science" are cowardly yielded as being above question. While the Bible is either ignored altogether or its language is tortured to fit the latest modern scientific theory; or the pitiable excuse is made that the "Bible was not intended to teach Science," and that the writers wrote —not as they were moved by the Spirit of God—but according to the general belief of a past and ignorant age! Shame on such "Christian" defenders of God's Word!

MODERN ASTRONOMY.

But the Bible *does* deal with the question of Creation, and it gives an account of the universe in harmony with natural appearances; and if its various writers were wrong in their harmonious teachings respecting God's world and this earth, they could not well have been inspired by the good Spirit of Him who created it. In fact, a brother in the faith gave it as a proof that the Bible is *not* inspired because it describes the earth as an "outstretched" and motionless plane, having "ends," "corners," "foundations," &c. It had never entered his mind to question the modern globular theory. Let us, however, be honest with sacred things, and venture to meet the sceptic on his own ground. Truth has nothing to fear from facts ? and only *theories* can be feebly opposed to it. Take for instance this popular theory, that the earth is a Globe. This is but a *theory* after all. It has never yet been *proved* by a single fact in nature. The Bible speaks of the earth as an "outstretched" plane, resting on the waters of the *great deep*; and *all natural phenomena can be explained on this basis*, without assuming as astronomers do assume, the sphericity of the earth. It can be round and flat too. "Scientific" theory says that the earth is not only a vast globe, but that it is whirling away and flying through "space" at the awful rate of over a thousand miles a minute in its orbital motion. This is not only contrary to Bible teaching, which represents the earth as being at rest on "foundations," and "established," so fast that it "cannot be moved" at any time; but it is contrary to the testimony of our senses, and the intelligence of the wisest men in the world for over five thousand years. This supposed motion, whether "axial" or "orbital," is neither felt nor visible; and if a thousand pounds were offered for *proof* of the earth's rotation or revolution, not a single proof could be given in support of it. If otherwise, let the proof of the earths' motion be forthcoming; Mr. Carpenter offers one hundred dollars for it, and we only wait to see it fairly put in print. It would be a literary curiosity! Yet, forsooth, we are called "cranks" for not believing this monstrous idea; while the poor deluded infidel *prides* himself for not believing the Bible! However, we invite him honestly to try his hand at the proof asked for; or at least to give up sneering at "Joshua commanding the *Sun* to stand still, instead of the earth." Yet the infidel who is unable to prove whether it is the motion of the Sun or of the earth which causes day and night, is far less inconsistent in his unbelief than the professed Christian, who, while equally ignorant with the sceptic in this respect, yet professes to believe that the Prophets were inspired of God, to write and speak as they did. Brethren, let us be consistent with ourselves and with our profession. If modern Astronomy is right, Joshua and the Bible are wrong. But let the proof asked for be given before we yield the contention against the Bible. This is only reasonable and fair.

WATER LEVEL.

The Bible is more scientific than many people are aware; and it cannot be overthrown quite so easily as some of our opponents imagine. Let them try it here, and *prove* that the earth has any motion, sidereal or orbital, to say nothing of the awful diurnal head-over-heels motion attributed to it. If it has not this motion—and we defy any man in the world honestly to prove it has—then the earth is not a globe at all, and the *natural idea* of a motionless and extended plane, in harmony with Bible teaching and ancient astrological belief, is after all right! We know that scores of other questions might be propounded here, *but let this be settled first*; for if the earth has no motion, then it is modern astronomy that is wrong, and not the system of the Bible. We can answer all other questions when time, space, and means are allowed us; but we here and now challenge these *fundamental* theories, or hypotheses of modern theoretical astronomy. The cleverest astronomers, Newton and Copernicus, admit that they are but theories, suppositions, not facts; mere hypotheses, not *science*, or knowledge.

On the other hand *we* can give proof, *to those who desire to know*, that the earth *is* a motionless and "outstretched" plane. This proof is found in connection with the grand fact that the surface of all *still water* is perfectly *level*, not convex as it ought to be if the earth were a globe with the sea all round it. The fact that water is level is at the basis of the Zetetic teaching; but many other facts besides, facts found in nature and outside Bible teaching, go to prove that the Bible view of the Creation is right, and that of the so-called "Scientist" is wrong. Our own senses too, tell us that people are never found in any part of the world living with their heads downwards and "their feet towards our feet," at some fancied "antipodes." It is those who believe, or rather who promulgate such absurd notions that ought to be considered cranky; not we who believe in the deliberate and intelligent verdict of our God-given senses, and in the teachings of his own infallible Word. However, I for one am not going to be scared out of my senses, and out of my belief in the Bible too, because some superficial sceptics ignorantly cry out "Crank"! I have generally found such people utterly incompetent to attempt even to give a decent so-called *proof* in favour of their own position. If I speak unadvisedly, let them attempt to give the proof asked for of the earth's supposed terrible motion.

While on the subject of "Cranks" I would commend those who call us such names, for believing in the evidence of our senses, to read what an American humorist is said to have written about them in the *Alliance News*. If "cranks," or "paradoxists," are such as are here described, we need not be ashamed of being compared with them. It runs as follows:—

A WORD FOR THE CROTCHETEERS.—Cranks, my son? The world is full of them. What should we do were it not for cranks? How slowly the tired old world would move did not the cranks keep it rushing along. Columbus was a crank on the subject of American discovery and circumnavigation, and at last met the fate of most cranks, was thrown into prison and died in poverty and disgrace. Greatly venerated now? Oh, yes, Telemachus, we usually esteem a crank most profoundly after we starve him to death. Harvey was a crank on the subject of the circulation of the blood; Galileo was an astronomical crank; Fulton was a crank on the subject of steam navigation; Morse was a telegraph crank; all the old abolitionists were cranks. The Pilgrim Fathers were cranks; John Bunyan was a crank; any man who does not think the same as you do, my son, is a crank. And by-and-bye the crank you despise will have his name in every man's mouth, and a half-completed monument to his memory crumbling down in a dozen cities, while nobody outside of your native village will know that you have ever lived. Deal gently with the crank, my boy. Of course, some cranks are crankier than others, but do you be very slow to sneer at a man because he knows only one thing, and you can't understand him. A crank, Telemachus, is a thing that turns something, it makes the wheels go round, it insures progress. True it turns the same wheel all the time, and it can't do anything else, but that is what keeps the ship going ahead. The thing that goes in for variety, versatility, that changes its position a hundred times a day, that is no crank, that is the weathervane, my son. What? You nevertheless thank Heaven that you are not a crank? Don't do that, my son. Maybe you could not be a crank if you would. Heaven is not very particular when it wants a weathervane; almost any man will do for that. But when it wants a crank, my boy, it looks about very carefully for the best man in the community. Before you thank Heaven that you are not a crank, examine yourself carefully, and see what is the great deficiency that debars you from such an election. *(From the "Banner of Israel.")*

Again, another writer in the *San Francisco Truth*, well says:—

"If a man is too much for you in argument, or so much better informed than you are, that you do not enjoy his conversation, call him a crank. If his conscientious devotion to principle makes you ashamed of your own loose morality, political or otherwise, just call him a crank, and get even with him. It is really quite an honor to be called a crank (under such circumstances) but the fools who use the term so freely have not yet found it out."

OUR BELIEF.

But we must, for want of space, briefly summarize our position as follows:—

We believe, for reasons which we are able to render, when the opportunity or the means are forthcoming, that,

1. The earth is a stationary and "outstretched" plane, resting, as the Bible teaches, on "foundations."

2. That the sun and moon are simply "two great lights," circling around and above the earth, not more than about three thousand miles off; and only large enough, by reason of our thick atmosphere and the laws of refraction, to illuminate about one half of the world at one and the same time.

3. That the stars are still smaller bodies of light, electrical or magnetic, but not "worlds like ours," nor *suns*, but "lights," intended to subserve this the only world known to the Bible or to *fact;* and as the Creator himself said, intended "for signs and for seasons," "to rule over the day and over the night," and "to give light upon *the earth*." Gen. 1: 16-18.

Now, all the phenomena of Nature, the seasons, tides, eclipses, &c., are explainable in harmony with the above propositions; while we challenge any scientist, infidel, or christian, to give us a single reasonable "proof," or a fact not otherwise explicable, to prove that we are living on a vast rotating globe, flying through "space" two or three times faster than a flash of lightning! Only *one* good proof of this is wanted. Who can give it? Christian friend, which will you choose to believe? The Bible or "Science"? The Holy Scriptures which are able to make you wise unto salvation, or "Science" which an inspired Apostle intimates may be "falsely so-called," and which is fast leading men into doubt and infidelity? "Prove all things: hold fast that which is good," is the precept addressed to you. If you are not able to prove these scientific theories true, then stand by us and your Bible like Christian men, as you ought to do, at least until some one can do it for you. And, unbeliever, if you think you can prove Joshua was wrong in attributing motion to the sun rather than to the earth, set to work and put your proof in print, that we may view it, and *review* it, for your benefit and ours. But let us have argument not assumption, reason not ridicule, science not sneers. We are willing to abide by the result. *Are you?* If honest you are.

An Inspired Warning.—"Beware lest any man spoil you through philosophy and vain deceit, *after the tradition* of *men*, after the rudiments of the world, and *not after Christ.*" Paul.

Another—"If ye believe not his (Moses') writings, how shall ye believe my words." *The Christ.*

SCIENTIFIC BLASPHEMY.

The following is proof, for those innocent minded Christians who still need it, that "Science" has something to do with the question of Salvation, inasmuch as it is leading men not only to deny the truths of the Bible, but, as a consequence, to deny the Christ and the God of the Bible. The first paragraph is from a weekly paper with the very suggestive title, *Lucifer*, published in America; and from a number dated "December 23, E.M., 287." This is instead of calling it the year A.D. 1887. Why do they refuse to acknowledge the A.D.? The editor himself shall tell us. He says:—

"We date from the First of Jan. 1601. This era is called the Era of Man, (E.M.) to distinguish it from the theological epoch that preceded it. In that epoch the earth was supposed to be flat, the sun was its attendant light revolving about it. Above was Heaven where God ruled supreme over all potentates and powers, below was the kingdom of the devil, hell. So taught the Bible. Then came the *new* astronomy. It demonstrated (?) that the earth is a globe revolving about the sun; that the stars are worlds and suns; that there is no "up" and "down" in space. Vanished the old heaven, vanished the old hell; the earth became the home of man. And when the modern cosmogony came, the Bible and the church, as infallible oracles, had to go, for they had taught that regarding the universe which was now shown (supposed?) to be untrue in every particular."

Gently, friend *Lucifer*, for you are somewhat in the dark here, notwithstanding you assume to be a great *light bringer!* It never has been "demonstrated" or "shewn" that the earth *is* a whirling globe, and that, therefore, the Bible cosmogony is wrong. It has been quietly *assumed* by the "new astronomy" and the assumption has been cowardly yielded by the "Christian" who ought to have challenged it. But it never has been *proved.* Never! If it has, kindly give us the name and the address of the man who "demonstrated" it. Newton and Copernicus, both, were candid enough to confess that the theory called by their names is but a theory, a mere assumption not based on known facts. Their disciples forget this.

And now, what is the result? It is seen in the above infidelity; and in the further fact that a correspondent signs his name under these words :—

"Yours without Christ, and no hope or desire of ever reaching the New Jerusalem." See *"Folly,"* March, 1890.

But, Christian friend, if the earth be a great globe, shooting and spinning away through "space" immeasurably faster than the deadliest cannon ball, how can the New Jerusalem, a city twelve miles square on every side, and twelve miles in pyramidical height; how, I ask, can such a city come "down from Heaven," as the apostle John shews it will, and rest, in a particular and a prepared locality, upon its grand and glorious foundations? Well might Thomas Paine say, as he did say, in his *Age of Reason*:—

"The two beliefs,"—Modern Astronomy and the Bible—"cannot be held together in the same mind: he,, who thinks he believes both, has thought very little of either."

Once more, in Reynold's Newspaper, (England,) Sunday, Aug. 14, 1892, under heading "Democratic World," we find the following blasphemous paragraph, which is also ignorantly based on the assumption of the truth of this much vaunted "New Astronomy." It is written by some one who very suitably signs himself "Dodo," and it runs as follows :—

"We are trembling on the eve of a discovery which may revolutionize the whole thought of the world. The almost universal *opinion* of scientific men is that the planet Mars is inhabited by beings like, or superior to, ourselves. Already they have discovered (?) great canals cut on its face in geometrical form, which can only be the work of reasoning creatures. (?) They have seen its snowfields, and it only requires a telescope a *little stronger* than those already in existence to reveal the mystery as to whether sentient beings exist on that planet (!) *If* it be found that this is the case the whole Christian religion will crumble to pieces. The story of the Creation has already become an old wife's tale. (?) Hell is never mentioned in any well-informed society of clergymen ; the devil has become a myth. *If* Mars is inhabited, the irresistable deduction will be that all the other planets are inhabited. This will put an end to the fable prompted by the vanity of humanity that the Son of God came on earth and suffered for creatures who are the lineal descendents of monkeys. (?) It is not to be supposed that the Hebrew carpenter Jesus went about as a kind of theosophical missionary to all the planets in the solar system, reincarnate, and suffering for the sins of various pigmies or giants, as the case may be, who may dwell there. The astronomers would do well to *make haste* to reveal to us the magnificent *secret* which the world impatiently awaits." Dodo.

Yes, yes, "Dodo"; you are evidently a very fine bird to be the "lineal descendant of a monkey"! Quite a *rara avis* in your way, no doubt. Although the famous astronomer Signor Schiapparelli has said, "The newspapers are wrong in attributing to me the idea of finding in the duplication of the lines on Mars a proof that that planet is inhabited, based on the *supposition* that the lines are the work of reasonable beings"; yet, "Dodo" as you confess the whole affair is still a great "secret," do push on the astronomers to make their glasses only a "little stronger," that we may know whether your progenitors were really monkeys or not. Do do, "Dodo" pray do do. If you can but prove that the earth *is* a "planet" at all, you will not only have the pleasure of overthrowing the Christian religion, which you evidently hate, but a friend of mine offers one hundred dollars towards making your telescopes a "little stronger" still, so that you may yet *see* the water in those parallel "canals" of Mars! Do push on the astronomers "Dodo," pray do do; and thus crumble to "pieces," if you can, this

BIBLE VIEW OF THE WORLD.

1. Heaven is *above* (not all round), earth *beneath*, and "water *under* the earth." Ex. 20, 1-4.

2. Heaven, the firmament : a semi-transparent structure, strong enough to divide the waters "above" it, from those "below" it. Gen. 1 : 7 ; Job, 37 : 18; and Psalms 19 : 1.

3. The sun, moon and stars, placed within the firmamental *vault*, are powerful "lights" only, some greater some lesser, electrical and magnetic, intended for "signs and for seasons," and to give light to this the only world. Gen., 1 : 16-18; Psa. 136 : 7-9; and Rev. 6: 13.

4. The earth is represented as being "outstretched" as a plane, with the "outstretched" heavens everywhere above it, like a circular "tent" to dwell in; to the great confusion of our so-called "wise" men. Isa. 40: 22; Prov. 8: 27; Isa. 44: 24-25 ; Luke 4: 5; and I Cor. 3:19.

5. The earth (or *land* portion of the world) is firmly and immovably fixed on "foundations," or "pillars ;" having "ends" and "corners" jutting out into the sea, like *Land's End, Cape Finisterre*, &c. Gen. 1 : 10; Job 38 : 4-6; 1 Sam. 2: 8; and Psa. 93: 1; and 104,: 5. R.V.

6 The sun, moon, and stars move around and "above" the earth (not more than a few thousand miles off) so that day and night are "ruled" by the motions of the heavenly bodies, or "lights," and not by the supposed axial motion of the *earth*, which contradicts the Holy Scriptures as well as our own God-given senses. Heaven is nearer to us than we have imagined. Josh. 10: 12-14; Psa. 19:4-6; Luke 24: 51; and Dan. 9: 21-23.

7. All that exists was created in six days (of the same kind as the seventh), and not slowly evolved, as infidels suppose and recklessly affirm, during "millions of millions of years." God said; " In six days the *Lord* made heaven and earth, the sea, and all that in them is, and rested the seventh day; wherefore the *Lord* blessed the Sabbath day, and hallowed it." Ex, 20: 11. SHALL WE BELIEVE THE CREATOR, OR THE CREATURE?

Christian, will you be guilty of so great a sin and enormity; and especially for a modern unproved and *unprovable* assumption? *See previous notes.* For, " *He that believeth not God hath made Him a liar.*" Dare you act thus, and deny the truth of His Word? which, in spite of what half-hearted Christians say to the contrary, *does* deal with the question of the Creation and the Universe, setting forth the wonderful works of God as the basis of our allegiance to Him as the Creator.

TO BE HAD FROM THE WRITER, AS BELOW. PRICE 1d.
OR POST FREE 1½d.

Further information, with a list of similar publications, to be obtained on receipt of a stamped and addressed envelope, sent to:—

ALBERT SMITH. PLUTUS HOUSE, ST. SAVIOUR'S ROAD, LEICESTER, ENGLAND.

Printed by R. W. COLES & Co., 68, Highcross St., Leicester.

TRUTH

•

THE EARTH IS FLAT

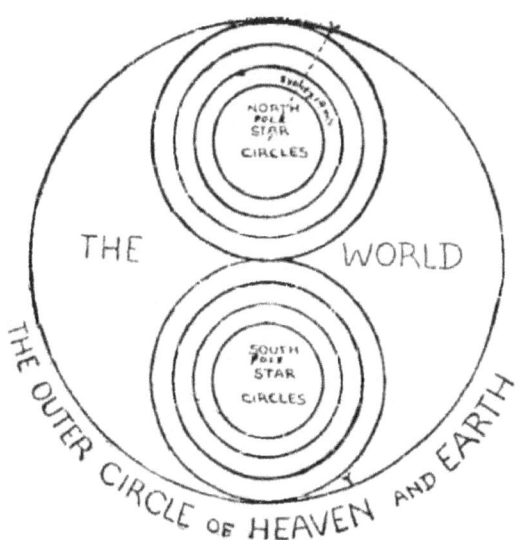

BOOKLET VI.

Printed by
E. & L. ATKINS
1 Lark Street,
Belmore

TRUTH

BOOKLET VI.

Foreword

The writer of this pamphlet feels almost shame that he is called upon to prove the flatness of the surface of the earth.

The subject is so elementary and the proof so obvious that he feels that his reader will be hurt by the simplicity of things.

It will be, as one of his listeners has already said, "The thing is too simple to be true." Or may I express it, in the words of another listener "Oh, how did they come to believe it?"

Reader, I will tell you how the Learned came to believe in a globular earth.

The ordinary and the commonplace find no place in the mental soul of the learned. Commonsense irritates him, his mental soul is wrapt in an inner conscienceness that his learning displaces commonsense and that it rides "rough shod" over the mind of the layman of the earth; and when the layman looked on the flat surface of the ocean, the man of Science could not let him believe in what was so obviously true so at the first opportunity, he did away with the flatness of the water surface and curved it into a globe; and "all the world wondered" and Professors and Pedagogues were given a free hand to debauch the intellect of the laymen as often as they wished.

Reader, I am forced to admire the Learned for his progress into the "unknown", and though I am forced to denounce his innate "conscience", yet, on account of my own conscience, I travel softly.

TRUTH

What has warped the mind of the learned is that his knowledge comes from "nowhere"; it has no background of reality, no Mother to bring it up, step by step, limb by limb, bone by bone, sinew by sinew, life by breath, mind by reason, and continuity into eternity.

Reader, behind all knowledge lies one little word "Wisdom", Wisdom the Eternal Mother of Mind and Law.

All knowledge is built up from A. B. C. building blocks, but the end of the Alphabet will ever be illimitably beyond the greatest of any living mind. However, the dead mind or state of physical nature can always absorb the living mind of Wisdom.

As expressed in Booklet III, "The finite (dead physical nature) encompasses the infinite (Wisdom, in her never-ending principles)", for the subject remains immutably existent through every mutation.

The end of the matter is expressed by God, as follows: "There is nothing new under the Sun." He continues to say: "Is there a matter whereof it may be said, 'See, This is New!'"

Whatever can exist has always been potentially existent.

Correspondence invited, for publication if desired.

S. G. FOWLER,
5 Graham Street, Seaforth.

TRUTH

THE EARTH IS FLAT

Reader: It would be a strange life trying to live on a Globe. Geometry knows no other law than Geometry, and the physical properties of a physical globe would create insurmountable difficulties for the being called "man" for man is a two-legged, smooth-footed, clawless-toed, and heavily-built creature.

Short of wind, slow-moving, full of fear and full of nervous re-actions. Picture him on the outside of a sphere in our popular 34° South latitude.

He has his boots on and his head is depressed in space 34° to his feet. Consider him magnetised through his boots to the centre of the globe, where the "big magnet" is located. Picture him looking down into the gaseous void, with his eyes gouging out of their sockets, and his heart in his mouth; and his prayer that his "hob-nailed" boots will not lose their magnetism.

There he stays "stuck-put" until he swallows his heart, and gingerly withdraws one foot towards his head to see if he can do the trick "one hand".

"Like a "Pelican in the wilderness" he finds himself standing on his magntic leg of gravity.

Recovering his consciousness and his mental equilibrium he finds he has no space below his head save the little bit between his head and his feet.

Becoming bold in the recovery of his common sense, he finds he is standing upright to his feet, and perpendicular to the earth's horizontal plane.

What a mental chimera he has been through; he has imagined himself to be sailing on the OUTSIDE OF A MOUNTAINOUS GLOBE, that is, of itself, rotating, revolving, gyrating and nutating; and yet, his "mountainous globe" is but a mere atom being whirled about in space by the diverse energies of "system upon system": it is but a toy balloon fastened to the wheel of the Solar System of

TRUTH

the fiery Sun, which system is but a flyspeck on the "udder" of the Galactic Cow-System, which in its turn is but a "lost sheep" of the Stellar Herds of space that sport with velocities too fearsome to utter.

No wonder the world's brain got addled! The reader will be presented with a series of diagrams and from these diagrams he should have no trouble in realizing that he has been hoaxed by the stupidest manifest hoax ever perpetrated on a being who was divinely granted logic, but who by his own flesh perversities has "tied himself into knots" to live according to the dictates of his fleshy nature in lieu of the grandeur of his high mental calling to be rational and upright.

But maybe, some reader is seeking to change his life and mind; is seeking to know his life and the purpose he can put into it.

We are handicapped from birth, we grow up as children, as other children grow up. We are by nature inclined to be mean and our dictates are mainly evil.

If we are to make any moral headway, we need a lot of self-disciplining. But who will, or who can discipline himself without the instruction that manifests itself in "golden glory and hope."

If we don't understand the "why and the wherefore" we will never make progression. Millions upon millions of men and women have lived and died with the darkness that has always been over the Kingdom of the World.

Why? because the man of the world is satisfied with his own blindness, because he loves evil.

Moral good blinds him, he shuts his eyes, and says, "Away with it."

It was the writer's experience to be told by a girl student starting out on her scholastic career, that she did not want "to be good."

And she is a true representative of the main part of boy and girl scholastic students, who are looking for a profession, and the indulgences of human nature.

TRUTH

If you want to know truth, you should want to know morality, for morality is the basis of Wisdom.

Could you call a man "wise," that was cruel?

Could you call an immoral man, moral?

If you wish to know the purpose of life, I will tell you:

The purpose of life is to give you one, and only one, chance of obtaining immortality; through seeking the will of God in His Book called the Bible.

For the purpose of making our subject a personal one, quotations will be given from various letters received from correspondents: we have not received much correspondence but what has come along has been above the average in mental tone.

One correspondent (a foreigner) wrote:

"You ought to be able to see why nobody 'arose in the Wilderness' to comment on your theses, which are not yours at all."

We replied:

"You are right in saying the theses are not mine, the theses are taken from the Bible."

A Melbourne correspondent wrote:

"It seems to me that you base all your opinions on the assumption that you are regarding the earth from a point in space instead of with your feet on the earth."

We replied by stating a few basic facts concerning geometric laws, viz.

(a) Geometry controls all space relationships, whether it is empty space, or space physically occupied.

(b) A fixed physical object in space fixes a datum location for all space. If the location could be identified without the object, the object would not be needed.

(c) A horizontal and a vertical plane are the shortest planes of space.

(d) A globe is set up in space by two axes right angled to each other, one vertical and the other horizontal;

TRUTH

an infinite number of great circles going through the poles of the polar circle form the meridians of the globe and an infinite number of small circles ascending and descending and diminishing to infinity from the equatorial circle form the parallels of latitude.

(e) No location on a globe can be found without reference to a datum, and all locations on the globe are subject to signed co-ordinates fixing distance in regard to height and depth as well as in regard to length and breadth.

(f) What you fail to note is this: If you were literally on the outside of a globe, you would have literal gaseous space below your feet level.

A country correspondent, seeking a high diploma in Physics, wrote seeking more information, which we supplied as follows:—

(1) On a globe two ships could travel from the same equatorial location; one traversing a horizontal circle while the other traversed a vertical one: in other words, one ship travels up a ladder, while the other ship keeps to the flat. Refer to diagram.

(2) A plane flying off the Equator should have as much gaseous space below it, as it has above it. Has it?

(3) That from your present location your head is 34° lower than your feet. Cf. diagram.

(4) That if you set a sextant on to the South Pole it would show an angle of elevation of 34° from your feet; and your heavenly pole would appear to be about a mile and a half away; and that if you went to get underneath it, you would have to travel 56 x 60 nautical miles, and as you travelled towards it, its altitude would rise. The law of optics does this.

(5) Your orthodox global South Pole is located some thousands of miles below you, and to get to it you have to travel over two thousand miles downwards in space and over three thousand miles northwards, and then above your head (90° in nautical measurement) is your heavenly South Pole. How can this be so, seeing that you perpetually see your heavenly South Pole

TRUTH

always constant in the one and same direction from your fixed location view-point. Nothing moves in space of the heavens and the earth to cause dislocation.

(6) If you climb a perpendicular ladder set up on earth, as you ascend each step of the ladder you raise the whole ocean one step in height, and if your ladder were 4000 miles high, you would have an optical globe of the same size as your bogus spherical globe.

(7) The farthest range of optical sight is approximately 4000 miles.

We will conclude our subject with two remarks made to our Melbourne correspondent:

(a) All locations on earth correspond to Greenwich Aries. All stars at that moment correspond with their meridian transit at places whose latitude and longitude correspond with the star's right ascension and declination.

(b) All physical bodies have four dimensions from their centre, viz., height, depth, length and breadth.

FINAL

Reader: The earth belongs to God: He formed it from eternal substance, which he claims to own; and which he claims he can hold against dispossessors.

The heavens also belong to God: he claims to have garnished them with stars. When did life arrive, how it arrived, the cause of its arrival, and the result of its arrival, being the resultant Jehovah are all told of in God's book, the Bible.

The mind of God is the one mind worthy to be known, the purpose of God is the highest purpose that can be known.

Eternity must continue, Substance must remain, Wisdom must triumph, and God claims that in that triumph He will be there, and so can you if you use your sense and seek His aid and shelter all the days of your life.

TRUTH

FOWLER'S FLAT EARTH SEA PLANE MAP

An example given on page 44, Norrie's 1944 Edition, is illustrated hereunder on Fowler's Projection. This projection proves the EARTH TO BE FLAT by merely showing how the law of perspectivity operates.

The example is: At what distance will a tower 200ft. high be visible to an observer whose eye is elevated 15ft. above the water?

At 15ft. high the horizon distance is 4.45 miles.

At 200ft. high the horizon distance is 16.2 miles.

At 215ft. high the horizon distance is 16.8 miles.

Note the difference in distance caused by breaking the Ocean at Horizon Point into 2 parts.

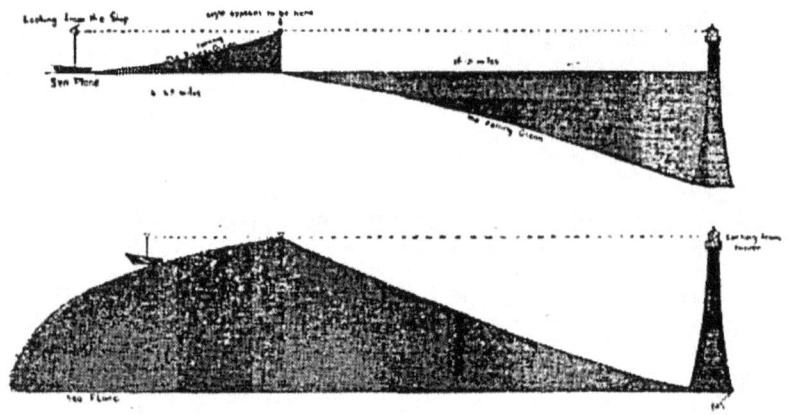

TRUTH

FOWLER'S SYDNEY MERIDIAN SKY-PLANE PROJECTION

The Eye creates the Dome of the Sky: the law of sight creates its mathematics. The Navigator measures all terrestrial distances on the Dome of the Sky and then assumes he is measuring on the dome of the Earth. The Stellar-plane is flat: the Earth-plane is flat; and an observer located at radius distance from these two planes would create our bogus globe. The actual distance to the Stellar-plane is 5,400 sea-miles from our earth plane. The eye pulls the sky down $(\frac{\pi}{2} - 1)$ radian $= 32°\cdot7$. The Lat. (dek) of a Columbae is S $34°\cdot1$.

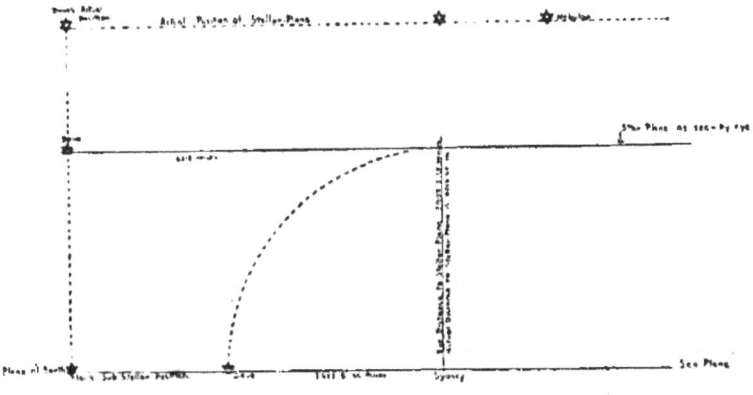

Actual distance to get under "Dove" (at lower transit) is 5,400 sea miles or 6·218 st. miles. What actually happens when "George" walks to get beneath "Dove."

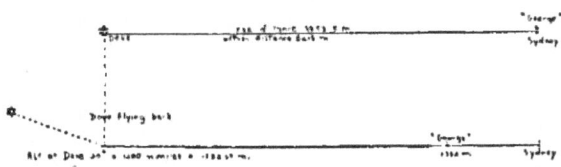

TRUTH

HOLD THIS SHEET VERTICALLY

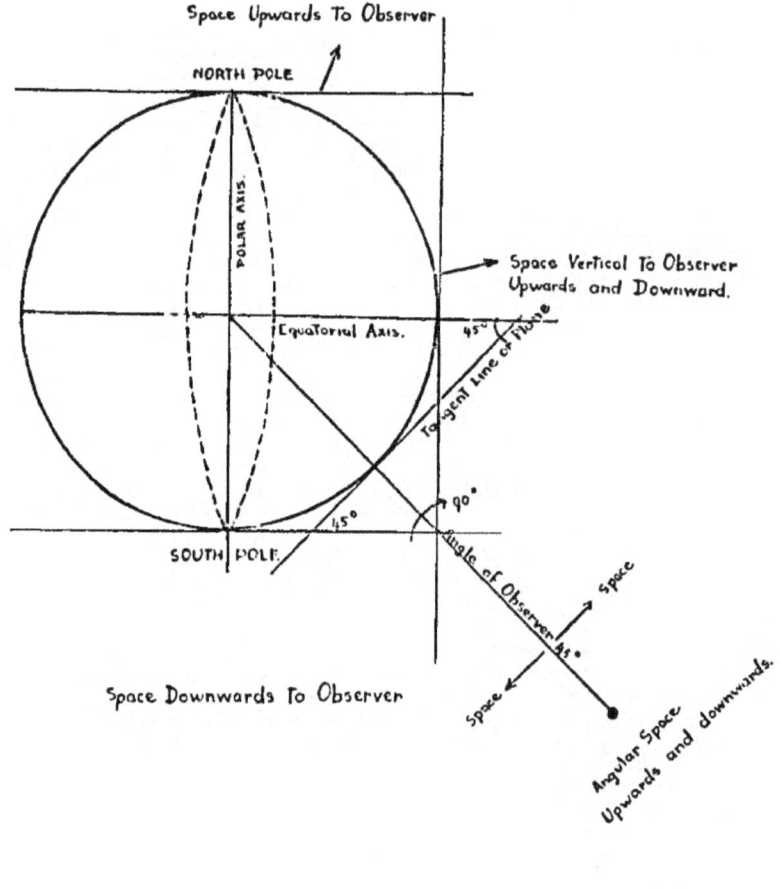

TRUTH

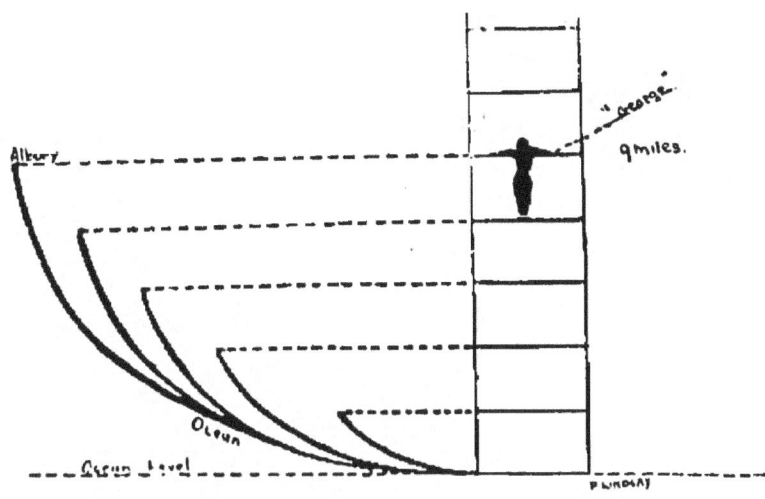

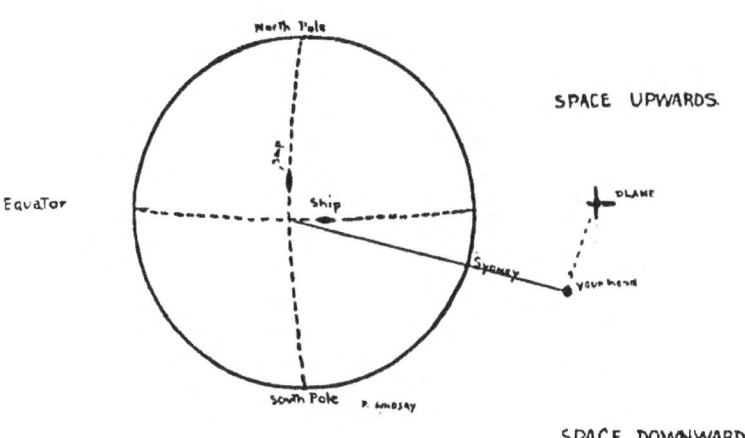

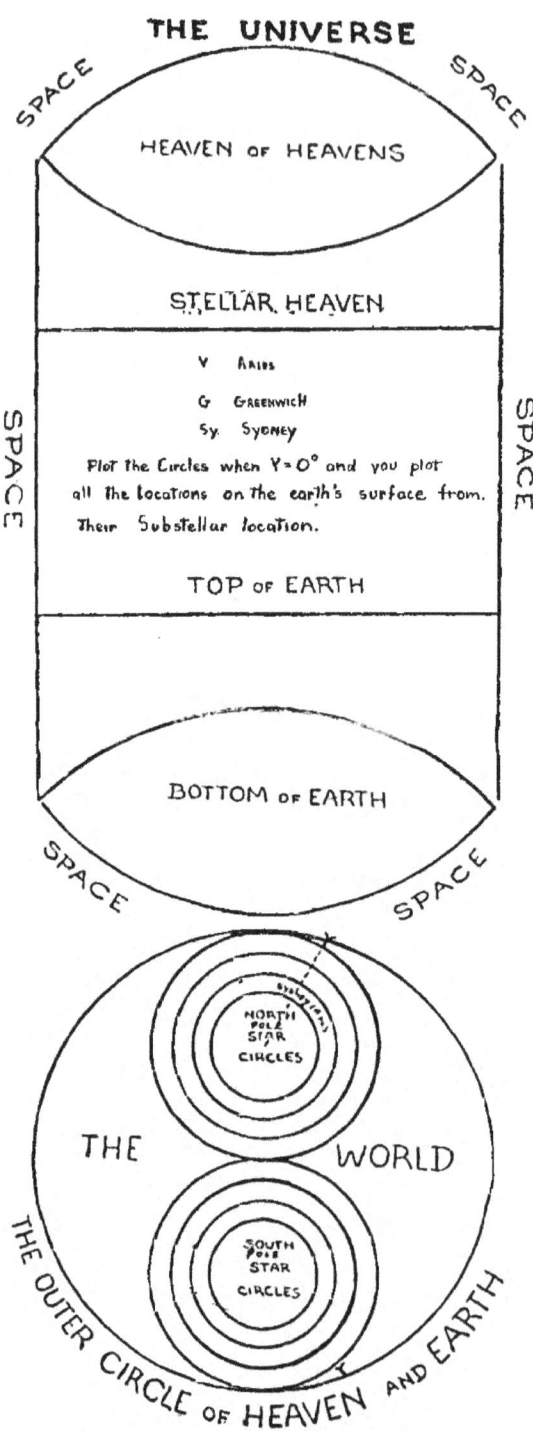

The Vanishing Ship.

By "Search Truth."

Proofs (so-called) of the World's Rotundity, examined in the Light of Facts and Common Sense.

PROOF I.—"If on a clear day we take our stand on a hill above "a seaport while ships are leaving, we shall see that the ship does not "become dimmer and dimmer, and is so lost at last to our view, but "that we first lose sight of the hull, then of the lower half of the masts, "and last of all of the top masts. In the same way, if we catch upon "the horizon the first sign of a ship, we shall find it to be the top "masts and top sails; then we shall next see the masts, the whole "masts, part of the hull, and, last of all, the entire hull. In both "cases it is as if the one ship were going down, and the other were "coming up, a hill. This is one proof that the earth is round," *i.e.*, a globe. The above is copied from "A Senior Geography," by John Markwell, M.A., corrected down to 1882, and used by the London University.

PROOF EXAMINED.—If a good telescope be used when the hull of a vessel has disappeared very frequently the whole of the vessel will be restored to sight, specially in calm weather. How then can the hull of a vessel have gone down behind a "hill of water"? One must either believe that the telescope enabled the observer to see through a "hill of water," or else that there is no "hill of water" at all. The writer has seen the whole of a vessel through a telescope when, with the unaided eye, only the top of a mast could be seen. The vanishing hull trick is thus exposed as a fallacy, for it is certain that, if the ship had gone down behind a hill of water, no telescope could restore it to sight again. Often, when at the seaside, the hull of a vessel has disappeared to one person, but to another, of longer sight, it can be seen quite plainly. This proves it is partly a question of optics, for if once a vessel had gone behind a real hill of water, no difference of sight could possibly restore it to sight again. The Laws of Perspective alone are quite sufficient to account for the way ships disappear at sea, and it is strange that in almost all geography books these laws are ignored, as the following sentence clearly shows:—"The ship does not become

THE VANISHING SHIP.

dimmer and dimmer." This is untrue, and is supporting a THEORY at the expense of FACTS. Let the reader watch for himself, and he will find that a receding vessel appears to become both smaller and more indistinct, until first the hull vanishes from sight and afterwards the masts, which gradually appear to grow less as the distance increases. The hull vanishes first partly because it is in and upon the water which forms a dark background to the observer. The following diagram will illustrate the Law of Perspective, and show that it is quite in accordance with those laws for the hull to disappear first upon a plane surface.

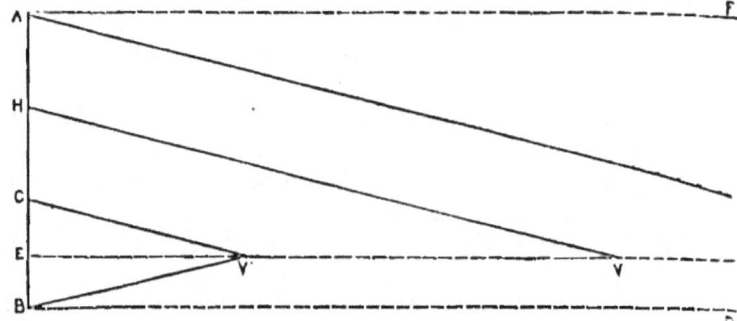

Let A C represent the mast 20 feet high, and C B the hull 10 feet high; E the line of sight 5 feet above the surface of the water B D. The horizon will be formed at V, where the sea appears to meet the line of sight E V. The hull C B will appear to vanish gradually and equally until it is lost at V, because its higher and lower parts are equidistant from the line of sight E V; but the mast which rises 20 feet higher will not vanish at the same time, but will do so at a greater distance on the line E V. Thus, besides being against a clearer background, it will be evident that in such a position the hull must disappear first, and the mast afterward, by the laws of perspective alone. Because a hull would disappear if it actually went behind a "hill" it is concluded that the world is a globe; but if the earth were a globe a ship's hull could *never* be restored to sight. As this can happen on a flat surface, it can only be regarded that the earth and sea form a vast plane. It can, however, be demonstrated and practically proved in other ways that the sea is a vast extended plane, and that the world is *not* a globe.

Price, 7d per 100.

JOHN WILLIAMS, 96 ARKWRIGHT STREET, NOTTINGHAM.

Is the Newtonian Astronomy True?

GLASGOW, 15th MAY.

SIR,—Your correspondent seems to think this a question entirely of flatness or convexity: whereas there are four sects of globists all at loggerheads:—(1) The Ptolemaists, represented by J. Gillespie, of Dumfries, who suppose the "earth" globe a centre for the revolution of the sun, moon, and stars; (2) The Koreshans of America, who suppose the "earth" a hollow globe for us to live inside; (3) The Newtonian Copernicans, who suppose the sun a centre, keeping the planets whirling in orbits by gravity; and (4) the Copernicans, who suppose the planets to whirl round the sun, without the necessity of gravity, Sir R. Phillips heading up this school. However, here are a few nuts especially for Copernican teeth :—Why are railways and canals constructed without any allowance for terrestrial convexity; and why do artists in marine views represent by a straight line the horizon, whether running east and west, or north and south? How can all the vast continents, with convexity only imaginary, along with the oceans, stick together to make a ball something like a little schoolroom globe, able to whirl on an axis only imaginary—that is, no axis at all; and though very many million tons in weight float light as a little cork in ethereal fluid found only in Copernican brains? How can gravity, which no one can describe, or prove, toss nineteen miles in a twinkling the great oceans and continents over the sun, and yet we are not accordingly killed outright, or even conscious of any such horrible motion? Is not this pagan Aristotelian gravity only a disguised theory of heaviness, representing the moon as falling 16ft. per minute towards the earth, but somehow deflected into an orbit; also the "earth" as falling towards the sun, but likewise deflected? Why do astronomers differ so much as to the size of the "earth" and as regards distances of sun and stars? Why believe antiquated fables devised thousands of years ago by stick worshippers, such as Thales and Pythagoras, who foolishly believed the sun a god to govern all, and hence the centre of whirling worlds, instead of the true God, who has declared that "the earth stands in and out the water," and is so fixed that it never can move.—I am, &c., A. M'INNES.

[All calculations of the earth's size, and therefore of the distances and magnitude of sun, moon and stars, depend wholly in the length of a terrestrial degree. The land and sea are first supposed to unite into a sort of ball, shaped like a turnip, orange or lemon, and then the circumference is divided into 360 parts called degrees, but not all equal, as is evident from Newton's supposition of ellipticity. Aristotle, about 300 B.C., said that mathematicians fixed the globe's circumference at 40,000 stadii (or 5000 of our miles). Fifty years afterwards, another Greek, Eratosthenes, first devised the plan of measurement still generally followed, that of determining by celestial observations the difference of latitude between two places on the same meridian, and then measuring the earth's distance between them. He calculated the earth's circumference to be 250,000 stadii (or about 32,000 of our miles). Various attempts have been made within the last three centuries to measure a degree, but with results so unsatisfactory, up to this hour, that the International Geodetic Association have lately resolved to hold a conference at Berlin during the summer to consider this much vexed question. The common method of measurement supposes the sky for the nonce a hollow globe corresponding precisely to the terrestrial one which it completely envelopes, and hence a degree as measured on the sky is believed to be the same as a terrestrial one; though again astronomers suppose the sky to be boundless space. Thus God's challenge to Job thousands of years ago may be repeated to the modern astronomer, "Hast thou perceived the breadth of the earth?" The terrestrial base line being therefore unreliable, all calculations of terrestrial magnitude, as well as distances of sun, moon, and stars founded thereon, are as fabulous as the monkey-man of evolution, or Lord Kelvin's third guess of past time at 4,000 million years.]

GLOBULARITY.

Sir,—Mr Harpur assures us that "surveys for canals and railways are made without mention of curvature, because the levels are taken by a succession of short tangents which overlap; so that, in surveyor's slang, "the backsight cancels the foresight." Now, we know that surveyors require back and foresights for uneven ground, and that their "datum line" must be parallel to the horizon, which is invariably level; nor have I ever seen it otherwise. Mr Harpur is challenged to prove that this cancelling allows for the fall of 8 inches per mile, increasing as the square of the distance. Nor can he prove his short tangents to be less imaginary than the globe itself, whirling on an imaginary axis, with an imaginary lurch of $23\frac{1}{2}$ degrees on an imaginary plane, driven along an imaginary orbit by the imaginary centripetal and centrifugal forces. Since the earth is alleged to whirl 1,000 miles an hour, how many billion tons of centrifugal force, according to mechanics, does Mr Harpur grant to pitch us off, seas and all movables, against the man in the moon? Again, if the lightning globe flashes over the sun about 19 miles every tick of the clock, how many billion tons of orbital centrifugal force will dash us, perhaps, against Neptune, whose imaginary inhabitants get only a 900th of the sun's heat and light compared to us? Now, isn't this sea earth ball a curiosity; nobody able to explain how all the great continents and oceans stick together to make it? Over its shape, size, distance, &c., how star-gazers squabble! Herschel will have it like an orange with two axes, but Ball with three axes, and Airy thinks it like a turnip. Herschel makes an astronomical degree 70 miles, but Airy 69, so that the globe's circumference may be either 25,200 or 24,840 miles. Again, according to Lardner, its distance from the sun is 100 million miles to Herschel's 95 millions, to Airy's 92 millions, &c.; but whilst to "Copernicans" the phantom's whereabouts is uncertain, common sense knows that it exists only in Newtonian brains.—I am, &c., A. M'Innes.

Sir,—"C. H.'s" supposition of the difference in levelling are surely exploded by the letter from the Manchester Ship Canal Office denying all allowance for curvature. It seems to be forgotten that the sun's distance is the astronomer's unit rod of measurement, and that seeing the astrologer Copernicus started with three million miles, now swollen up to some hundred millions, why, according to Newton's rule, we being now thirty times further away, have only a nine-hundredth of the sun's light, heat, and gravity formerly enjoyed. Further, the speed of our lightning ball is increased from less than one foot per second to nineteen miles. Indeed, as regards distance, speed, light, heat, gravity, the whole machinery of the solar system, without leaving out the millions of twinkling globes outside, has been for 300 years getting such a tinkering as must gladden the heart of "Topsy-

Turvy" to behold. Moreover, Brewster and Herschel, in calculating the distance of the nearest fixed star, differ by eight hundred thousand million miles; but, of course, millions of miles are as so many paltry inches to star gazers. A correspondent cites in proof of globularity the supposition of boundless space, which in turn is the usual inference from globularity; but to prove this assumption would require a boundless astronomer accordingly, without shape, centre or gravity, and bodily organs; or an astronomer of organic protoplasm, but endued with unending life, to explore creation through endless time, therefore, never to get at the evidence sought for. If our "globe" needs tangential force and gravity to spin it round the sun, just as does the moon to spin round us, why not the sun to spin round its vaster centre towards Hercules, whilst the sun's centre needs another centre still more monstrous? Thus there must be an infinity of globes. Then omnipotent gravity, operating everywhere, yet nowhere to be found or seen, a universal cause without beginning, to operate on uncaused creation, itself uncaused, is hence another god, yet to be everywhere opposed by a nameless, invisible, utterly mystical rival, equally omnipotent, tangential force; lest the infinitely many globes be mutually smashed into infinite molecules and atoms.

SIR,—Are not Newtonians logically bound to look fairly and fully in the face the contradictions, the assumptions, and the absurdities of the whirling, flying globe? They stoutly contend for curvature on land and sea, yet, *mirabile dictu*, they allow flatness at the poles, which no Arctic or Antarctic expedition has ever reached, claiming at the same time an extra amount of curvature for the equatorial region, where the earth, they say, bulges out, turnip-fashion. Thus the earth is flat and not flat, globular and not quite a globe. Well, if there be flatness at the poles, how do they know but the flatness may be of so wide an area as to make the earth's shape cylindrical, quite according to some old Greek scientists? Astronomers also suppose that there is no roof overhead called heaven, as if there could be blueness without a sky to be blue, where sun, moon, and stars may move: or space without a substance possessing the three dimensions of space. Still, the fact of heaven is again granted by them, but only as a celestial globe, whereon latitude, longitude, declination, and right ascension may be calculated—a heaven and no heaven! How can the earth globe, so like a wheel, turn on an axis only imaginary, the ends or poles of which, though imaginary also, nevertheless nod without causing fearful earthquakes, the north pole waving so as to describe a circle in 25,868 years? Since, too, a plane and an orbit merely imaginary cannot support a globe millions of millions tons heavy, won't astronomers mercifully scratch their heads for a new supposition, to save us from tumbling down into the horrible gulf of boundless space? Now, if we pour water on a school globe, the liquid runs off; then must we accordingly grant that all the oceans are frozen hard as steel, lest they spill, and that all the high mountains are flattened down level with the plains, so that all, holus-bolus, may unite into a smooth ball, turning round as nicely as a clock wheel, to make days and seasons? Next, there is the stereotyped trick of calling the real size, motions, and distance of sun, moon, and stars, apparent; whilst magnitudes, motions, distances, only supposed, are called real. Herschel, by persuading us that our eyes are nothing else but a cheat, would have us believe that what we see moving, stands still, and what stands still rushes faster by far than any express railway engine; but no astromoner has yet even attempted to prove the globe's exact whirl of 1,000 miles an hour, or the fling over the sun of 19 miles per second, any more than that the globe so knowingly preserves the parallelism of its axis at an angle of 23½ degrees. And since a Newtonian is accustomed to hang head down from the earth twelve hours in every 24 hours, may he not, by way of experiment, hang himself by the heels from the ceiling of his bedroom? Isn't a horse running 50 feet per second reckoned smart, as well as a whale swimming a mile a minute? But a man able to fly 19 miles per second when hooked on by gravity to a big globe, ought surely to be able to bear being tied to a cannon ball and shot from the mouth of an Armstrong gun. Our opponents are challenged to name anything outraging to common sense and reason more than the phantom globe of ancient heathendom.

To Editor of " Halifax Courier."

SIR,—" The distance," says Dr Rowbotham (Zetetic Astronomy, p. 102), " from London bridge to the sea coast at Brighton in a straight line is 50 statute miles. On a given day at 12 o'clock, the altitude of the sun at London Bridge was found to be 61 degrees of an arc; and at the same moment of time the altitude from the sea coast at Brighton was observed to be 64 degrees of an arc." With these data he calculates by the method called "construction." However, I shall here follow J. Layton's method, taking his diagram; the base A B being 50 miles, the angle at A 61 degrees, and the angle D B C 64 degrees. Then multiplying the sine of 61 degrees, or ·87,462 by 50, and dividing the sine of 3 degrees, angle A D B, or ·052,336, we have about 835 as B D. Next multiply the sine of 64 degrees, or ·898,794 by 835 and divide by the sine of 90 degrees or 1, and the result is approximately 750 miles.

Will your Halifax correspondent kindly tell us, if he can, where he found his data of 151 miles as a base line, with the altitudes of 55 and 53 degrees? Does he really accept the astronomy of Pythagoras, who, as a worshipper of the sun, imagined it to be the centre of worlds depending on it as a god for light, heat, and rain, and every blessing? If so how does he calculate the sun's distance by parallax so as to reconcile the conflicting opinions of Pythogorean astronomers? Accordingly, he must believe himself tied by the gravity of the infernal regions to the end of a globular wheel with spokes 4,000 miles long, to be unceasingly tossed thousands of miles either upwards or downwards, and at the same time pitched over the sun, 19 miles a twinkling. Moreover, he is liable to be tossed off his big ball by a centrifugal force of about 116,000 billion tons, due to diurnal motion; by another of 24,000 billion tons due to annual motion; by a third force due to an imaginary flight of his solar system towards Hercules, 46 miles per second; and by how many more forces is a wretched globist tormented in the hell of "endless space?" Further, there is an atmospheric pressure of 24,000 million tons per square mile, to squeeze the poor globe into a monster jelly; and how many million tons of coal are daily shovelled into the sun to keep up its heat and light?—Yours, etc.,

SIR,—Your Halifax correspondent fires his Copernican popgun, then runs away. Yet the book from which I get my data for calculating the sun's height is by "Parallax," the *nom de plume* of the late Dr Rowbotham, with whom I corresponded before his death; and I am not aware that zetetics, according to J. Layton, believe that the sun's distance is approximately 4,000 miles, far less 6,000. We must calculate by plain trigonometry, seeing that the surface of water is level, whilst the Bedford Level, the Salisbury Plain, &c., are what their names imply, and not arcs of a globe. Canals and railways are constructed without any allowance for a convexity which necessitates the rule of mechanics,— "The difference between real and apparent levels is equal to about eight inches in every mile, and increases as the square of the distance." As one example among many, the Suez Canal is 100 miles long, and therefore there ought to be a difference in deviation of 1¾ miles, whereas it is a dead level from end to end.

J. Layton would have us believe without proof and in opposition to sense, that the sun which we see moving westward, is in reality moving the opposite way, and that the moon, with Jupiter, is moving east and west at once, therefore standing still! These contradictions, he assures us being explained (and how?) by modern astronomy taught by the idol worshipper Pythagoras 2000 years ago. He asserts, however, that zetetics assume that the earth (meaning all the vast oceans and continents) has neither axial nor orbital motions; whereas we believe our senses, which testify that the continents do not fly through the air, especially with the awful speed of 19 miles a twinkling. Copernicus himself confessed that the whirling lightning globe of heathendom was not even a probability; Herschell, that we must take it for granted; Professor Wodehouse, that it cannot be proved, &c. Further, as John Wesley, founder of Methodism, long ago pointed out, astronomers prove the distance of the stars by their great magnitude, and the magnitude by the great distance. Zetetics hold according to common sense, sound argument and God's revelation, that the earth or land floats in the great abyss of waters fixed there by the Creator so that it cannot move, the ocean being surrounded on all sides by the Antarctic icebergs (Psalm xxiv, 104; Job xxxviii, 10, etc.), and that the sun, moon, and stars move in the vault of heaven always at the same altitude above the earth, neither rising nor setting, but as the Hebrew Scriptures say going forth and going in, from horizon to horizon (Gen. xix., 23; Eccles. i., 5).

Here are nuts for Copernican teeth :—(1) How can the continents extending thousands of miles with vast mountain ranges, great plains, and rivers, along with the immense oceans (the Arctic and the Antarctic unfathomable), be rolled together into something like a little schoolroom globe, and the whole mighty mass be tumbled over and over, and heaved about the sun with more than lightning speed, without the earth and all its inhabitants being at once destroyed thereby? (2) How is gravitation or attraction proved—first taught by the idolater Aristotle of ancient Greece—which supposes almighty power not up in heaven but down in the infernal regions or heart of the earth? (3) How is the imaginary globe proved to be 24,000 miles or so in circumference, without supposing the sky, otherwise called "infinite space," after all a vault for the sake of measuring degrees; and how is it possible to reconcile the conflicting calculations of distances of sun and stars, differing by millions of miles? (4) Are the Copernican diagrams, mathematical "proofs," the schoolroom globe and maps of a turnip-shaped earth; the technical terms, such as parallactic motion, spheroid, terrestial axes, plane, orbit, equator, poles, &c., anything else than the tricks of a disguised atheism now misleading multitudes? (5) Is not the parallelism of the globe's axis with the lurch of $23\frac{1}{2}$ degrees on an imaginary plane, as well as centripetal and centrifugal forces to account for orbital motion, polar nutation, infinite ether, infinite space, stars with men hanging from them by the heels, globes of many million tons floating on nothing, light as feathers, moonshine being sunshine, &c., mere suppositions, unsupported by one fact or solid argument, but making up one huge anti-Christian fiction for simpletons to swallow?—I am, &c.,

---

Sir,—Mr Layton still dissatisfied, yet too nervous for your arena, writes to me confessing an "Atmospheric" squeeze for the poor globe sides of only 27 millions tons. Then by mensuration, multiplying the supposed circumference of 25,000 miles by the diameter of about 8,300, we have a surface of something like $207\frac{1}{2}$ millions square miles, which again must be multiplied by 27 millions to get the whole pressure, so that we may well wonder why we are not balancing our bodies on something like a jelly inconceivably large. But the "atmosphere" (a word from the Greek, meaning "smoke of the ball") is said by physicists to be just as high as their theory will allow, 40 or 50 miles, and as a kind of outside ball for the earth whilst squeezing it so terribly, it whirls as fast or 1,000 miles an hour, and flashes along with it quicker than forked lightning round the sun. Now, why does the air press down and not up, though astronomers conveniently deny anything to be either up or down, there being therefore neither east nor west, north nor south, nor, indeed, any direction for the big ball with its air envelope to turn? What makes the "atmosphere" or the sea-earth globe whirl top-fashion, astronomers can't tell, and whether the "centripetal and centrifugal forces," which push the "Earth" along the imaginary rails of the annual orbit, do the same friendly job for the air, astronomers don't say, perhaps thinking nobody would ask. Proctor, with 22 figures for the "Earth's" weight, or six sextillion tons, makes his dupes scratch their heads utterly dumfounded; and yet the monster mass floats light as a little cork on a boundless sea of unknown ether found only in Copernican heads. Then as the globe flashes along faster than a red thunder-bolt, according to Herschel it nods and waves its enormous head as if seized with St. Vitus's Dance. Yet what common sense can't see is, how with all this constant tumbling head over heels, we don't choke or get pitched against the man in the moon. Ah! but don't we always carry on our head and shoulders an aerial pillar 50 miles high, some tons in weight, enough to crush us outright, whilst the devil with his gravity far down in the infernal regions hauls us towards himself? Still that we may bear all this squeezing, tugging, and tossing, ought not astronomers to say men are not flesh and blood but malleable iron?

Moreover, the scientists say the stars are not stars, any more than the sun and moon are what they are, but tremendous globes as many million miles off as they please to put them, with men hanging from the heels in danger of being burnt by flying comets. Also in "words of learned length and thundering sound" the evolutionist tells his "tale of wonder to gull the mob and keep them under,"—that our great, great, grandpapas were monkeys, baboons, apes, or gorillas—what a beastly pedigree! So we have a new edition of Genesis 1.: "In the beginning, somewhere (says Lord Kelvin) between 20 million and 400 million years ago, was fiery gas; and gas after many years hardened somehow into solid rocks, which partly softened at length into cabbages, &c., whence sprouted long after tadpoles, and the tadpoles begat a donkey, and the donkey begat a monkey, and after many ages the monkey begat Adam." These tales are great theories, of course, because fathered by men great because of their theories   Yes, but there are mathematical " proofs "—aye, and as mystical as Egyptian hieroglyphs, entitling the scientists to a monopoly of mutual squabbling.—Yours, etc.,

To Editor of " Hebdon Bridge Times."

Sir,—Since a blundering translation answers a pagan astronomy, Job 26, 7v, is rendered, " He stretcheth the north over the empty place and hangeth the earth upon nothing." The Hebrew is *Notheh tsaphon ab tohu toleh erets al-belimah*; the correct rendering being, " He stretcheth out the north over desolation, he hangs the earth on its fastenings." The north was called by the Romans, Septentriones, rendered by Max Muller " the Seven Stars," which are in Ursa Minor, that part of the heaven being spread over a desolation of snow and ice. The Hebrew verb talah, to hang, excludes the idea of motion, applying to the suspension of shields, harps, vessels, dead bodies, as in Psalms 137, 2v; Ezekiel 15, 3; 27, 11. Can lightning be hung up, *a fortiori* the globe of Copernicus flashing 19 miles every tick of the clock? The verb balam signifies according to Parkhurst, and Bresslau, to fasten, bind. Col himself declares the earth fastened in the mighty abyss of waters, Job 38, 6. The cuckoo cry that the Bible is unscientific needs also a reply. The other day in the University here the fact was mentioned to the mathematical class, that the value of the ratio of the diameter to the circumference of a circle was given in the temple of Solomon by the molton sea ten cubits from brim to brim, and thirty cubits in circumference (2, Chronicles, ii, 9); Solomon, according to Gould, of America, having measured the diameter from the outside and the circumference from the inside of the circle. Let " Leo " tell the breadth of the rim, and he solves the problem that has baffled mathematicians for 2000 years. This ratio is the key to the harmonies of all plain figures. Then the solid geometry of the temple itself, which was built without the sound of a hammer or workman's tool, the stones having been " made ready in the quarry," may well astonish the moderns, who cannot even restore the lost books of Euclid. Next look at the arithmetic of Noah's ark, the model for shipbuilders, fitted to accommodate every species of land animals zoology can mention; also the deep significance of numbers as used in Christ's parables, the Apocalypse, and Moses' tabernacle; the seven lamps, the five virgins, ten talents, the New Jerusalem, a perfect cube whose side is 12,000 furlongs &c., &c. Solon and Lycurgus were bunglers compared to Moses, by whose wise legislation every Israelite was made a land-owner, paying no rent or taxes except a voluntary tithe in kind to the Levites, there being no need for jails, police, navy, standing army, coinage, paid judges, or king. The wondrous summaries of history in Daniel and the Apocalypse, particularly the metallic image of Nebuchadnezzar's dream, supply us with the " philosophy of history " which historians have failed to construct. What glowing tableaux of the future universal restoration are in the prophets, contrasting with the beggarly emptiness of the pagans Homer and Virgil! The Puritan Fleming was able 200 years ago to predict with the dates of the Apocalypse the fall of the French Monarchy in 1792. Descartes 300 years ago proved the ancient syllogistic logic a fraud; whereas the Scriptures identifying speech with life and mental light (John 1), lays down the great rules of proving all things and calling everything by its right name, and supplies the models of reasoning in Paul's infallible dialectics and Christ's dilemmas that silenced all opponents. What ancient sage, pagan, or Jewish, ever conceived the Christian ethical precepts of peace, love, self-denial, perfection? For definition of soul and spirit I turn to Levi. 17, 11, I. Cor. 2, 11, &c. The structure of the universe, the aqueous origin of the earth, &c., I learn from Gen. 1; the fact of the earth's arrangement in layers from Job 38, 5; the universal deluge accounting for geological remains. Let the agnostic name a true science which is not discussed in the Great Book. Even electricity, inexplicable to Lord Kelvin, is explained in Ez. 1. By means of the Bible, Mr Dimbleby, of London, has rectified the calendar and reduced chronology to an exact science. The true use of astronomy, he contends, is the measurement of time. By the classification and enumeration of eclipses (giving cycles of 18 years) and transits, they have been reduced to a system of metrical indicators proving the whole or any fractional part of past time. "All Biblical history is astronomical, and hence all the dates of Scripture fall with precision on the lines of scientific time like the cogs of a wheel." The book of Genesis supplies that point of time which chronologists and astronomers had long been desirous of obtaining in order to rectify all subsequent periods of history and celestial phenomena." "All Past Time," p. 143.—Yours, &c.,

To Editor of "Lincoln Chronicle."

SIR,—Two queries by C.D. may be answered with the help of Solomon's proverb, "A wise man carries his eye in his head." We see, as the Hebrew prophets reveal, "the earth stretched out upon the waters," and the vault of Heaven spread overhead, whilst, on looking all around, we find a limit to our range of vision, called the horizon, towards which the earth appears to slope up, and the Heaven to slope down. Now, whatever point of the horizon we look at becomes a point of convergence, being on a level with the eye, and all lines above the eyeline must descend, whilst all lines under it must ascend to that point, where, indeed, we see an object vanish. Hence a ship receding from view appears to ascend to the horizon, but approaching to descend towards us, though when about to disappear the hull (on account of the obscurity) becomes confounded with the watery base; yet, with a good telescope we see no "hull down." What artist would represent a ship "hull down" by a mast like an indiscoverable north or south pole, sticking up behind a hill of water? Also when a ship has disappeared we have but to climb some eminence to see it again, and whilst it goes away it appears to go up, but never goes over the imaginary hill. Likewise the sun appears to rise and go down, but the Hebrew Scriptures, with scientific accuracy, speak only of the sun going forth in the morning and going in at night. In page 20 of "Herschel's Astronomy" there is a cut representing a man looking down on the horizon. The pillar he stands on is supported by a curve of the "earth," fully 60 degrees, and, therefore, one-sixth of an entire circle, or 4000 miles in extent; whilst the man with the pillar equals in height one-fourth of the curve, so that, according to the scale, the man is 300 miles high and the pillar 700! A line is drawn from his eye, as a tangent to the circle, representing him as elevated thousands of miles above the horizon, which, according to true perspective, can never be below the level of the spectator's eye! The third question by C.D. is seemingly about the "phases of the moon," which he wishes "fully explained." Herschel, in his "Astronomy," by way of explaining, gives a diagram representing the moon in eight different positions circling round the "earth" globe, but shining only by the reflected light of the sun. Yet he only presents new mysteries even more inexplicable than the lunar "phases." How can moonshine be sunshine, or Moses a liar when he called the moon a light, his writings being endorsed by Christ himself? How could all the oceans and continents adhere to make up a whirling lightning globe, or a globe of such an unconceivable weight need no support, and whirl on an axis only imaginary, that is, no axis at all? Further, what causes the big globe to lurch over at an angle of $23\frac{1}{2}$ degrees on no plain at all, and preserve the parallelism of its imagainary axis with more than the nicety of a clock wheel, or by what power does it revolve and pitch itself over the sun with more than the speed of a thunderbolt? If we are told that gravity explains some of these mysteries, we ask what is gravity and its cause, but receive no reply. We are only treated to old Pagan fables set forth in big words, and spiced with lying diagrams and mathematical jargon. Now, I challenge any professor or scientist to prove that the geography and astronomy of the Bible was ever known to the ancient heathens; but I am not aware that God has revealed, through His prophets, the causes of eclipses and lunar "phases," which I hold, therefore, to be inexplicable. Have not self-styled philosophers, esteeming themselves gods in omniscience, vainly sought to explain how they saw, heard, perceived, &c., instead of being content with the wise use of their senses and faculties? In like manner speculation is rife as to the causes of "celestial phenomena," whilst the use of the sun and moon as the clocks of the universe is entirely over-looked, with the results that scientists cannot agree as to the date of creation—chronology and the calendar are in helpless confusion. Yet, by star-transits, eclipse cycles, lunar and solar years, lunar seven days' phases, all past time may be reduced to a most wonderful order, in accordance with the date of creation as given by Moses, and other scientific data supplied by the sacred Scriptures generally.—I am, etc.,

---

Sir,—Isn't it as hard to open the eyes of a bigotted Copernican as to convince a Bedlamite that he is not butter? In vain we show the lunatic that his body has none of the qualities of that dairy produce, for he stubbornly persists in denying his senses. Also we show the globist that the earth has none of the qualities of a globe, and that not one of his senses witnesses the horrible motions supposed by astronomers; yet he persists in believing the old pagan delusions, in opposition to all his senses, to all facts, reason, and God's own revelation. He believes himself glued by gravity to his big globe, flying with it among the stars of Heaven, far faster than a rifle bullet, and tumbling head over heels once every 24 hours. He gapes in wonder at Herschel's lying pictures, mathematical gibberish, and long-winded outlandish words—all of them to him infallible unanswerable proofs. Nay, he even thinks some tons of atmosphere are pressing on his shoulders lest he jumps too high, that his great-great-great-grandpapa was an ape or some such beast, and that all creation began to evolve from involved eternal gas, somewhere, says Professor Thomson, between 20 millions and 400 millon years ago! The mesmerised simpleton must be right, forsooth, for everybody around believes as he does, great men have taught the great delusions, and infinite self-created nature, with self-imposed omnipotent laws, is virtually his god, but with no commandments to keep nor judgment to come. Poor Mr Fieldsend, over flats and sharps, is as funny as a clown in a Christmas pantomime, so that children may laugh;

but were he sharp and not a flat at reasoning, he might see a logical connection between the general belief in Copernican fables and the latest saying of Carlyle, his theological teacher, that this is "a generation of conceited fools," aye, too, with engineering inventions to fill the world with accidents, and warlike engines to destroy mankind.—Yours, etc.,

To Editor of "Dundee Weekly News."

SIR,—Mr Gillespie, whose book I have long ago read, supposes with Thales, Ptolemy, and other pagans of remote ages all the oceans and continents rolled into something like a monster turnip or apple, and whirling mysteriously round itself upon nowhere 1000 miles an hour, though he denies the thunderbolt dash of 19 miles a twinkling round the sun. But even if to the Pacific and Atlantic along with the unfathomable Arctic and Antarctic Oceans we add all the islands and continents, how can they unitedly have the quality of solidity, without which there can be no globe? unless the astronomers want us to imagine all the oceans frozen into a huge iceberg and somehow soldered to the land. Yet, again, how can we all endure the awful daily somersaults caused by what is quietly called axial motion, to be pitched as often heels uppermost as heads thousands of miles down into a fearful gulf, and after being hurried under, to be tossed aloft to a giddy height of thousands of miles, being all the while bound as to a vast wheel like the fabulous Ixion? Then as for gravity, that ties us on, and moves the sea-earth wheel, 8000 miles high: what is it? Hobgoblin or fiend, that Newtonians suppose everywhere, yet is found nowhere, unless in astronomers' brains, though guessed by Mr Gillespie to be lurking far down in the earth's centre or infernal regions, and hindering our escape into the Copernican boundless space where world's fly more plentiful than cannon balls in battlefields? Moreover, as drums, poles, pennies, cups, &c., are round without being globes, let Mr Gillespie show (though not yet done anywhere) that everywhere on surface of land and sea, there is, according to mechanics, a fall of eight inches per mile, the increase being as the square of the distance. Hence, our island being about 700 miles long, John o'Groats should be at least 60 miles higher or lower than Land's End in Cornwall. Also, the Atlantic cable was laid over 1656 miles without any allowance for rotundity, and yet there should be at the centre, according to globularity, a hill of water 150 miles high.—Yours, &c.,

ECLIPSES.

To Editor of "Birmingham Mercury."

SIR,—A "Curious" correspondent thinks globularity proved by lunar eclipses; the difficulty, however, being to prove the darkness on the moon to be a shadow, and that shadow a terrestrial one; unless, as usual, after proving globularity by the shadow, he prove the shadow by globularity, just as the sun's great distance is generally proved by its size, and its great size by its distance. A hat, a saucer, a cheese, &c., have round shadows, and to prove roundness is not to prove globularity. The tendency of everything not to the centre of a globe, but to fall off, except at the top, &c., is conveniently unnoticed by Newtonians. More than 2,000 years ago the Chaldeans presented to Alexander the Great at Babylon, tables of eclipses for 1,993 years; and the ancient Greeks made use of the cycle of 18 years, 11 days, the interval between two consecutive eclipses of the same dimensions. The last total eclipse of the sun occurred on Jan. 22, 1879, and the preceding one on Jan. 11, 1861 Now, have not mere theorising about the sun and moon—the great unerring clocks of time—thrown chronology and' the calendar into confusion, and hence scientists cannot agree as to the world's age; and the year absurdly begins on Jan. 1 instead of at the vernal equinox, the months consisting of 31 or 30 days, one of 28? However, it can be shown that with eclipse and star transit cycles, the greatest accuracy as to dates may be attained.

Going back, for example, from Jan. 11, 1861, through a period of thirty-six eclipses, or 651 years, we find that a total eclipse occurred also on Jan. 11, 1210; and, continuing backwards, by such cycles we arrive precisely at the date of creation as given by Moses in Genesis. Also, as related by Josephus, the moon was eclipsed in the fifth month of 3998 A.M., when Herod the Great died, and Christ being then two years old, His birth occurred 3996.

GRAVITY.

SIR,—Your Newtonian correspondents have only one string to their tuneless fiddle, curvature, so that they may prove the "earth" a big drum, a barber's pole, a Cheddar cheese, or the fixed globe of Ptolemy; but they carefully shirk the dynamics of gravity borrowed by Newton from the ancient pagan Aristotle. Well, how could the old serpent deceive Sir Isaac with the fruit of an apple tree? Our philosopher did not believe anything up or down, light or heavy; though sitting once in his garden, he saw the apple hanging up, whilst he sat down below. Then, he saw the apple fall or go down, and if stooping down he lifted it up, he would have seen it was too heavy for the branch to hold. Still, might it not have fallen either up or down, since his science supposed it neither light nor heavy? Yet, down it must fall after all, obedient to the power far down in the earth's centre. Again, if there is nothing up or down, why should there be north and south, right and left, or any direction whatever; then why any space, especially infinite, where unnumbered worlds of apple shape may fly? And since apples are so obedient to gravity, why do feathers, smoke, steam, balloons mount on high in defiance of gravity's almighty downward pull? Further, if the apple was neither heavy nor light, how could it be hard or soft, large or small, round or coloured, or, in fact, an apple at all? Now, take away Newton's apple, and where can scientists hang up their wonderful tales of astronomy, geology, and evolution? Accordingly, Sir Isaac announced that "every particle in the universe attracts every other with a force whose magnitude is proportional to their masses divided by the square of their distance from each other."

SIR,—"C. H." supposes a lunar eclipse proves moonshine sunshine, just as a solar eclipse might prove sunshine moonshine. Does "C. H.," with his eyes open, see himself by his unsolid globe whirled heels over head, whilst securely hooked on by friendly gravity? Surely the mystical ether theory needs cobbling, as Lord Salisbury lately hinted. Newton affirms of the undiscoverable gravity, denied 2,000 years ago by Lucretius, that "every particle in the universe attracts every other." From this it follows that the smallest crumb between "C. H.'s" wireless teeth pulls every drop of oceans, streams, &c., all the particles making up continents and islands, as well as the numberless globes of "boundless space." Whence the wee crumb has got the infinite power, or by what *modus operandi* it pulls the whole universe, are surely difficulties not less inexplicable than the mysteries meant to be explained. A " Lover of Nature " believes with Hutton, though contrary to Lord Kelvin, that there is as much hot lava under our feet as could make all the seas boil; and that as Copernicus, with many others, thought, not the moon, but the daily whirling of the globe, causes tides, though why the whirling does not empty all the ocean beds is another mystery of scientific Babylon.

Sir,—" C. H.," believing Newton's assumption that " every particle in the universe attracts every other," grants consistently enough that the crumb between his teeth pulls the whole universe or all the unmembered globes of " countless space ;"—the crumb, however, being accordingly omnipotent and a god! Then, if the crumb really draws, we ought to see all creation move towards it; and as " C. H." confesses he sees the pulling, shouldn't he also see the motion towards the crumb? Further, according to gravity the whole of creation in turn combines to pull the crumb—and yet contemptibly small as the plucky little bit is—in spite of the awful tugging it sticks as lovingly as ever to " C. H.'s " teeth. Moreover, a living flea no bigger than the dead crumb ought *à fortiori* also to move the whole universe; consequently the drawing of a 100-ton gun should be a very easy job for the clever little insect. And wouldn't the wee crumb require, moreover, to be infinitely large so that, in order to pull, every one of the particles infinite in number may get a hold of it? But how does the crumb pull all in turn? When a man, a horse, or an engine draws, it must be by hands, ropes, claws, with some sort of couplings; therefore, how could the crumb pull all creation except with an infinite number of hands, ropes, &c.—all, of course, being invisible, that is, as imaginary as the hauling itself. " C. H." would have us believe in the omnipotent crumb without any proof whatever, though he refers to gravitation as a " common element (!) among many observed facts," not one, however, of the " many observed facts " being named. The Latin name " gravity " or " gravitation " has its Saxon equivalent —weight or heaviness, which certainly our senses witness as a fact; but that there is a power miscalled gravity down in the infernal regions drawing us and all on the surface to itself is evidently an absurdity as destitute of proof as the belief of Aristotle, father of gravity, that a painted stick was a god.

THE GLOBE'S MOTIONS.

Sir,—Surely " Copernicus " knows as much of mensuration as to be aware that a globe is entirely solid, and though the earth or land is solid, yet the oceans are fluid, with three times more surface than the land, the Arctic and the Antarctic being unfathomable; so that the broad solid continents, with unfathomable oceans still broader, uniting as one vast solid called a globe, must be opposed to common-sense. Also, when a lunar eclipse is supposed to be caused by the earth's shadow, how much more by the shadow of the oceans—the proof being of the *circulus in probando* kind. Then, if the circumnavigation of the earth proves globularity, Ireland, and all other islands ought, as well as the continents, to be globes, requiring axes, poles, planes, orbits, with the rest of the astronomical gear. In vain have I sought a discussion of the dynamics distinguishing the Copernican globists from the Ptolemaic, because the former, fearing an exposure of the gross absurdities involved, cobble away at curvature—an *ignoratio elenchi* argument.

Who, since the idolater Aristotle, the father, has ever proved gravity; or what Newtonian, tangential force, so that the rolling globe may flash along an orbit elliptical?—elliptical forsooth, not because the parallelogram of forces has been proved, or any other reason exists why the resultant velocity should not be in a straight line; but because the Pythagorean sun-worship requires the globe's course to be circular, yet, after all, not quite circular, but elliptical to a trifle of six million miles. Of course, all the terrible nodding, tossing, whirling, flashing of the big ball (because imaginary), the Darwinian progeny of apes may easily endure, but could they believe if they were men? Yet dare they dispute the fables taught by scientific gods with titles so lofty, gowns so learned, and words of such thundering sound?

CONFESSION OF COPERNICUS.

Sir,—Can "Novelist" show that it is not disgraceful for so many reputed educated and Christian people to believe, contrary to sense, reason, and God's revelation of nature, the globe fable devised by ancient Egyptian priests and Chaldean astrologers; denied in modern times by the theologians, Luther, John Wesley, Bacon the philosopher, Goethe the German poet, &c., nay, confessed by Herschel and Professor Wodehouse as incapable of demonstration? What said Copernicus himself, according to Humboldt? "*Neque enim necesse est eas hypotheses esse veras, imo ne verisimiles, quidem sed sufficit hoc unum, si calculum observationibus congruentem exhibeant;*" or, "For neither is it necessary that these hypothesis (globular) be true, nay, not even probable; but this one thing suffices, that they show the calculations to agree with the observations." Moreover, none of your correspondents have yet given one proof of globularity. Since "Novelist" is so much at home in starology, let him prove terrestrial motion (earthquakes excepted), hitherto a mere *petitio principii* with astronomers; also Newton's dynamical laws and Kepler's planetary laws; the nebular hypothesis of La Place; the alleged distances and magnitudes of the sun, moon, and stars by trigonometry; the velocity, refraction, and aberration of light; the atmospheric pressure of 2,160 lb. per square foot; that the stars are inhabited, or that there is a man in the moon; the reality of ether and infinite space; that the stars called "fixed" are not stars, but suns, and that the planetary stars are not stars, but globes, with axes, planes, poles, &c.; that moonshine and starlight are sunshine; that the moon causes tides and solar eclipses, whilst the earth causes lunar eclipses, and the sun lunar phases; that the sky is no sky at all, yet a celestial globe; that flesh and blood can bear to hang head downwards twelve hours daily, and tied to a big ball, to be flashed along the sky quicker than a red thunderbolt, &c. I engage to discuss with "Novelist" all these various points in turn.

ATMOSPHERIC PRESSURE.

Sir,—"Novelist" grants that the earth is not proved a globe by circumnavigation, any more than Ireland itself, though it again being surrounded by water, cannot be part of a solid called a globe. Then, if we start, as he suggests, from Cork in a direct northern course, we come at length to the impenetrable central region of desolation haunted by the north pole, where we must make a detour of 180 degrees to regain our straight line; and, continuing, we go south through the Pacific until stopped, as were Ross and other navigators, by the antarctic walls of ice. Again we must make a detour, but east, passing Cape Horn into the Atlantic, thence on to Ireland once more.

"Novelist" thinks that the ability of Newtonians to calculate eclipses, transits, &c., is a proof of the solar system; not knowing that the practical astronomy, as a science of observation, is independent of all theories, whether of Aristotle, Ptolemy, Tycho Brahe, or Copernicus. Does "Novelist" not know that the interval between two consecutive eclipses of the same dimensions is 18 years, 11 days, 7 hours, 42 minutes, 44 seconds; so that a schoolboy, knowing the eclipse of March 10, 1876, could easily calculate its return in 1894, March 21, at 2 hours, 3 minutes, 46 seconds p.m.? Hence the great utility of eclipse, transit, and other astronomical cycles in chronology. Has "Novelist" never read that 4,000 years ago the mean notions of sun, moon, and stars were known to a second, just as at present; that Alexander the Great was shown Chaldean tables of eclipses for 1,900 years; and that the ancient Hindoos knew trigonometry and sexagesimal arithmetic?

In turn, I ask "C.H." how does the air-pump prove an atmospheric pressure of 2,160 lb. per square foot, according to my original terms? If he can prove that men hang by the heels from that twinkler Mars, I shall grant him the man in the moon for his cleverness. Further, since "C.H." asks me to prove that the sun does not somewhere dip, surely he believes it does dip or drop itself down into something—salt water, perhaps, as a refresher after the day's race; but how could Newton's ghost permit the bath?

Sir,—Is not atmospheric pressure another name for weight of air—after all, the old gravity pulling to the centre? The air being supposed to be forty or fifty miles high, an ordinary-sized man carries continually on his head and shoulders fourteen tons of oxygen and nitrogen, though how he can is a nut yet to crack; and why, when inside a house, and, therefore, under an aerial pillar only a few feet high and about one pound weight, can't we jump with ease to the roof? Strange, too, the pressure is down, though Newtonians deny up and down; but if, as with water, the pressure is perpendicular to the surface, our sides ought to be sore as well as head and shoulders. A slate, 10 inches by 5, allowing 15lbs. to the square inch, sustains 750lbs. of air, and yet I can balance the slate on my finger's end. Moreover, scientists to account for trade winds say the air lags behind the globe as it spins on its axis, but is there no danger then of the atmosphere altogether slipping off the globe as it flashes round the sun, so as to leave us all choking with incurable asthma? Again, there being a pressure of twenty-seven million tons of air upon every square mile of sea and land, multiply the supposed circumference of 25,000 miles by the diameter of about 8,300, and we have a surface of about 207½ millions square miles to multiply by the twenty seven millions—and what a terrible squeeze for the globe's poor sides! —a squeeze endured, too, during how many million years of evolution? Lastly, what pressure does "C. H." allow for undulating ether, which, filling infinite space, ought to give an infinite squeeze to the globe, reducing it to an invisible atom infinitely small?

Sir,—For experimenting purposes, "C.H." would have us swallow the cunning bolus that "heaviness is 1-289th part greater" at the poles than at the imaginary equator—poles and equators being essential to his ideal earth. He prefers, of course, to calculate by metres—because the metre, as elementary arithmetics say, is supposed to be a ten-millionth part of the distance between the pole and the equator, being calculated on a circle of longitude." Now, is not heaviness, according to common sense, due chiefly to bulk and density; and has "C. H." not seen feathers and smoke mount on high in spite of resisting boundless ether, millions of tons of air above, and gravity's omnipotent pull below?

LIGHT

Sir,—Will "C.H." explain how all the oceans, mountains, and flat plains can unite two or three times a year to cast their shadows up to the sky, making after all such a small shadow on the moon? Surely "A Lover of Nature" ought to believe we may have daylight without sunshine, that is, the sun does not cause day; though if we stand on a globe of fire with a very thin crust, our feet ought to be uncomfortably hot, and all the seas to disappear in a cloud of steam. However, if light is a fluid, or consists of material particles, with a velocity of 192,000 miles per second, every ray, as Franklin confessed, ought to crash down upon us more terribly than a 24-pounder from a gun. Then, the rays being without number, what awful showers of cannon-balls to wreck the earth! And yet light does not move the smallest speck of dust. What a tug-of-war there must be between ether and gravity? An ancient sophist used to argue—"Whatever body is in motion must move either in the place where it is or where it is not; but neither of these is possible, therefore there is no such thing as motion." Some logicians have answered, *Solvitur ambulando.*

MYSTERY

Sir,—If we grant "C. H" his sea-earth globe, because, as he says, it is scientific, then—since a globe or ball is made to roll, we may also grant axial motion; and with the rolling ball, the sun's distance by parallax is calculated as immense. Hence that light must be a million times larger than the earth. Again the sun being vastly larger, is surely more likely to be a centre of revolution for the earth than the earth for it; and so it is easy by arithmetic to calculate our orbit, and consequent rate of motion with the globe lurching 23½ degrees to account for the seasons. Next, we may grant the moon's distance to be 60 times the earth's radius, and a lunar diameter one-quarter that of ours; and surely, when sun, moon, and stars, as they go merrily round, happen to be in a straight line, there ought to be lunar or solar eclipses! However, when the globe is shown to be nothing but an old pagan myth, the gigantic astronomical bubble bursts.

Life still defies our imitation and scrutiny, its origin being yet unaccounted for by Darwin's "Natural Selection;" whilst the mathematicians, led by Lord Kelvin, grant only 100 million years for the period of organic life to the evolutionists, whose theory needs almost an eternity. But beyond our little oasis of knowledge, bounded by common sense and Divine revelation—what but clouds and darkness? On sun, moon, stars, clouds, and winds, on every drop of stream and ocean, on every blade of grass, on every leaf and tiny insect thereon, on every bone, muscle, member of fowl, beast and man, has not the all wise Creator written the humiliating word—MYSTERY? And do not our hearts in consciousness beat responsively a life-long Amen?

SHIP ON THE HORIZON.

Sir,—Allusion has been made to the disappearance of a ship below the horizon, whereas that phenomenon exposes their fable. Standing on the shore, we see the sky appearing to slope down to meet the ground plane, which seems to slope up, the union making an angle in which the ship vanishes; whilst the horizon is on a level with the spectator's eye. Could the earth and sea possibly make a globe, we should look down at the horizon; hence a vessel in approaching would come up and in receding go down; but the very reverse is apparent. Whilst the masts, standing up against the clear sky are visible, the hull in the distance is readily confounded with its watery base, though made distinct with a good telescope. Also if, after the ship has disappeared, we mount an adjacent hill, the hull with its masts is again visible, always seeming to go up, but never actually to tumble over the marine hill imagined by prejudice. If, again on the shore we look in quite the opposite direction, the surface we stand on seems to slope upwards, indeed doing so in every direction we may look, just such a perspective as a level surface presents. Then if, fixed in the same spot, the spectator looks all around, he finds himself the centre of a circular plane, the boundary of which is entirely due to his eyes. Next, if in a balloon we mount a few thousand feet into the air, the horizon, according to Glaisher, the aeronaut, is still on a level with the eye, whilst the earth beneath appears not convex, but concave.

COMMON SENSE.

Sir,—Newtonians from a blundering perspective infer curvature, and from that, again, globularity, just as bottles, hats, &c., having curvature may be globes. Then, objects seen to move are said to be fixed, and *vice-versa*, objects apparently small to be incredibly large, and the vast oceans with continents to shrink into the bulk of a little star, with the sky as a nonentity. Now, last century the metaphysician Berkeley, with much smart sophistry, gained dupes for his idealism, similarly Hume for nihilism. Herschel would have us believe the opposite of what our senses declare, saying that "in the disorder of our senses we transfer in idea the motion of the earth to the sun, and the stillness of the sun to the earth." But had Berkeley and Hume met the astronomer, they would with subtle reasoning have persuaded him that the earth, ocean, sun, moon, stars are only apparently outside realities, yet, after all, but ideas inside his head, and his head also a mere idea inside a brain nowhere; that there is globularity as well as motion, but no globes to be round or roll; stillness and brightness, but no sun to be still or to shine; distance, but no stars far away; vastness, but no worlds to be vast; seeing, but no eyes to see; sensations, but no senses to deceive, as astronomers say. Hume by himself, would have added, "You imagine, Sir John, that there is everything and everywhere, but I can prove that there is nothing and nowhere." Yet Hume and Berkeley would not have walked into fire and water believing these only ideas, any more than Herschel would have hung himself by the heels at a butcher's door to prove the whirling of his big globe.—A. M'Innes.

"Earth not a Globe Review." London: 32 Bankside, S.E.
"Debate on Moses and Geology." 1d. W. Love, 221 Argyle St., Glasgow.

THE BIBLE AND SCIENCE.

THE following letter, an invitation to discuss being addressed to the Agnostics of London, appeared in April last in the "Birmingham Weekly Mercury":—

SIR,—A discussion recently arose between Lord Kelvin and Professor Perry over the earth's age, the former Professor, assuming that the earth is a homogeneous body cooling at a fixed and uniform rate, and, therefore, somewhere between 20 million and 400 million years old. The latter regarded these numbers as insignificant, assuming the earth's centre to be in a highly molten state. His lordship spoke as confidently of the world's primeval state as if he had witnessed the creation; whilst Professor Perry seemed as familiar with the infernal regions as if he had been down there making a personal inspection! What a wide gulf of 380 million years Lord Kelvin makes light of!—a period of time to count which at the rate of 60 per minute, twelve hours daily, would consume 24 years of a man's life. Finally, to humour Professor Perry, the Glasgow scientist is willing to widen the gulf enormously from 20 million even to 4,000 million years, thus confessing a possible blunder of 3,600 millions. What of Moses' chronology making the earth's duration about 6,000 years, as defended by the London Chronological Society with tables of eclipse and transit cycles?

In Dr Dick's "Natural History" we have a specimen of the geological method of calculating. He supposes (no proof whatever) that God did not make the bed of the Niagara, but that the river cut for itself a passage of six miles below the Falls. Supposing it cuts at present about one foot yearly, then it must have been so working for 31,000 years; or, if it cuts, as others think, only one inch yearly, the period is 300,000 years. Then, the rocks of the quartary, or present period, being 500 feet thick, and those of the previous, or tertiary, period being six times thicker, we have six times 31,000, or 300,000 years, to add for the earth's age! Next, the thickness of the secondary rocks being 15,000 feet, we have thirty times the duration of the quartary period. The primary rocks, also, are three times, and the primordial rocks five times, thicker than the secondary. The duration is thus somewhere between eight-and-a half and eighty-three million years! Now, further, we must take into account the incalculable period of the igneous rocks, or chaotic state.

Now, according to Genesis i., God made heaven and earth, with all therein, in six days, all the rocks on the third day, and in strata, according to Job xxxviii. 5. Finally, we believe Moses's writings because we believe Christ, who said, "If ye believe not his (Moses') writings, how can ye believe My words?"

ALEX. McINNES.

Now, the publication of all my opponent's replies would be inconsistent with the limits of a cheap pamphlet. Besides, they only contain the usual cunning sophisms along with ample quotations from that shallow-pated apostate Colenso, and Huxley, the recognised pope of the many wrangling infidel sects to which he applied the fitting name Agnostic (of Greek derivation) signifying Ignoramus. Here, however, is the first of the series, of course full of impudent bounce and profane sneers.

———

Sir,—Theologians, generally, have become very anxious to make their peace with science, and to convince us that the palpable contradictions between its teachings and those of the Bible are not real, but only apparent. It is quite refreshing, therefore, to come across an honest, old-fashioned believer like Mr McInnes. He sets the scientists at defiance, and, taking his stand upon the Hebrew cosmogony, asserts his conviction that "God made heaven and earth, with all therein, in six days; all the rocks on the third day, and in strata, according to Job xxxviii. 5." He concludes his argument by saying, " Finally, we believe Moses's writings because we believe Christ, who said," If ye believe not in his (Moses's) writings how can ye believe My words?" Now, I do not propose to enter into any lengthened discussion with Mr McInnes; I simply ask to be enlightened on one or two points that are not quite clear to me. If God made all the stratified rocks on the third day, how came He (since animal life did not make its appearance until the fifth day) to put in them, in the shape of bones and skeletons of countless animals, what He knew would come to be regarded as demonstrable proofs that the earth existed millions of years before it really did? Fancy the Almighty playing a trick like that upon earnest inquirers after truth! How does Mr M'Innes propose to prove that the account of creation contained in Gen.1.which he speaks of so confidently as Moses's, was written by Moses? Does he know nothing of the "higher criticism?" How does Mr McInnes know that the Book of Job, to which he appeals, is the Word of God? Who was the author? When was it written? Canon Driver and Dr Dale have said that it is a dramatic poem. That being so, how can it be used as a scientific textbook? Again, two Hebrew commentators, Aben Ezra and Spinoza, say that the Book of Job carries no internal evidence of being a Hebrew book, that it has been translated from another language into Hebrew, and that the author was a Gentile. I defy Mr McInnes to prove that Christ used the words which he attributes to Him, and which, in the dogmatic way of the orthodox apologist, he says Christ did say. How can we possibly know with certainty what Jesus said and did, when, as Matthew Arnold says, "The record, when we first get it, has passed through at least half-a-century, or more, of oral tradition, and through more than one written account?"

AGNOSTIC.

———

But is the Bible responsible for the bungling attempts of theologians to reconcile it with some cunningly-devised fable, tricked out in big words and mathematical jargon for simpletons to swallow? Moreover, are not Dale and other enemies of Moses infallible oracles for Agnostics who strut as Biblical Critics, though more ignorant of the Greek New Testament than a schoolboy, and unable to tell a Hebrew yod from a vav?

I answered, first, as to
"THE HIGHER CRITICISM."

SIR,—In turn I challenge "Agnostic" to prove his old-fashioned fable of geology, which supposes fire and water the creating gods, and was taught more than 2,000 years ago by Thales and Pythagoras. As for the Gospels, let him prove that the early Christians held the modern theory of canonical Scriptures and a peculiar inspiration to write them; and surely as to authenticity he will grant the same fair play to the Gospels as to other ancient books? However, let it be noted that in Apostolic times historians wrote on wax tablets, a copy being then made on papyrus by an attendant, which the bibliopolist could afterwards multiply as ordered. Hence, who can find the autographs of ancient authors? The four oldest MSS. of the Gospels in Greek are those of Sinai, the Vatican, Paris, and the British Museum (4th and 5th centuries), and contain Christ's saying as to Moses' writings in John v. 46; that of Wolfenbüttel being fragmentary. In all there are forty-five codices, some entire, others fragmentary, extending down to the 10th century. Then we have 661 of the cursive sort, ranging from the 10th to 14th century. From any MSS. of the entire Gospels can "Agnostic" prove the disputed passage excluded?

Further, as to versions in languages other than Greek, there are the Old Latin, or Vetus Italia, of 2nd century; the Coptic, Peshito Syriac, and Thebaic (fragmentary), of 2nd or 3rd century; the Vulgate, Ethiopic, Armenian, Jerusalem Syriac, and Bashmuric (fragmentary), of 4th century; the Georgian, Slavonian, Frankish, Arabian, Anglo-Saxon, &c., of 5th and 6th century. Again, the epistles of Barnabas (companion of Paul), Clement of Rome (97 A.D.,) Ignatius (80 A.D.,) and Polycarp, disciple of the apostle John, contain facts and sayings found in the Gospels. Of 2nd century, Quadratius, Justin, Apollinarius, Athenagoras, Melito in apologies to Emperors, refer to Gospel facts; Papias to Matthew and Mark; Theophilus and Tatian, having written harmonies of the four Gospels. Irenæus (177 A.D.,) mentions the partiality of particular sects for a special Gospel. Moreover, we have the testimonies of opponents, Pliny, Lucian, Tacitus, and Celsus, of the 2nd century. We have, too, the internal evidence of the Gospels, the heavenly teaching and Christ's perfect character, of which the world had previously no conception.

Thus the denial of Moses is also that of Jesus, whose Christhood can be proved by incontrovertible evidence. Next, as regards the "higher criticism," I ask "Agnostic" if he has read near the end of Deuteronomy how Moses declares himself the author of the book of the Law, requiring the Levites to lay it up in the Ark? Has he heard of the ancient Samaritan Pentateuch, of the Alexandrine and other Greek versions: of the Talmud, the Gemara Targum saying that Moses wrote the Pentateuch and Job? In Arabia, where Job dwelt, did not Moses tend cattle forty years? No traces of a later era are to be found in the book, but an identity of style is evident between the poetry of Uz and that of Moses. Has "Agnostic" heard of the testimonies of the Hebrew prophets, of the Apocrypha, Josephus, New Testament writers, and Christian Fathers, as to the genuineness of the Pentateuch? Has he heard of the labours of the early Rabbins; and of Christian scholars, from Origen and Aquila down to Michaelis, Kennicott, Havernick, &c.,

over the text of the Law? Can he distinguish the peculiar Hebraisms of the three periods, of the Law, David, and Ezra? Long ago, in opposition to German Neology, now smuggled into our midst as "higher criticism," Havernick, of Konigsberg University, pointed out the characteristics of the Pentateuch, its lofty poetry and concise prose, the peculiarities of its grammatical forms, noun forms, verbal suffixes, pronouns, expressions, and phrases, not found generally in later writers. He pointed also to the absence of Greek, Chaldee, and Persian words, which occur in the age of Ezra, as due to the intercourse of Jews with the nations speaking those languages. Strange that Moses was credited with his own Torah all along until modern times, when Spinoza and Hobbes denied Moses that they might deny Christ.

SIR,—In reply to "Agnostic," permit me to touch briefly on some internal evidences of the Pentateuch's authenticity. If the events narrated there did really occur, surely, then, we have the *a priori* argument of tradition, oral and written. The Creation and Fall as related in Genesis is somewhat confirmed by the Greek and Roman traditions given in Ovid's Metamorphoses. The Hindoos and Chinese still believe that all nature is contaminated and the soil under a curse; the Ceylonese, that from Adam's Peak the first man took a farewell view of Paradise; and the Christian Fathers, that he was buried at Calvary, in the very spot where the Saviour was long after crucified. The fact of the universal deluge is yet to be dealt with. Josephus quotes from the Egyptian Manetho and Hieronymus, Nicolaus of Damascus, Berosius the Chaldean, &c., to show that the ancients lived about a thousand years. As regards the Tower of Babel, the ancient Sybil says:—"When all men were of one language, some of them built a high tower, as if thereby they would ascend to heaven; but the gods sent storms and wind to overthrow the tower, and gave every one his peculiar tongue, and for this reason the city was called Babylon." The old historians, Berosius, Nicolaus, and Hecataeus, supply facts regarding Abraham remarkably confirmatory of the Pentateuch; and to this day the Patriarch is honourably mentioned all over the East. Again, Strabo reckons thirteen cities, Sodom the chief, as once standing on the spot now occupied by the Dead Sea. Justin Martyr quotes ancient Egyptian traditions regarding the wise government of Joseph. Manetho, who quoted from the records of Egyptian priests, and is confirmed by the translation of hieroglyphic inscriptions on existing Egyptian monuments, likewise Diodorus, Siculus, give accounts of the exodus of Israel from Egypt under Moses, Diodorus also referring to the drying up of the Red Sea. Quotations might here be made from Herodotus, Eratosthenes, Strabo, and other early writers in corroboration of the Pentateuch.

SIR,—Can "Agnostic's" vigorous crowing cover his utter inability to grapple with my historical evidence for the genuineness of the Pentateuch? To bring him to bay I now challenge him: (1) To give what he considers proof for the genuineness of any book of antiquity, such as that ascribed to Herodotus, Xenophon, or Cæsar; and I undertake to produce better evidence for the Torah; (2) To point out in the Hebrew of the Torah any words, phrases, grammatical forms, anachronisms,

&c., betraying a post-Mosaic date; (3) To prove that any other than Moses wrote the great national book of the Israelites, regulating so long the civil and religious institutions of the Israelites, and still read in synagogues; (4) To show that before this age of neologists and agnostics there ever was any serious doubt of the Torah's genuineness; when infidels are forced by their very position to attack the sacred records; the friends of Moses having formed an unbroken line, from the ancient rabbins, through the New Testament writers, the Christian Fathers, onwards to the famous Hebraists, Havernick, Hengstenberg, Kennicott, Stuart, &c.; (5) To name beyond a very few learned works denying Moses's authorship, and written by Christians; (6) To show why the Jewish targums, the Samaritan Pentateuch, the Alexandrine Version, Origen's Hexapla, the Apocrypha, Josephus, Herodotus, Diodorus, Strabo, &c., as already cited by me, are of no account, as well as those ancient traditions so convincing to Humboldt, who believed the great fact of the Deluge despite geological prejudice.

Next as to

MOSES AND GEOLOGISTS.

SIR,—" Agnostic " wonders how I can account for broken strata and fossils, out of which geologists make up so much capital. I answer by the universal Deluge, respecting which the learned Humboldt says :—
"The ancient traditions of the human race, which we find dispersed over the surface of the globe (earth) like the fragments of a vast shipwreck, present among all nations a resemblance that fills us with astonishment. There are many languages belonging to branches which appear to have no connection with each other, and those all transmit to us the same fact. The substance of the traditions respecting the destroyed races and the renovation of nature is about everwhere the same, although each nation gives it a local colouring." Moses, supplying the very date, represents the water as falling from the sky and rushing in from the ocean, so as to destroy mankind and the earth (or strata), the flood covering the highest mountains, and continuing a whole year, surely sufficient to petrify the dead animals and plants imbedded among the rent and piled-up masses. Besides the references in the New Testament, Josephus describes the flood, quoting from Berosius the Chaldean, Hieronymus the Egyptian, and Nicolaus of Damascus. Further, Plutarch, Plato, Diodorus Siculus, show that the ancient Egyptians believed the Deluge to have been universal; and the Flood of Deucalion, as believed by the ancient Greeks and Romans, is narrated in Ovid's " Metamorphoses," in striking harmony with the Mosaic narrative. Accounts are quoted by Sir W. Jones, from Hindoo poets, as well as from Confucius and other ancient Chinese writers. Nor are traditions of a universal flood wanting among African tribes and the Celtic Druids. Humboldt found legends among the ancient Mexicans and other American aborigines; Ellis and Sir A. Mackenzie, too, among the Polynesians. Finally, does not the denial of the Flood involve the denial of all history, so that we may be as ignorant of the past as Darwin's monkeys?

SIR,—The following is the humiliating confession of Skertchley ("Geology," p. 101) :—" So imperfect is the record of the earth's history as told in the rocks, that we can never hope to fill up completely all the gaps in the chain of life. The testimony of the rocks has been well compared to a history of which only a few imperfect volumes remain to us, the missing portions of which we can only fill up by conjecture. What botanist but would despair of restoring the vegetation of wood and field from the dry leaves that autumn scatters? Yet from less than this the geologist has to form all his ideas of past floras. Can we wonder then at the imperfection of the geological world?" Accordingly it is before a geological tribunal forced to confess its extreme imperfection and consequent incapacity to judge that agnostics would drag the Saviour Himself for immediate condemnation because of his endorsement of Moses. In opposition to the Huttonians, who suppose the earth a ball of fire with a thin rocky crust, Lord Kelvin believes the earth-ball as rigid as steel, with only as much internal fire as may cause volcanic eruptions, &c., allowing for its age only a few trifling hundred million years, though evolutionists require almost an eternity for their theory. Hutton fancied that from the ruins of old worlds new ones are being made. But Kelly, vice-president of the Royal Geological Society of Ireland, holds that this the only earth was made during six successive periods corresponding to the six days of creation recorded by Moses, and to six different systems of rocks; also that particles of mud and sand decomposed from rocks and carried down by rivers to be deposited in sea bottoms could only become rocks of a heterogeneous mixture, but never such as the primary with sub-divisions having each its own marked peculiarities. "Neither the brown gueiss, nor the primary red sandstone, nor the yellow quartz rock, nor the gray mica slate, nor the blue limestone. Not one band out of all these could be formed out of the river sediment coming down from the pre-existing continents, because not one of them has mixed particles. The quartz rock has no lime, the limestone is purely crystalline, &c." (Errors of Geologists p. 15.) He wonders how a continent composed of a variety of rocks could send down at one time particles only fit to make 2,000 feet of the gueiss in the West Highlands of Scotland, then yield only particles fit to make primary red sandstone, 3,500 feet thick, as in Ross-shire; next the yellow quartz beautifully homogeneous and *stratified in thin*, smooth, hard flags, &c. Kelly further denies the existence of a wondrous order of fossils among rocks, and that fossils are a guide to strata, narrating various personal examinations of fossils which he found common, for example, to two different systems, Carboniferous and Devonian. Moreover in classifying rocks, Kelly divides the systems into six—primary, Cambrian, transition (*including* Silurian and Devonian), carboniferous, secondary, tertiary. Dr. Page gives five—primary, transition, secondary, tertiary, recent; next Dr. Dick—primordial, primary, secondary, tertiary, quartary; then Wylde names—palæozoic, secondary, tertiary, &c. How many more squabbles will Agnostics tell? As regards the accommodation of Noah's ark, zoologists allow that there are in all nearly 1,700 species of mammals, 1,242 lizards, 10,000 birds, 980 reptiles, and 100,000 insects. Then the ark was 300 cubits long, 50 cubits broad, and 30 cubits high. The length of the cubit has varied from 1½ft. to 1¾ft., and therefore the cubical extent was from 675,000 to 787,500 cubic feet. The ark had a window (which was 1 cubic square) rather as an outlook, and the three stories must have consisted of one deck and two galleries above. Multiplying 300 by 50, length by breadth, we find the area of the deck to be about 4,500 square cubits, or from 33,000 square feet to 45,540. Then allowing 20 square feet for an ox, as do the New York steamers, there was room on the deck alone for 1,650 oxen or 2,272. But the average size of a mammal is allowed to be less than one-fourth of an ox, therefore there was accommodation for, instead of the 3,400 mammals, actually from 6,600 to 9,088, so that ample space was left for stores of food, "Agnostic" asks why "human remains have never been found in the primary and secondary rocks, which abound in those of lower forms of life?" I deny the classification, but at Predmost, in Moravia, a few years ago, Hert Mascha found in a cave alongside mammoth bones the entire skeletons of six human beings, a discovery which, says the *Standard*, "contradicts the assertion of those specialists who deny that the mammoth was contemporary with man." Lastly, can "Agnostic" account for his creating gods, fire, and water; also life and organisation, natural laws, human speech, reason, morality, and his own freedom from the monkey's skin and tail?

Sir,—In reply to "Agnostic's" last letter I now invite him to prove—(1) That the granite and trap rocks are, as affirmed, of igneous origin, and that the metamorphic are of aqueous, and by heat and pressure are being changed, sandstone into quartz, clay into slate, &c.; (2) that the sedimentary systems (chalk, clay, &c., being manifestly not rocks, though called so) were ever formed by the alluvial deposits of rivers in sea bottoms; (3) that geological systems and groups, or even the natural strata, can represent different vast periods of past time, or that they can tell the earth's age; (4) that the classification into primary, secondary, and tertiary systems are not mere moonshine, or that the boundary lines drawn between them can be logically justified. I have already proved man and the mammoth to have been contemporaneous, contrary to geology, which assigns them to different periods. Of the tertiary Dr. Page says, "Even yet the limits of the system may be regarded as undetermined." (p. 355.) As regards the line between secondary and primary, what was formerly the new red sandstone group is now divided into the permian beds to be driven down among the primary, and the triassic to be pushed up among the secondary, all because of a new discovery of fossils. Where, too, are the transition rocks that used to lie between secondary and primary? Can we forget that the twelve sedimentary groups "are not everywhere found," as says Dr. Page, "all lying above each other; but on the contrary, only *one or two* of the groups may be developed, and that very scantily? "All that is meant by order of succession is that where two or three systems occur together they are never found out of order; that is, the chalk is never found under the oolite; or oolite beneath coal, or coal beneath the old red sandstone." Further, let it be proved (5) That all the pieces of coal, limestone, etc., called fossils (over which professors themselves squabble), because of a fancied likeness to leaves and branches, or bones, were, instead of being so formed originally by God, portions of plants or teeth and limbs of animals living millions of years ago; and too, despite such a confession of great imperfection in knowledge by Skertchly (Geology, T. Murby, London), we are accordingly to reject the Pentateuch and our Christianity, though both are so fully established! Forty years ago, Hugh Miller, in his "Testimony of the Rocks," regarded Dr. P. Smith's reconciliation of Genesis and geology, attempted eighty years ago, as inconsistent with the scientific progress since then made; but now Hugh Miller's six periods for creation are in turn left behind by palæontological advance, and what a cobbling the rocks must have for the next forty years! According to the "Testimony of the Rocks," fossils prove the primary to have been pre-eminently the period of immense forests and gigantic plants; the secondary that of sea monsters, flying serpents, and mighty reptiles; and the tertiary, the period of huge quadrupeds, such as the lordly mammoth; so that the present ought to be a degenerate age. My figures for Noah's Ark were taken from Geikie's article on the Deluge, in Kitto's Dictionary; and let "Agnostic" prove by mathematics insufficiency of room for the necessary food. As for Noah's skill in gathering and feeding the animals, we know he was assisted by God, without whom how can "Agnostic" account for the fire and water, supposed to have created the earth?

THE TORAH OF MOSES.

SIR,—"Agnostic" now tells us he cares not a straw whether Moses wrote the Torah or not; though, after all, he does care, as he promises to do what Colenso vainly tried—to disprove Moses' authorship. "Agnostic," overlooking the philological and historical tests considered as essential by critics, believes the works ascribed to Herodotus, Thucydides, and Tacitus, in so far as they relate what he judges probable, because those ancient writers tell who and what they are, with their purpose—as if Moses did not do as much, and with far more fulness. But why should "Agnostic," smiling at the supernatural, scorn the many accounts of Egyptian gods in the Euterpe, since ancient Paganism is only the worship of nature, above which does he believe any possible existence? Yet does not this very word "nature" signifying by its root "nascor" that which is produced, imply the higher and antecedent producer? and so the Mosaic record of creation necessarily recognizes the supernatural—the personal living God, without whom how can nature be accounted for? Can "Agnostic" find in the Clio any other claim to the authorship than the passage, "Herodotou Halikarnesseos histories apodesis," or, "The publication of the history of Herodotus of Halicurnassus"? Also, is he aware that the critics doubt the 8th Book of Thucydides; and can he quote the passage in the "Annales" proving Tacitus the writer?

That the "fire and faggot" of sectarian zealots hindered an earlier outburst of infidelity is only the supposition of "Agnostic," who ought to know that since Christ forbade the sword, enjoining the return of good for evil, persecuting sects are no more Christians than the French atheists who a century ago reddened the Seine and Rhone with human blood. I have already quoted Humboldt and Hugh Miller concerning the Deluge. The Caithness geologist adds, "Sir W. Jones, perhaps the most learned and accomplished man of his age, and the first who fairly opened up the great storehouse of eastern antiquities, describes the traditions of the Deluge as prevalent also in the great Chinese empire with its 300 million inhabitants." Likewise Dr P. Smith, in his book on "Geological Science" (p. 74), "The historical traditions of all nations, ancient and of recent discovery, furnish ample proof that this great event (the Deluge) is indelibly graven upon the memory of the human race." Besides, is not Christ's own reference to the Deluge, Noah and the ark (Matt. xxiv. 36), found in the three oldest Greek MSS. of the New Testament? "Agnostic" wonders how Noah managed the ark with its inmates, as if the Deluge so well attested is less a fact because we do not know all the mechanical arts of antiquity. How can he explain the building of the pyramids, or the morning cry of Memnon's statue? Were those ancients only stupid Darwinian apes that built Babylon, surpassing in splendour Paris, or any other modern city, and executed those works of art we can imitate, but cannot equal? I do not expect "Agnostic" to do more than either touch or overlook entirely the following questions :—The ancient national records of the Jews being divided into (1) the Law or Torah, (2) the Hagiographer (Job, Psalms, &c.), (3) the prophets—all in Hebrew excepting some Chaldee passages in Daniel, Ezra, &c., where is the evidence that the first was divided originally into five books? At the end of the Torah is appended a brief notice of the lawgiver's death, as says the Gennara Targum, by Joshua. Now, are there not traces of the Torah in every other Jewish National book, Joshua and Judges naming the Book of Moses, the Psalms giving abstract of the Mosaic narratives ; the "Kings" relating how Hilkiah the high priest found a temple copy of the Torah, Nehemiah as an eye witness relating how Ezra read the Book to all the people, &c. Then why believe the mere supposition of the Atheist Spinoza, relating to a matter 2,200 years before him?

What is proved by the Torah written in Syriac characters, not in the square Chaldee letters, along with an ancient Arabic version, both possessed by the Samaritans, who claimed descent from the ten tribes that revolted from Rehoboam four centuries before Ezra, separating for ever from the Jews? Do not the Apocryphal books and the Septuagint existing centuries before Christ, Josephus' history (first century), the Talmud Commentaries on the Torah with the Targums or Chaldee versions, Origen's Hexapla, and other Greek recensions, the Vulgate, the labours of the Massoretes, and rabbins, &c., form an unbroken chain of literary evidence extending through thousands of years? Are not the chief events of the Mosaic history the Creation, the Fall, the Deluge, Tower of Babel with dispersion, Abraham's piety, the march out of Egypt under Moses into the desert, confirmed by ancient writers and others—Hieronyous, Strabo, Manetho, Berosius, &c? Do not the abundant allusions to primitive geography, customs, &c., in the Torah prove the writer to have resided (as did Moses) in Arabia and Egypt? That is evident from the absence of Chaldee and Greek words in the Torah, as well as the following peculiarities not found, or rarely, in the later national Jewish records :—The Hebrew text makes no distinction in gender in the use of the third personal pronoun singular, preferring the older form of the demonstrative, as well as a peculiar abbreviation of the imperative, with original forms of certain verbs ; also certain strong noun forms, phrases, expressions, &c., which might be quoted.

With the date of Creation found in Genesis, along with eclipse and transit cycles, is not the earth's age known, rather than by the vague conjecture of Lord Kelvin? Do not the following predictions prove a Divine origin?—(1) That of Jacob, the Messiah to spring from Judah ; (2) that of Moses, the Jews, as we now witness, to be a hissing and a bye-word on the earth, also to be scattered among all nations; (3) that regarding Ishmael, fulfilled in the greatness of the Caliphate empire, also in the independent and predatory life of the Bedouins; (4) that of Noah, African slavery, also the dwelling of Japhet in the tents of Shem, fulfilled in the colonisation of America by Europeans, and the division of Asia chiefly between Russia and Britain, the Buddhist empire of China now becoming the prey of the sons of Japhet.

Can "Agnostic" show from the Hebrew any other rendering of Gen. xlix. 10, than—"A sceptre shall not fail from Judah, nor a lawgiver from between his feet, until Shiloh come?" (Shiloh signifies Saviour). Is it not a fact that the servitude of the sons of Ham to those of Japheth has occurred only in modern times? As regards Noah's curse (Gen. ix. 25). Negro slavery followed the discovery of America four centuries ago, and was vigorously maintained by the American colonists, Portugal, Spain, France, Holland, and England. This kingdom alone exported 300,000 slaves from Africa between 1680 and 1700, carrying during the next 80 years 610,000 to Jamaica. Why does "Agnostic," whilst believing without sufficient evidence the books ascribed to Thucydides, Tacitus, Herodotus (the supernatural excepted), reject every fact in the Torah, though he dares not grapple with its evidence, particularly its peculiar Hebrew? Again, though Christ taught his followers to love their enemies, does not "Agnostic" plead that the saying, "I came not to send peace but a sword," was a command to shed blood? though the apostle explains that "the weapons of our warfare are not carnal, but spiritual." Where in the New Testament do we find that after the Apostles began to disciple the nations any Christians ever used violence, even in self-defence? Does not Hebrew iv. 12 show the spiritual sword to be truth—Christ's wondrous utterances (never before known to Jew or Gentile) with which a spiritual war has been waged against the polytheism of the world? I have quoted from Humboldt, Miller, and Smith by the way of *argumentum ad hominem*, showing how conclusive is the fact of the Deluge, because incapable of being denied even by geologists, whose theory it overturns, the fossils found in every region having been regarded generally as the evidence of a universal flood until last century. But has not the device of Miller and Smith been to plead that the Deluge was confined to a level district near Ararat, where they say all mankind originally dwelt? And their ingenuity was taxed to explain away the universal terms applied by the Torah to the awful catastrophe. Yet all the ancient traditions so convincing to geologists speak of an ark, the need of which is clearly absurd for a local flood, since Noah, with family and animals, would rather have emigrated in due time out of the doomed locality. In answer to Miller and Smith's supposition of the ark's inadequacy to lodge all the animals, I have already given a calculation with Geikie's estimate of species However, according to Humboldt, there are 500 kinds of mammals, 4000 birds, 700 reptiles, and 44,000 insects. Now, the ancient cubit being 1ft. 9in., the ark was 525 ft. long, 87½ broad, 52½ high, with three storeys (each 15ft. high) or decks, the area of each being 45,937½ square feet. The New York steamers allow 20 square feet for every ox, and the average size of a mammal is one-fourth that of an ox, so that 5ft. for each of the 1000 mammals would leave 40,937ft. of area for the 4000 birds, as well as tanks, troughs, stairs, waterpipes, &c. A second deck could accommodate in separate rooms, as in the first, the 1400 reptiles, the tiny insects, as also the odd clean beasts (seven of every kind) for food to ͏ ͏ ͏ nivora. Thus a third storey was left for the eight persons, and a year's provisions. As regards the feeding and the watering of the animals, the cleaning of decks, etc., how do the ocean steamers arrange that carry regularly thousands of cattle? Also, the window in the roof of the ark being used as an outlook; surely the builder would see the necessity of ventilators, and the insertion of transparent substances in the sides to give light. And why do zoologists differ so much in their estimates of species, unless they confound mere variety with essential difference? Humboldt gives 500 kinds of mammals, Philips 1200, and Geikie 1660. According to Humboldt there are of reptiles 700 kinds, and to Swanson 1500; whilst of birds, according to Humboldt 4000, but to Geikie 10,000; the species of insects varying from 11,000 to 100,000. Lastly, is not the Deluge inseparable from the miraculous, as Moses reveals, God Himself causing the Flood, Noah miraculously forewarned, and directed to make the ark? As is related, Noah with his family entered the ark first, then all the animals were led in by God's word; and after being shut in a full year, the eight persons went out first, God by His word causing all the animals to follow.—ALEX. McINNES.

GEOLOGY.

SIR,—How do we know that all those pieces of chalk and limestone called fossils, instead of having been originally so created, were the bones of animals, and the leaves of plants living millions of years ago? Let me next ask "Agnostic" if he has ever heard of "simulative structures?" 'The student," says Dr Page, "should be informed that its (*Ozoon Canadense*) organic structure has been called in question by some who regard it as a peculiar mineral structure mimetic of the organic examples, such simulative structures being well known in other formations besides the Laurentian. Those who take an interest in this matter may refer to the Journal of the Geological Society for 1865-6." Ought not, then, the announcement of a discovery of new species, however pompous the names, to be received *cum grano salis?* As formerly quoted, Skertchly (F.G.S., H.M. Survey) confesses geology to be a history of which only a few volumes remain, the rest being mere conjectures. Indeed, is not the geological argument, without one living witness of the fabulous past ages, precisely that of Scott's antiquary who imagined he had discovered the remains of an old Roman camp? Eddie Ochiltree, however, suddenly appears to prove that twenty years before he was present at the rearing of the supposed camp, the "remains" of which were around. And how many fish scales, dry leaves, bits of coal and limestone, cannot the antiquarian geologist produce wherewith to fill up the fauna and flora of ages as vast as Lord Kelvin pleases to say?

Are not the remains of sea monsters satisfactorily accounted for by the existence still of the "sea serpent," of which we have ample evidence in the August magazines? And what is the Mammoth but a huge elephant, the bones of which were lately found in a Moravian cave along with six gigantic human skeletons? That beasts and birds have degenerated is in accordance with the fact shown by the Mosaic and other ancient records, that men in longevity and stature are far inferior to the ancients, especially the antediluvians. "Agnostic" infers from the discolouring of the sea for 300 miles at the mouth of the Amazon that strata were originally formed by the sediment of rivers. But where is the proof that all that soft sand sinks far down to the bottom of the sea to harden into rock? Even granting the hardening process, the mud particles coming from a variety of rocks, and therefore of a heterogenous kind, could never form separate beds of the so-called primary or secondary red sandstone, blue limestone, yellow quartz, gray mica-schist, brown gueiss, white chalk, &c. That strata were formed from pre-existing continents Hutton could have learned from the pagan Greek Pythagoras, 2,300 years before him; but how can we account for the first continents unless we assume an infinite series of them, and call the earth a god?

The tyros in geology imagine that the twelve groups composing the metamorphic rocks, the primary, secondary, tertiary, and recent, are all piled above each other like so many shelves, with their peculiar fossils; whereas, according to Dr Page, there are generally only two or three groups together; whilst the only order is, that chalk is never found under the olite (limestone), nor olite under coal, nor coal under old red sandstone. Further, Dr Page confesses the tertiary limits to be yet undetermined, and when will the tinkering of the boundaries between secondary and primary cease so long as geologists use two principles of classification, fossiliferous and lithological? Where now are the seven zones once so orthodox, each, as affirmed with its peculiar fossils; and what of six creation-periods upheld by Hugh Miller and Kelly? Strange, too, that though we can dig down only a few hundred yards, and the greater part of the earth's surface is yet unexplored, acquaintance is claimed with the whole earth's "crust" even to 4,000 miles downwards. Now, all that is meant by huge beasts being of the tertiary rocks is, that their bones were found in caves or in the sand of the sea-shore, and by marine monsters being of the secondary strata is meant that pieces of stone shaped like peculiar bones were found in quarries. Of course, the miner also finds down in the pit bits of coal shaped like shells, &c., hence of the primary.

SIR,—Professors Keil and Delitzsch of Germany, in vindicating Moses, point out (vol. i. 42) the contentions as to the origin of rocks among the Neptunian, Plutonic, and other geological sects; though Gen. i. shows that on the third day of creation the earthy atoms in a chaotic state within the primeval waters, at God's word united into one great solid mass of land with a dry surface. The argument of Keil and Delitzsch may be given in the following questions:—(1) Is not the order of systems and sub-divisions of rocks as laid down by geologists often found reversed, crystalline primary rocks lying upon transitional (once put between primary and secondary) stratified and tertiary formations (granite, syenic, gneiss, &c., above both Jura limestone and chalk); (2) Do not the various systems and sub-divisions frequently shade off into one another so imperceptibly that no boundary line can be drawn between them, and the species distinguished by oryctognosis are not sharply and clearly defined in name, but that instead of surrounding the entire earth, they are all met with in certain localities only whilst whole series cf intermediate links are frequently missing, the tertiary formations especially being universally admitted to be only partial; (3) Are not the following mere assumptions :—(1) That each of the fossiliferous formations contains an order of plants and animals peculiar to itself: (2) That these are so totally different from existing plants and animals that the latter could not have sprung from them; (3) That no fossil remains of man exist of the same antiquity as the fossil remains of aminals; (4) That in the strata there is progressive development of plants and animals, or that the transition rocks contain cnly fossils of the lower orders, that mammals are first met with in the Trias, Jura and chalk formations, and warm blooded animals in the tertiary systems; (5) That the fossil types are altogether different from existing families, though all the fossils can be arranged in the classes of existing fauna and flora, there being only specific differences, but no essential ones as regards genera, and many existing types being smaller than those of the old world; (6) Even if the old species differed from those now existing, which, however, can by no means be proved, would there be any valid evidence that the existing plants and animals had not sprung from these that have passed away:—remembering that the origin and formation of species is still a mystery to human science?

THE HEBREW AND GREEK,

SIR,—Is it not evident that throughout this discussion " Agnostic " has failed, (1) either to prove Moses not the author of the Torah or to name another instead; (2), to grapple with my historical and literary evidence for that book in particular and the Bible in general; (3), even to defend the popular theories of geology; (4), to prove the earth's age more than six thousand years? Now, at last, when he approaches the philology of this question, according to him the Hebrew of Moses had twenty-two letters without any vowels. However, according to the grammars of Parkhurst and Gesenius it had only nineteen letters; four being vowels—Aleph or A, Vau or U, Yod or I, and Ayin or O. The Chaldee square letters, not used by the Jews till after the captivity, amount to twenty-two, the three superfluous letters being Teth or T, Samech or S, and Zain or Z. Our (Roman) alphabet has nine redundant letters, Q, X, Z, being double consonants, C having the sound of K or S, I and J being identical, as well as V and U, the E as pronounced in Latin is the diphthong ai, and W is twice U, whilst Y is the tripthong uai. The a and i ought to be sounded as in Latin and the continental languages. The Hebrew vowels have now acquired a consonantal force owing to the innovation of vowel points after the language ceased to be spoken, to preserve the pronunciation. Hebrew, however, is still read without points with Prkhurst's Lexicon, the meaning of words depending on the stems, biliteral or triliteral; but as with all languages there may be ambiguity with an isolated word or phrase. In Latin, for example, the isolated phrase 'jus naturale' may mean either natural law or natural soup. Moreover, can linguists name a language more original, pure, picturesque, logical, than that of Moses, which bespeaks an age earlier than the eras of David and Ezra?

" Agnostic" refers to the various readings of the Greek and Hebrew

MSS., as if he could name any ancient book with MSS. of different ages without such variations. Do not the many readings and translaitons of the Bible prove that, being of divine origin, it has been copied, edited, criticised, read and circulated more than any book in existence? Scholars, of course, collate the various MSS. to make a *textus receptus* for translators; and, after all, can any essential differences between the MSS. be pointed out? Taking the three oldest codices, Vatican, Sinaitic Alexandrine, and opening at Mark i., we find the title is in the first two merely, "According to Mark"; but in the third, "The Gospel according to Mark. Again, at v. 4 the first two MSS. read "I will send," but the third "I send," etc. Thus, thousands of such petty differences may be found among MSS. Professor Gaussen, of Geneva, mentions that in 1817, Claudius Buchanan found among the black Jews of Malabar, who had no intercourse with the west, a Hebrew MSS., 48 ft. long, 22-in. broad, containing all the Old Testament except Leviticus and part of Deuteronomy, in thirty-seven sheepskins, which he deposited in the Cambridge Library. It was compared, letter for letter, with the Hebrew edition of Van der Hooght, and only forty small differences were detected. The extreme reverence of the Jews for the ancient text is well known. Besides, was Moses responsible for the errors, unavoidable or intentional, of all future copyists of the Torah, or of any possible translator? Moreover, a teacher of languages knows that if any two pupils give precisely the same rendering of a Greek or Latin author that there has been copying or cribbing; as verbal variations prove independent work. Yet, let "Agnostic" compare the old and new versions of the Bible, and tell how far they differ in sense.

As regards Gen. xlix. 10, Dr Young, reckoned the best oriental scholar of this century, gives it—"The sceptre turneth not aside from Judah nor a lawgiver from between his feet until the seed come"; whilst the new version is, "The sceptre shall not depart from Judah nor a lawgiver from between his feet until Shiloh come,"—both the same in sense. The Greek of the Septuagent is—" Ouk ekleipsa arhon ek Iouda kai hegoumenos ek ton meron autou, eos ean elthe ta apokeimena auto"; or, "A ruler shall not fail from Judah, nor a leader from between his thighs until the things stored up come;" a rather loose translation, the version having been executed under the pagan Ptolemy (3rd cent. B.C.) which Theodotus and Origen failed to improve. If "Agnostic" knew Hebrew, I could here give the original of Gen. xlix. 10 from the text of Van der Hooght, that he might justify Arnold's rendering. Then as regards the slavery of Ham's progeny foretold by Noah, did Plato, Aristotle, or any ancient sage ever condemn slavery before Jesus said, "He that would be greatest among you must be servant of all"—the death-knell of a then universal evil? Further, "Agnostic" repeats what all the sects of infidelity assert without a shadow of proof, that there are "glaring inconsistencies and contradictions in the Bible;" yet what says the poet? "Better had they ne'er been born: Who read to doubt, who read to scorn." The Mosaic record has come down to us through the long march of thirty-three centuries, during which how many a generation has come and gone, how many an empire with all its royal and priestly pomp has passed away! But the Torah still stands the monument of God's imperishable truth.

NUTS FOR AGNOSTICS.

A few weeks ago, this letter was inserted in the *Newcastle Chronicle*, but as yet no answer has appeared:—

Can any of your Huttonian readers answer the following queries? —Is it not mere supposition (1) That the granite and trap rocks are of igneous origin or that the metamorphic change by heat and pressure from sandstone into quartz, clay into slate, &c.? (2) That the sedimentary systems (clay, sand, chalk, &c., being manifestly not rock, though called so by geologists) were formed by alluvial deposits of rivers, which no one ever saw harden into rock, and being. besides, confused heaps of particles from various rocks could ever become separate beds of red sandstone, yellow quartz, blue limestone, white chalk, &c.? (3) That, living witnesses being wanting, the different systems and groups of strata can represent vast periods of past time, or can indicate the earth's age? Is not the division into primary, secondary, and tertiary mere moonshine, the limits of the tertiaries, as Dr. Page confesses, being yet undetermined, and the boundary between secondary and primary but lately changed again? Is it not confusion in classifying occasioned by sometimes arranging according to fossils, and, again, according to lithological character of strata? (4) If as Hutton taught, and as did the pagan Pythagoras 2,300 years before him, that the present continents were made out of antecedent ones, how were the first continents made, unless we suppose an infinite series of them and the earth without beginning? (5) Did not the earlier geologists, such as Dr. P. Smith, teach that the Mosaic six days of creation were literal, but with previous creations. Also did not Hugh Miller and others afterwards teach that the rocks exhibit six divisions answering to six ages symbolised by the six days of Genesis? whereas are not the strata now in three classes, and pray what next? (5) Is not the mammoth the only entire animal found as a fossil, all other fossils consisting of decayed bones, dry leaves, pieces of stone, coal, etc , geology being likened by Skertchley ("Geology," p. 101) to a history mainly conjectural, a few volumes only being left, so imperfect are the fossils? (7) And do we know that all the pieces of stone called fossils shaped like bones and leaves were the parts of animals and plants living millions of years ago, especially since, as Dr. Page says, there are structures mainly simulative or mimetic of fossils over which professors have had their squabbles? (8) Were not fossils and broken strata, up till last century, regarded as evidence of the universal deluge of Noah, attested too, by the traditions of all ancient nations? (9) Are not tyros apt to imagine that the twelve groups or rocks making up the systems are all piled upon one another like so many shelves stored with their peculiar fossils; where as no more than two or three groups are generally found together without being even contiguous, the only being, as Dr. Page confesses, that chalk is never found under oolite nor oolite under coal, nor coal under Old Red Sandstone? (10) When the bones of elephantine mammoth (beasts and men having degenerated since the Deluge) are found in caves or in sandy beaches, do not the geologists talk of finding extinct species of mammals in tertiary rocks? or, if in a quarry, is found a lump of stone shaped like a big bone, do not geologists rejoice over the discovery of the remains of ichthyosauri or plesiosauri, though the existence of the sea serpent (see August number of the "Strand Magazine') may suffice to account for marine monsters?

A. McINNES,

MORE NUTS.

(1.) Did any of the pagan sages of antiquity ever devise an ethical system so perfect as that of Moses; or can the legislation of Solon, Lycurgus, or any other ancient lawgiver be compared with that of Moses, which made every Israelite a landowner, without need of taxation, coinage, police, jails, paid judges, royalty, navy, or standing army? (2.) Did not the tree of knowledge of good and evil (Gen. ii.) symbolise man's mixed condition till now? But why cannot we have justice and truth without crime and fraud, peace without war, food without labour, life without death, and the whole earth one paradise of joy? Moses foretold a Deliverer to give the good without any evil, and whom can the Agnostic name? (3.) Do not the doctrines in Genesis of the Fall, the curse of labour, the command to multiply, man's power to subdue, the confusion of Babel, etc., harmonise with the facts of prevalent crime, the toil of millions, the advance of populations, the triumph of human art, etc.? (4.) Moses applied to God two names, Elohim or Power, Jehovah or Self-existence. Now, "nature" is either self-existent or not, that is, it either had a beginning or not. If it had a beginning, as Moses records, then it is rightly called "nature," which signifies that which is produced, and its existence is therefore originally due to a higher and antecedent existence, underived and possessed of all power—God. But how (and it cannot be so proved) could nature have had no beginning and therefore to be called Jehovah or God, seeing it is not a living personality with speech, intelligence, a moral will, &c., all of which man has, though said to be the product of nature, yet higher by far, though with a beginning; and whence then has he derived all his powers? Not from the fabulous, dead, impersonal, dumb, useless soulless, inorganic primeval gas of evolution.

A CHALLENGE.

I hereby Challenge any infidel scientist, professor, editor or lecturer, to debate in writing on the genuineness of the Bible or the truth of the Christian religion, and undertake to publish the correspondence in pamphlet form, granting my opponent equal space with my own letters.

ALEX. McINNES.

204 Dumbarton Road, Glasgow.

SCIENCE VERSUS CHRISTIANITY.

THE Universal Zetetic Society will pay, through their Secretary, 96 Arkwright Street, Nottingham, the sum of £100 to any Newtonian sending satisfactory answers to the following questions :—

1st.—Is popular science anything else than a disguised pantheism or deification of nature, expressed in all the pomposity and verbosity of technical terms, along with cunning and elaborate mathematical formulas and diagrams to mystify the multitudes and delude University students -- a system of knowledge which originated 4000 years ago with the priest Hermes, Thoth, or Misrain, grandson of Noah, who founded the Egyptian monarchy? Is not, therefore, popular science necessarily opposed to the true religion of one God the Creator of Nature, and Jesus Christ the only Saviour?

History.—The German Ennemoser identifies science with heathen mythology (vol. ii., p. 17) remarking, "All the more profound modern inquirers say that the ancient myths had a physical foundation." Then after quoting from various authors, he adds, " We have already seen that the ancient philosophers treated theology (paganism) as a part of physical science, and that it is openly declared that the primeval doctrine of the gods was founded on natural philosophy. We have the propagation and the connection of the secret knowledge from Egypt and the East descending from the traditional period through the Greek and Roman mysteries, and that the ancient forms of the gods are not the ideas of the poet but of "the natural philosopher." Schweigger has shown "that the most ancient and influential mythic circle, viz, that descending from Phœnicia and Samothrace certainly reposes on a basis of natural philosophy, and that it was regarded as a hieroglyphic record of electricity and magnetism." It is well-known that the scientific mysteries of the Egyptian priests were imported into Greece some centuries before Christ by Thales and Pythagoras, the founders of the Ionic and Italic schools of philosophy, which subsequently branched out into many sects, whose doctrines are now taught in all our schools and colleges. The astronomer's globe was an object of veneration in the temples of ancient Egypt, and the Egyptian priest Manetho maintained that the world was moved by the magnetic attraction of the sun. The Greek Thales taught that the earth was a fixed globe round which sun, moon and stars revolved ; but Pythagoras later on held that the sun was a fixed centre for the revolution of earth, moon and stars, yet a third theory is now held of an infinite number of suns with circling worlds. Baal, Helios or Apollo, the sun god of antiquity was believed to rule the universe, dispensing all possible blessings ; and do not modern astronomers invest the sun with the same divine powers without plainly calling it a god? Aristotle, Plato, &c., believed the stars to be moved by the gods inhabiting them, for which the moderns substitute mysterious forces and laws.

2nd.—Did not that astrologer Copernicus, in reviving the Pythagorean astronomy, plainly confess it incapable of proof and useful only for certain calculations? Then amid much whining that science was the handmaid of religion, and with the countenance of certain clergymen thirsting for popularity, did not Hutton and other self-styled geologists, further maintain, with the ancient magician Pythagoras, the self-creative power of the earth? But the latest pagan deception has been that of evolution—a modification of the old Egyptian and Buddhist myth of the metemsychosis, according to which man's soul after leaving his body enters that of a beast and continues to migrate through the whole brute creation for ages; whilst Darwin and his followers teach that the human soul after going the round of the animal creation for many fabulous ages, has at last entered a human body.

3rd.—Do not scientists now openly stigmatise the rational belief in God as superstition, and craftily invest nature with the Divine attributes, assuming without shadow of proof that creation is self-existent, without beginning and indestructible, or incapable of an end, with unchangeable, omnipotent powers called laws; as also infinite in extent, having evolved from itself all organised, living and intelligent creatures, without the aid of the one personal, all wise, and Almighty God, the Creator and Lord of all?

Are not these scientists by their inventions which enrich only a small class of capitalists, shareholders and speculators, filling the earth with accidents and woes; at the same time busy devising rifles, guns, explosives, and many murderous engines of war, that threaten to exterminate the human race?

4th.—All our youth by stern coercive laws being educated in the belief of this atheistic science, and crime now prevailing to an extent unexampled in any former age, is it not evident that a reign of terror is impending—the "time of trouble" predicted by the prophet Daniel and our blessed Saviour?

5th.—Is it surprising that the idolater, the atheist, and the scientist agreeing to deify nature, necessarily deny the Word of God Himself, who has revealed that at the creation He called forth the earth from the waters of the abyss, created on the fourth day two great lights (not vast globes) the sun and moon, as well as the stars (twinklers, according to Hebrew); causing them to move in the expanse or roof overhead called heaven; that He founded the earth upon the seas, so fixing it that it should never move at all; that Satan from a high mountain showed Christ *all* the Kingdoms of the world; that the sun and moon stood still at Joshua's order; that Enoch, Elijah and the Saviour ascended to heaven, &c., &c.? Is it not a fact that Christians have from the first opposed all worldly or pagan science, until these last days of apostasy?

6th.—Is not the Bible a most wonderful cyclopædia of universal history, perfect juris-prudence, the sublimest poetry and all true science, physical and spiritual? Scientists are challenged especially to dispute the Bible's unerring chronology, cosmogony, astronomy, geography and geology, the universal flood accounting for the broken strata on which modern geologists build their fable. Additional questions are contained in the following letters which have been inserted in newspapers:—

ASTRONOMY.

Sir,—In "J.H.W.'s" letter there is no lack of Copernican bounce and mathematical brag; yet, could telescope or microscope detect in it one authentic fact or logical argument? What lunatic ever believed anything more opposed to common sense than the old Pagan globe of Pythagoras—that all the vast oceans and continents are really and somehow rolled together into something like a little football, where we all crawl as midgets, and whilst daily whirling heads over heels, and yearly tossed over the sun quicker than lightning, we are held securely on by the invisible chains of gravity, a mysterious power that lurks far down in the infernal regions? Now, can "J.H.W." calculate how many million tons of centrifugal force are due to orbital and axial motions, and can he explain why (gravity pulling us in an opposite direction, lest we be pitched off among the stars), we are not torn asunder by the contending forces? Or ought not the resistance of infinite ether filling boundless space render the globes' motion impossible? Also, let him explain why the globe, with all on it, is not squeezed into one monster jelly by the many million tons of air that press all around, 15lbs. per square inch. Moreover, let him prove globularity by showing that everywhere on surface of sea and land, there is, according to mechanics, a fall of 8 inches per mile, the increase as the square of the distance. Hence, our island (a globe like the earth because circumnavigable!) being about 700 miles long, John O Groat's in Caithness ought to be 60 miles lower or higher than Land's End in Cornwall, neither railways nor canals allowing for such a curvature.

Then how cleverly "J.H.W." begs the question by supposing a tangent to the earth's surface, as if there could be a tangent without a curve. How cunningly a circle is defined as "A polygon, whose sides number infinity; accordingly the sides or straight lines being so many are not straight lines at all, whilst infinity that cannot be numbered is numbered! And why is that old blind idolater Euclid infallible any more than the pope, that so many silly flies should be caught in those spider-webs called mathematical proofs? And though no mathematician has yet been able to tell precisely the ratio of diameter to circumference, yet the secret was revealed 3000 years ago in Solomon's temple (1 Kings vii. 24). Further, have not astronomers since the days of Eratosthenes who, more than 2000 years ago, devised the modern method of terrestrial measurement, squabbled over the length of a degree? Herschel now making it 70 miles, but Airy 69, so that the globe's circumference may be either 25,200 miles or 24,840. So, by parallax which supposes globularity, the sun's distance is, according to Lardner, 100 million miles, to Herschel 95 millions, to Airy 92 millions, etc. Moreover, the sun's distance, about which there is no agreement, being the unit rod of calculation, there can be no definite agreement as to the distances of stars, as also to the velocities, heat, light, gravity, etc., involved. Herschel reckons the distance of the nearest fixed star at 19,200,000 million miles, Brewster differing by the trifle of 800,000 million miles!

Moreover, there are brawls over the globe's shape, whether like orange, turnip, apple, or lemon; as to axis whether, 1, 2, or 3; as to motions, whether the sun wheels round the earth, or the globe flashes round the sun; or, again, if sun with planets dash like thunderbolts towards Herculus round a greater centre, as gravity requires—aye—even an infinity of centres and motions. Proctor, too, makes the globe's weight to be as many million tons as must dumfound simpletons, yet, supported by an orbit only imaginary; but the measuring rod transforms the monster into a mote in bulk compared to the sun, a million times larger, and, since a mote dancing in sunshine has no dark side, how can "J.H.W." account for alternation of day and night? Next, we are to imagine the mote changed into a school-room globe with a lurch (no proof) of 23½ degrees, and with intelligence enough to preserve the parallelism of its imaginary axis, as it is dashed by fictitious forces round the sun, about 100 million miles away; and how then can there be any appreciable difference between sloped and perpendicular rays, so that the mystery of the seasons may be explained by "J.H.W.?" Still, do not astronomers calculate eclipses very precisely? How, let Professor Philips tell :—" The precision of astronomy arises not from theories, but from prolonged observations and the regularity of the mean motions, and the ascertained conformity of their irregularities." Did not the Babylonian astronomers, 22 centuries ago, present to Alexander the Great table of eclipses for 1900 years? Also, the interval between two consecutive eclipses of the same dimensions is 18 years, 11 days, 7 hours, 22 minutes, 44 seconds; the last total eclipse of the sun occurring January 22nd, 1879, and the preceding one on January 11th, 1861. With eclipses and transit cycles, together with the date of creation given in Genesis, the earth's age is proved to be nearly 6000 years, in opposition to Lord Kelvin's very vague conjecture of its being somewhere between 20 million and 400 million years. Navigators, such as Ross, have sailed through the unfathomable Antarctic Ocean round the earth thereby proving it to be a vast island ("founded on the sea," Ps. xxiv.), but they found all round (according to Job xxxviii. 10) doors and bars of ice, hindering progress southward. Moreover, if from any point of the southern boundary we go north, we arrive finally at the Arctic Region, round which the land is massed, thence extending in irregular triangular forms to the Southern Ocean, and terminating in capes (Horn, Good Hope, &c.) called in Scripture "ends of the earth." The sun, according to Moses, a light and set in the sky ("strong as a molten mirror," Job) performs with the seven stars of Ursa Major as a centre, its daily circle, contracted in summer, but expanded in winter, not going down or rising; but "going in and going forth," according to the Hebrew of Moses, lighting up the regions it visits, whilst those behind are left in darkness. Now whilst the ancient Pagans thought the earth and sea an extended plane UNDER which the sun travelled by NIGHT, or else a GLOBE governed by the SUN GOD, the Bible alone has revealed the true science. Again, on travelling north or south, we, by-and-bye, lose sight of the stars (Hebrew "twinklers") in the roof of heaven, because that roof is not as astronomers suppose millions of miles high, but a few miles as proved by the ascent of Our Lord to heaven.

Here is a distich by the Ayrshire ploughman, which I cannot help thinking is a *multum in parvo* shoe for the moderns to put on :—

"The ancients tauld great tales of wonder
Tae gull the mob and keep them under."

Now, Sandy, Paddy, John Bull, &c., if not wide awake, are mightily astonished when told by the scientist that, instead of standing on level ground, they are perched on the top of a big globe, off which they are apt to tumble, more especially that they are whirling and flying faster than lightning; but their alarm is ca'med down by the assurance of a mysterious power, "gravity," far down in the lower regions, holding them on secure. They are still more bewildered on learning that their shoulders carry about day and night an atmospheric column, some tons heavy, and fifty miles high; that the sun is a globe a million times bigger than the one they are on, whilst the moon and stars are also tremendous globes with men hanging from them by the heels, yet all hindered from dashing against one another by powers too mystical to be explained; that our own globe is found, without being put on scales, to be of an inconceivable but exact number of tons weight, yet needing no support; and that the stars are so many millions miles off that their light takes thousands of years to reach us. Lastly, amazement is at a climax when we are told that our great-great-great-grandfathers were apes, gorillas, or some such beasts. Accordingly, evolution calls for a new edition of Gen. i :—" In the beginning, somewhere (says Lord Kelvin) between twenty millions and four hundred million years ago, was fiery gas; and gas, after many ages, hardened somehow into solid rocks, which partly softened at length into cabbages, &c., whence sprouted long after tadpoles, and the tadpole begat a donkey, and the donkey begat a monkey, and after many ages the monkey begat Adam." These tales are great theories, of course, because fathered by men who are great because of their theories. Yes, but there are mathematical "proofs"—aye, and as mystical as Egyptian hieroglyphs, entitling the scientists to a monopoly of mutual squabbling.

———

SIR,—The mystification complained of by J. B. may be due to his ignorance of the Newtonian theory, especially the numerical calculations; but ought paunch-bellied parsons, even moderators of church assemblies, to stand hat in hand before Darwin and Hollyoake's tribunal humbly apologising for the writings of Moses and the prophets as unscientific, unphilosophical and begging credit at least for the spiritual doctrines, at most a small percentage? Ingersol, the agnostic Goliath, throws down the gauntlet, saying—" If it shall turn out that Joshua was superior to Laplace, that Moses knew more about geology than Humboldt, that Job as a scientist was superior to Kepler, that Isaiah knew more than Copernicus, then I will admit infidelity may be speechless for ever." Now Humboldt, in his "Cosmos," quotes the confession of Copernicus, "For neither is it necessary that these hypotheses be true, nay not even probable, but this one thing suffices, that they show the calculation to agree with the observations." Then these very "hypotheses," not necessarily true, or even probable, are the essentials of popular astronomy, first taught by the Greek idolater, Pythagoras, contemporary of Daniel the prophet, which supposes, proves nothing, that the universe is governed not by God, but by self-existent, eternal, omnipotent laws (a mere atheistic necessity), and that space, or, virtually, creation, is infinite, with an infinity of globes or inhabited worlds flashing about central suns, all originally evolved from gas or vapour, itself uncaused or unaccounted for—thus deifying nature as invested with the attributes of God himself.

Must we believe, forsooth, that at the very time the ancient idolaters were luxuriating in the light of globose science the great Architect of creation should speak ignorantly or falsely of His own works and so as to keep in darkness His own worshippers? Of course simpletons believe that because the Bible writers did not use the pompous verbosity of technical terms intelligible only to the few, giving celestial magnitudes and distances such as those of modern astronomers, of whom scarcely two agree, or the lying diagrams of Herschel, &c, that therefore the Bible is unscientific. In the good old Greek and Hebrew book, so long and so carefully preserved, I find the true scientific creed. The heathen astronomy, to cause mental confusion, craftily supposes ocean and land one impossible ball called earth, yet a planet or star shining in the heavens. Moses, however, carefully distinguishes the trinity of creation, saying that God called the dry land arranged in layers (Job xxxviii. 5) out of the abyss of waters, which are surrounded on all sides by the Antarctic icebergs; and made the expanse of heaven, or roof, setting therein the sun and moon with stars to regulate day and night moving always at the same altitude from horizon to horizon above the earth and round the central north. The date of creation, as supplied by Moses, along with the fact that the sun and the moon are the great clocks of the universe, has enabled Mr Dimbleby, of London, to rectify the calendar, and reduce chronology to an exact science.—I am, &c.,

---

Sir,—Mr Layton's perpetual motion bubble to be full blown begs three assumptions, instead of as many facts, with a logic granting nothing without proof. That the earth is a whirling globe shaped like an orange, lemon, or turnip, could we believe without proof, any more than that a lunatic, though himself credulous enough, is butter or glass; for the earth or land made up of the great continents would then be supposed revolving in the ocean like the paddle of a steamer?

But if by "earth" is meant all oceans and continents rolled into a solid called a globe, then as sure as the earth is solid and ocean fluid, such a globe is only the phantom of a disordered imagination. Next, that the earth's surface has a fall of eight inches in the first mile, the increase being as the square of the distance, is only an inference made by mathematicians from the assumption of the sea-earth globe, but totally disregarded by surveyors in the construction of canals and railways. Thirdly, elementary books on mechanics assume, but without proof, that a stone irrespective of its weight falls through 16 feet in the first second of its descent from any height. Yet cannot Mr Layton calculate how much faster a stone 4lbs. in weight (not a body any more than a soul) falls in any number of seconds than a stone weighing 1 lb?

However, behind the three beggarly assumptions is a fourth, exhumed by Newton from the grave of that ancient idolater Aristotle—

attraction or gravity, denying that stones fall because of heaviness, though the fact is undeniable that smoke, steam, &c., ascend because of lightness or lack of heaviness. Hence, heavy substances falling not because heavy, it is next assumed there must be a mysterious power far down in the infernal regions, pulling down us and all on the earth's surface, though what pushes up smoke and feathers is not explained any more than gravity. Surely, too, Mr Layton didn't need by the subtleties of Algebra, even with his own assumption of 16 feet in the first second, to infer that 8 inches answer to one-fifth of a second. A schoolboy of ten or twelve years knows that a stone taking one second to fall through 16 feet, would take only one twenty-fourth of a second to fall through eight inches; but if it takes one-fifth of a second to fall 8 inches, it must take four and four-fifths seconds to fall 16 feet.

Newton assumed, but of course without proof and in defiance of common sense, that "every particle in the universe attracts every other"; hence, the smallest crumb sticking to Sir Isaac's teeth would pull every drop of ocean, streams, &c., all the particles making up continents and islands, as well as the numberless globes of "boundless space." Whence the wee crumb could get this infinite power, or by what *modus operandi* it could pull the whole universe, are surely difficulties as inexplicable as those meant to be explained. Moreover, if the crumb did draw, all creation ought to have moved towards it; and, according to Newton, the whole creation should have combined to pull the crumb, and yet contemptibly small as the plucky little bit is, in spite of the awful tugging it would stick to his teeth. Why, even a living flea no bigger than a dead crumb ought *a fortiori* to move the whole universe; consequently, the moving of a 100 ton gun should be a very easy job for the clever little insect. And wouldn't the wee crumb, after all, require to be infinitely large, so that, in order to pull, every particle might get a hold of it? Again, how does the crumb pull all in turn? When a man, a horse, or an engine draws, it must be by hands, ropes, or some sort of coupling; therefore, how could the crumb pull all creation unless with an infinite number of hands, ropes, &c., all of course being as imaginary as the hauling itself?

Lastly, Mr Layton, in calculating perpetual motion round his phantom globe, must consistently take into account the centrifugal force due to axial rotation, also that due to orbital motion, not forgetting the resistance caused by boundless ether, which, being infinite, ought to make both globe and stone perfectly motionless.

SIR,—Can your correspondent, "R.E O," amazed and amused as he is, by his reply so full of specimens of the petitio principii type, lend even a crutch to his confrere, Mr Layton, now in distress? "R.E.O." confesses the wish to notice only two points in my letter, and why? unless from a painful consciousness of his inability to defend the heathen globe, about whose shape and size astronomers have all along so much squabbled, the length of a terrestrial degree being to this hour as great a mystery as it was 2,200 years ago to those pagan globists Aristotle, Eratosthenes, &c.; and hence the forthcoming meeting of the International Geodetic Association at Berlin to crack the geodetic nut. The calculations therefore of terrestrial size and weight, as well as distance and magnitude of sun, moon, or stars being dependent on a precise knowledge of a degree, are as mythical as the man monkey of pantheistic evolution.

As regards falling stones (generally called bodies, though without any qualities of either body or soul), if "R E O" thinks that the *Evening Courier* and a ton of lead when let fall from a house-top simultaneously would reach the ground at the same time, let him try the experiment. Or let him say if he ever saw two stones, respectively one and four pounds weight, along with a feather, all drop through 16 feet in one second, in what is called a vacuum, remembering that Newton denied the vacuum, because, not squaring with gravity which requires a material medium. But, of course, gravity must be true, because Newton, i's advocate, was a great man, though his greatness is also due to that of gravity—the usual circulus in probando argument. Yet what avails the assumption of a vacuum since popular science supposes also an atmospheric pressure of 15lbs. per square inch of surface, and let "R.E.O." now calculate the globular surface and tell how many million tons of air are squeezing the poor globe's sides—explaining how in spite of this wonderful superincumbent mass as well as almighty gravity, a little feather can mount so triumphantly on high. Then when the magical air-pump removes the awful downward pressing weight, ought not the feather to soar still higher?

"R.E.O." crows over a supposed analogy between gravity and "magnetic attraction." Sir R. Phillips, though a globist, says:—"La Place invents gravitating atoms, and gives them a velocity of 6,000 times that of light, which in some way (known only to himself) performs the work of bringing the body in; others imagine little hooks. As to drawing, pulling, &c., it becomes them to show the tackle, levers, ropes," &c., "Million of Facts," p. 383. Proctor, after imagining all the oceans, including the unfathomable Arctic and Antarctic, rigid as steel and soldered somehow to the vast continents, so as to make a big ball 8,000 miles high, puts them on imaginary scales, and gives the weight in as many million tons as can make simpletons gape with wonder. Another inexplicable wonder is that mystical gravity, everywhere and nowhere to be found, can, without ropes, hooks, or any coupling, drag the mighty load swifter than lightning along an imaginary orbit, that is, no orbit at all; and let "R.E.O." say how many tons the magical magnet can pull or drive, especially along nowhere. Yet another wonder; the globe, 8,000 miles in diameter, with the enchanted rod of Proctor collapses suddenly into something like a little pea before a turnip-shaped sun a million times larger, which in turn contracts into the bulk of an apple as it flies towards Hercules, driven by gravity round a mightier centre than itself! Accordingly, how many worlds like cannon balls on battlefields are flying through the "boundless space" of astromers' brains, all the while multitudes of men and other living creatures being wheeled, tossed, and flashed up, above, around, along, without choking outright or tumbling off? Lastly, is it not the greatest wonder of all that men outside lunatic asylums should swallow down those silly fables invented by Egyptian priests and Chaldean astrologers in the heathen darkness of past ages.

IS THE SCIENCE OF ELECTRICITY AFTER ALL ONLY A REVIVAL OF THE ANCIENT MAGIC?

Three hundred years ago certain physicians and university professors in Europe began to observe what had been known to the Greeks 2000 years before, that amber, when rubbed with the hand, attracted light substances, as feathers, paper, etc. Subsequently, many other substances were discovered to have the same property, and with them even circular motion could be caused. It was also found that sparks could be extracted from the human body, as well as fire from ice and water. Moreover, by rubbing in the dark a light was made to shine. The supposition was then started that these phenomena were not due to the substances rubbed, but to a fluid within them, which was accordingly called electricity, a word derived from the Greek—elektron—meaning amber. Since, many instruments have been made for collecting, communicating, etc., the electric fluid. It was found, too, that shocks could be given to the human body, pains and convulsions produced; whilst attempts at healing were abandoned as a failure. About the end of last century Franklin, by sending up a kite into the air, brought fire down from the clouds, as had been foretold in Rev. xiii., 13, "He brings fire down from heaven in sight of men." Accordingly electricity has been identified with attraction, light, fire, lightning, etc. This present century has witnessed the construction of electric telegraphs over land and sea over thousands of miles, for the transmission of intelligence; and, lastly, of telephones, by means of which persons may mutually converse, though separated by many miles.

Dr Lardner in his "Museum of Science and Art," makes the humiliating confession:—"The world of science is not agreed as to the physical character of electricity. According to the opinion of some it is a fluid infinitely lighter and more subtle than the most attenuated and impalpable gas, capable of moving through space with a velocity commensurate with its subtleness and levity. Some regard this fluid as simple, others contend that it is compound, consisting of two simple fluids, having antagonistic properties which, when in combination, neutralise each other. Others, again, regard it not as a specific fluid which moves through space, but as a phenomenon analagous to sound, and think that it is only a series of undulations or vibrations." Lord Kelvin also confesses that what electricity is he cannot tell. However, Dr Lardner adds, "Happily, these difficult discussions are not necessary to the clear comprehension of the laws which govern the phenomena,"—that is, it matters not though electricity is itself unknown, seeing we possess its powers, welcomed by us even if got from the devil himself. Indeed, Dr Lardner, whilst exulting over the wonders of telegraphy, stumbles upon a solution, saying, "The genii of Aladdin's lamp yield precedence to the spirits which preside over the *battery* and the boiler." For electricity let us substitute the word *spirits*, and the mystery is solved. Here is a quotation from Ezekiel's vision of the cherubic spirits:— "Behold a stormy wind came out of the north, a great cloud with a fire unfolding itself, and a brightness round about it. And out of the midst thereof as the *electrum* to look upon. And out of the midst thereof came the likeness of four living creatures, and their feet sparkled like the colour of burnished copper; as for the likeness of the living creatures, their appearance was like burning coals of fire, like the appearance of torches, and the fire was bright, and out of the fire went forth lightning, and the living creatures ran and returned as the appearance of a flash of lightning " (i. 4 to 15). Here we have all the essential electrical phenomena—the sparkling, the fire, the lightning, the lightning velocity of the spirits; above all, the very word electrum, as given in the new version. We read also in the sacred scriptures of the spirit appearing to Moses in the burning bush, of the flaming sword guarding the tree of life, of the chariots, horses and horsemen of fire; of God descending in fire on Mount Sinai; of Elijah bringing fire down from heaven; of fiery serpents and the great fiery dragon (Rev. xii); of the fiery pillar leading ancient Israel, of God's throne of flames (Dan. vii); of the spirit Gabriel, whose face was as lightning and his eyes as lamps of fire, of the fiery tongues on the day of Pentecost, of the baptism of fire Christ received and gave to his followers, of the spirit or angel having power over fire, of the fire unquenchable, etc. Further, we read in Rev. ix. of the spirit horses that in these last days should send forth from their mouths fire, smoke, and brimstone, and having riders with breastplates of fire; and of God's two witnesses (xi.) sending forth fire from their mouths to burn up.

As regards electric light, we read that on the Mount of Transfiguration Christ stood in the white light, His face shining like the sun, and His clothes assumed a dazzling white appearance; whilst Rev. i. describes His face bright as the sun, His eyes as a flame of fire, and His feet as fine brass burning in a furnace. Moreover, by night, a spirit suddenly appeared to the Apostle Peter when he lay in prison, and enveloped him in the electric light.

Telephonic phenomena according to the Bible. Thirty-five centuries ago, God, from the lofty summit of Mount Sinai, spoke to ancient Israel, and was heard distinctly by the trembling multitudes far down in the valleys below. Saul the persecutor, when on the road to Damascus, heard distinctly the voice of Jesus speaking to him from *heaven*—which is so high above the earth, and in which God has set the sun, moon and stars. Scientists differ enormously in their calculations of the sun's distance from us, from 1000 miles to 100 millions. Evidently the height of heaven, like the breadth of the earth, the size of a terrestrial degree being unknown, is yet as great a mystery as ever. Indeed the difficulty is to find any two scientists who agree as to one essential point. However, we read in the Biblical narrative, that Saul's voice traversed the great space between earth and heaven ;. for he asked " Who art Thou, Lord," receiving for answer, " I am Jesus of Nazareth."

Ballooning—Scientists profess to explain this mystery by asserting that the hydrogen gas, said to inflate the balloon, being so very much lighter than the surrounding air causes the ascent. But does the difference between the alleged weights of air and hydrogen at all equal the weight of the attached car with its occupants? Besides, the ascent of the balloon with the car ought to be an impossibility, there being a downward pressure of several tons of air, or of 15 lbs. per square inch of surface, according to Pneumatics. Nor are we to forget the downward pull of gravity. Near y 3000 years ago fiery spirits bore up Elijah higher than any balloon ever could ascend, even to heaven itself.

The Saviour, in presence of more than 500 witnesses, went up to heaven without the assistance of any spirit. The prophet Ezekiel was carried by the hair of his head through the air, a distance of some hundred miles, viz., from Babylon to Jerusalem. Again, Satan once carried Christ through the air all the way from the wilderness to Jerusalem, and even to a mountain so high that from the summit were seen all the kingdoms of the world.

Important query :—It is evidently not astonishing that a mere child should talk to a doll or little idol, as if it were a real baby; but how could so many millions of men and women ever be persuaded to believe a stick or a stone a god? Not merely because the wood or stone was, by the skilful artist, changed into a beautiful representation of a man or animal, to be then placed in a temple of wondrous splendour where there was no lack of exquisite music and gorgeous processions of priests. No, but from the very beginning of idolatry, magical arts were employed by the priests to deceive. Also, is it amazing that in our day men reputed wise and learned should indeed believe themselves daily tossed and whirled head over heels round about the sun, whilst tied to a large globe by the invisible rope of gravity; that their great great, great grandfathers are apes, gorillas, or some such beasts, &c., &c. True, scientists issue mystical volumes full of beautiful pictures along with delusive mathematical "proofs;" and professors flaunting lofty titles strut in academical robes through the lofty and elegant halls of universities. But, just as the nations of antiquity were lured into idolatry by the magical arts of priestcraft, so now the world rejecting the true God, and the only Saviour Jesus Chris', gapes in wonder at the miracles of electricity; and how soon according to Rev. xiii. shall men worship the talking idol and receive on right hands and foreheads the mark of the last antichrist! The living Eddison can make dead heads speak, as did the friar Bacon of England some centuries ago. In ancient Egypt the colossal statue of Memmon uttered a cry every morning at sunrise, and wept audibly as the sun appeared to go down. Further, are not marine telegraphic cables, batteries-electrometers, &c., &c., an elaborate pompous disguise, and as unessential for magical purposes, as were the magical circles, incantations, rods, &c., of the professed magicians of antiquity.

The following letters on magic by the author of this pamphlet have appeared in the "Birmingham Weekly Mercury."—

SIR,—The word magician is derived from the Hebrew "mag" a priest—hence the Persian magi—and "mog" in modern Persian signifies a high priest. The oldest books on magic are the Zend Avesta of Zoroaster, the Jewish Caballa, and the Hindoo Laws of Menu—sources of knowledge, perhaps to the Pythagoreans, Apollonius, and the magicians of the Middle Ages. According to the Caballa, you may with the magnet ("magic stone") walk unharmed through legions of reptiles; and, according to the German Ennemoser, the magnet is the key to unlock the science of magic. The Roman Lucretius speaks of the negative and positive electricity, and the magnetic rings worn by priests in the mysteries of idolatry. Pliny calls the loadstone the Herculanean stone, and the magnetic needle used by Phœnician navigators the arrow of Hercu'es. References are also made to electricity and magnetism by Pausanias, Lucian, Claudian, and even Hesiod. The magnet was worshipped in Egypt thousands of years ago, whilst the inextinguishable fire of Vesta was but an electric flame, King Numa, the founder, having brought fire down from heaven (Revelations xiii.), as did the ancient kings of Babylon long before, and Franklin in our day. In ancient times the electric light was worshipped, shone on altars, whilst idols hung in the air by magnetic force. With scientists electricity is still a mystery, their knowledge being limited to obtaining, storing, and using the unknown power. Christ Himself came baptising with fire, which we must distinguish from the demoniac fire used by ancient priests and modern scientists. Christians must be energised with Divine fire to cast out demons, lay hands on the sick, &c., according to Mark xvi. From the hands, mouths and bodies of Christ and His Apostles flowed the Divine energy to do signs and wonders; the condition of Spirit Baptism being the surrender to Jesus in order to a pure and holy life. In Xenephon's Memorabilia we read that the ancient Greeks called their gods demons, Socrates claiming to speak because of one dwelling within him.

Those early Fathers—Clemens Alexandrinus, Origen, Gregory, Justin, Lactantius, &c.— say that demons maintain magical arts, are the the founders of idolatry, seeking to be worshipped as gods, and try to "injure men in every possible way by public calamities, death, disease, and all kinds of accidents." We know that what is now called natural science was formerly called magic. Moses (Lev. xx., 27 —Deut xviii., 10), refers to the magical practices of idolators, denouncing death as the penalty for such crimes. Lucretius (6th Book) says :—

> "Men see the stone with wonder as it forms
> A chain of separate rings by its own strength,
> Borrowing their binding strength from the strange stone.
> Such power streams out from it pervading all.
> But sometimes it doth happen that the iron
> Turns from the stone flies it, and is pursued.
> I saw the Samothracian iron rings
> Leap and steel filings boil in a brass dish
> So soon as underneath it there was placed
> The magnet-stone, and with wild terror seemed
> The iron to flee from it in stern hate."

Thus Lucretius knew both kinds of electricity. Claudian in his Idylls on the Magnet speaks of a temple where a magnetic image of Venus held suspended in the air an iron one of Mars; and Lucian says that he saw a very old image of Apollo lifted aloft by the priests and left hanging without any visible support. According to the German Ennemoser (Bohn's edition), Schweigger proves that the fire brought down from above by the ancient Samothracians was electrical, referring also to the Hermes-fire, the Elmes fire of the Germans, the lightning of Cybele, the torch of Apollo, the fire of Pan's altar, the flame of Pluto's helm, the fire in the temple of Athene in the Acropolis, &c. What, too, of the myth of Prometheus stealing the fire of Jupiter? Thus far electro-magnetism now astonishes the world; in opposition to which we have now many allusions to a holy flame throughout the Bible, from the flame revolving about the tree of lives in Genesis iii. to the crystal light of the New Jerusalem seen by John in Patmos?

SIR—Does not the claim of scientists to have made discoveries in electricity and magnetism merely exemplify the saying of Solomon that there is nothing new under the sun? Nor is the use of metals for supernatural (scientific) purposes a novelty. Three thousand years ago, by means of the metals on the high priest's ephod, King David conversed with God; and five hundred years previously Moses healed with a seraph or serpent of copper raised on a pole. Elijah, after building an altar of twelve stones, brought down the electric flame, which the priests of Baal (Bel, Zeus, or Jupiter) could not do, though confident according to ancient practices—praying earnestly from morning till evening. Whilst Egypt and Assyria gloried in the magnet, Isaiah's lips were touched by a seraph with a burning stone that he might prophecy; the word seraph signifying a serpent as well as to burn. The Hebrew Nachash applied to Satan in Genesis iii. signifies, variously, serpent, sorcerer, diviner, a spell, to view sharply; and well may electricians use copper wire emblematically, to attract the old serpent. Mesmerists now have a theory called animal magnetism, whereby they maintain an analogy between the action of the magnet and that of animal energy, professing to divine prophesy, and virtually to bewitch; whilst spiritualists seek after the dead and demons, performing some of the wonders of the scientists. Scientists are believed not because of the reasonableness of their theories but because of the marvels they display before the world. What avails now the mere traditional evidence of apostolic miracles done eighteen centuries ago to oppose the advancing flood of unbelief mainly due to false science? Rather a present living faith with the fire baptism. The miracles of true religion are those of mercy and kindness.

TO THE EDITOR OF THE TIMES OF RESURRECTION.

C. G. Cook's Evolution pills, in the absence of facts and argument, smack strongly of scholastic dogmatism. He certifies, with all the weight of his own authority, that science is really alive; though he immediately after puts Evolution into its coffin by likening it to a corpse with theorists squabbling over the cause of death. He confesses that science—every idea of which, according to Humboldt, one of its late popes, was known to the Pagan Greeks two thousand years ago—is still dragging at its infancy; that is, scientists are only squabbling bodies, as much as ever they were in the time of Daniel and the other prophets, and when Pythagoras, Plato, and Socrates worshipped the sun, moon and stars as gods. He is certain that Evolution is a fact, because voted true by all scientists; whereas it is but a lying blasphemy, and therefore a putrid carcass for the dunghill, and because opposed to the living truth which no scientific fool can overthrow:—that God, about six thousand years ago, created heaven and earth with all therein within six days by his almighty word.

Now, is not the denial of Moses' record of creation the denial of Jesus Christ? "If ye believe not Moses' writings, how can ye believe my words?" Then if C. G. Cook denies that Jesus is the Christ, let him deal with the evidence stated in my former article.

We are further told by Mr Cook, without one item of proof that "The Bible is the result of evolution: the Hebrew mind evolved it in course of ages," etc. As if we Christians were as silly as the bodies of scientis's who with their eyes shut gulp down these antiquated fables of Egyptian and Chaldæan priests, now called astronomy, geology, and evolution. Why not boldly say with Thomas Paine, that the Bible is full of lies; because it professes to give so many revelations from the God of creation, while it scorns all man's wisdom as the loud long-winded braying of an ass?

How the old dogs of heathendom did growl and howl over the dry bones heaped up in their sepulchres of philosophy! Stoics snarled at Epicureans, Platonists at Peripatetics, Electics at Cynics: dreamers, augurs, magicians, astrologers, fought over theories of the soul, eternal matter, space, atoms, morals, dialectics, metaphysics, the earth globular or cylindrical—movable or fixed, the universe—if of aqueous or igneous origin, etc. Then did not mediæval times abound with astrologers dreading unlucky stars, astronomers imagining themselves midgets on Ptolemy's motionless sea-earth globe, mathematicians vainly trying to square the circle and mechanicians to find out perpetual motion, alchemists in hot haste after the philosopher's stone and the elixir of life, schoolmen in the tight jacket of Aristotle's sylogisms; Realists and Nominalists, Thomists and Scotists, scientists, &c.? Among the moderns what a Babel of t) Newton's gravity, Descartes' vortices, Leibnitamonas; as to jurisprudence and political economy; as to ideas, sensation, perception, the will, matter and mind—whether real or illusory, logic and language; as to the classification of plants and animals, light—whether fluid or vibration, heat—whether fluid or atomic motion; sound—whether vibratory or undulatory or substantial, chemical elements and atoms, electricity—whether fluid or witchcraft; also over the sea-earth globe—whether movable or fixed, like an apple or an orange, with two or three axes; as to the distances of the sun and moon and stars, comets—whether igneous or aqueous, the sun—whether fixed or flying toward Hercules; among geologists, as to the earth's origin from fiery gas—whether a ball of fire or rigid as steel, how many millions of years old, as well as to the classification and formation of rocks, &c.

"In the study of science," says Dr Dick (Nat. His., p. 10), 'one is permitted to *suppose anything*, if he will but remember and acknowledge to others that he makes suppositions; will give reasons to show that what he supposes may be true," &c. Such then is the tree of knowledge with which the old serpent deceived the world!

God asked Job of old: "Who has set its (the earth's) layers? Or who stretched out a measuring line upon it? On what are its bases sunk? Or who laid down its keystone rocks?"—Job 38. 5. Thus it is revealed that the earth's strata were originally arranged by God himself, and (Gen. 1) within a few hours, with the regularity of the stones of a house, and as if the builder's measuring line had been used. The unstratified or keystone rock, whether basalt or granite, is lowest, and above are the various beds according to density—such as sandstone, limestone, coal, chalk, clay, with sand, gravel or soil on the surface. Geologists confess that the age of a fossil is not determined by the degree of its petrefaction, but by the age of the rock in which it is imbedded, and the age of the rock by its position among the strata; accordingly how the strata were originally formed and the age of the world are mere supposition, geologists being but of yesterday and therefore knowing nothing about the creation, unless they learn it from the Great Architect himself. The whole mighty mass of rocks appeared (Gen. 1) out of the abyss at God's command on the third day of creation, and was made to float in the unfathomable waters, yet as securely fixed as a ship in a Liverpool dock; the bases of the earth being so sunk as to make it immovable forever, and man is challenged to tell how.

An iron ship can be made to float, though the metal is seven times heavier than water; while the heaviest rock is only three times heavier than water. Then consider the tremendous buoyancy of the ocean; causing some substances to float on its surface, and others to sink only to a certain depth. The earth, its density decreasing from the foundation rock upward to the soil of the surface, is sunk to a depth of several miles in the sea, yet so as to have a dry surface and shores on a level with the surrounding waters It consists of four continents of an irregular and somewhat triangular shape, stretching out from the central north for thousands of miles towards the icy barriers of the far south against which winds and waters rage in vain. The continents are connected by submarine rocky beds of various depths; while the Arctic and Antarctic oceans are found to be unfathomable.

Christ mentions the fact of a universal deluge; and we learn (Gen. 7) that when Noah had entered the ark, the waters rushed from heaven and the abyss to fulfill God purpose to destroy the earth with its inhabitants. Hence, the rending of rocks, the shattering of hills, the breaking up of the earth's strata, the piling up of mass upon mass, wherein were buried animals and plants to be dug up many centuries afterwards. All lands were filled with the wreck of the old world—a terrible warning to all future ages against the commission of unrighteousness. And are we surprised at the petrefaction of fossils, seeing that the earth was covered by the deluge for a whole year!

Also on the third day of creation (Gen. 1) God made "the tender grass, the herb sowing seed, and the fruit-yielding tree, whose seed is in itself, *after its kind*," so that transmutation of species was impossible. On the fifth day He formed the great sea monsters and every living creature which the waters have teemed with *after their kind*; while on the sixth day the earth brought forth every living creature *after its kind*, cattle and creeping thing and beast of the earth *after their kind*; transmutation of species being thus impossible. Lastly Adam was formed fr m the *adamah* or soil, not "evolved," or as that word means, "tumbled out" of a monkey.

Of course scientists bark and bite over the definition of Evolution, making it a Babylonian mystery to themselves and everybody else : but the idea upon which most people are generally agreed is manifestly borrowed from the grand old Book. God's revelation and the common sense of mankind testify against the monstrous assumption of higher species of plants and animals being evolved or tumbled out of lower; every species or kind having been at first created incapable of transmutation.

Then as regards the world lying in wickedness and deceived by Satan, we have it on God's authority and agreeably to all history, that since Adam's disobedience there has been a continual devolution or tumbling down into fearful depths of misery, folly and crime, out of which only Jesus can and will deliver. The inspired writers show that God, according to a wondrous plan, has been advancing through ages and dispensations in the work of man's redemption. Beginning with the choice of individuals, from righteous Seth to faithful Abraham, He next chooses the family of Israel, which grew into a nation producing in fulness of time the great Deliverer. During the Gospel Age this Deliverer, Jesus, has been choosing from among all nations a body or ecclesia, to rule with Him in the coming age over a renovated world and a resurrected race consisting of a multitude that no man can number. From the present visible typical creation, the six days of which foreshadowed the six thousand years now closing, God has been advancing to the new creation—spiritual, heavenly, everlasting—of which the Son of God is the Head.

Mr Darwin, with all the confidence of an eye-witness, says in his "Descent of Man:" "The early progenitors of man were no doubt covered with hair. Their ears were pointed and capable of movement and their bodies were provided with a tail. The foot was prehensile and our progenitors were no doubt arboreal in their habits." This means in plain English that Adam and Eve were a couple of dumb, brutish, irrational apes skipping through forests or swinging by their tails from the branches of trees. Also, 'At an earlier period the progenitors of man must have been aquatic in their habits." Sharks or lampreys, perhaps? Then he traces our pedigree further back even to the jelly fish. which he thinks (contrary to embryology) the link between vetebrate and invertebrate animals. But Professor Haeckel, with an entire absence of proof, begins evolution with animalcules (which, having no organs or fixed shape, cannot be animals at all) originating by spontaneous generation; that is, they willed their own existence before they had either a will or any existence at all. Next these animalcules grew into worms, which grew into ascidians; and the ascidians into fishes, fishes into frogs, frogs into mammals, birds, and reptiles; lemurs into monkeys, lastly monkeys into men. Mr Darwin tries to account for transmutation of species by "natural selection." Of course, variations occur among plants and animals in a state of domestication, but according to him they also occur, though imperceptibly slow, in a state of nature, and are transmitted to the offspring of the individuals thus varied; these variations accumulating in course of long ages give rise to new species. Then he points to the excessive fecundity of animals and plants. with the consequent necessity of the destruction of many of them, and points out that there is accordingly a struggle for life going on resulting in the 'survival of the fittest." However, if natural variations do occur, he fails to prove that they accumulate, or that, notwithstanding the sterility of hybrids and the absence of all transitional types transformation of species has ever happened. Lyell denies that the geological record gives Mr Darwin any support; whilst Professor Agassiz (Natural History, pp. 51) maintains that the identity of the animals preserved as mummies by the Egyptians 5,000 years ago with animals of the like kind now living, is a proof of the stability of species. Besides, what struggle is there for life, unless in human wars and the contentions of wild beasts? Does not God make the fields fruitful for man's sustenance, and give cattle to us as food? And notwithstanding so many varieties of dogs, horses, and pigeons produced under man's care from the earliest times, is not a dog still a dog, a horse a horse, and a pigeon a pigeon? In the "struggle for life," why, too, should the race invariably be to the swift and the battle to the strong, the modern elephant now surviving the ancient mammoth, the weak sloth the extinct megatherium? Mr Darwin comparing us with apes, overlooks their beastly hide, want of legs and feet, as well as their incapacity to walk erect, but argues from "physiological analogies" got by comparing the bones, &c., of dead men with those of dead apes. Might he not as plausibly have contended that apes are degenerate men and give birth to lemurs, which in time degenerated into jelly fish, &c.? Moreover, Mr Darwin does not prove or account for the natural laws supposed to produce the variations; and whilst writing only as a naturalist, he prudently avoids the mental evolution so essential to the question, failing to bridge over the impassable chasm between mere beastly instinct and the mind of man divinely revealed as the image of his Creator—also the subjugator of nature and lord of beast, fowl, fish, and reptile.

ALEX. M'INNES (Glasgow).

Part I. PRICE 6d

Entered at Stationers' Hall.

ZETETIC PHILOSOPHY.

PATRIARCHAL LONGEVITY,

ITS REALITY, CAUSES, DECLINE AND POSSIBLE

RE-ATTAINMENT.

THE DIET ORIGINALLY ORDAINED, SUBSEQUENTLY PERMITTED, AND NOW BEST ADAPTED.

"Their food was then fitter for the prolongation of life; and God afforded them (the Patriarchs) a longer time of life on account of their virtue, and the good use they made of it in Astronomical and Geometrical Discoveries."—*Josephus.*

BY

"PARALLAX,"

(Author of "Zetetic Astronomy" and other Works.)

LONDON:
HEYWOOD & Co., (LATE J. CAUDWELL),
335, STRAND.
1867

The right of translation is reserved by the Author

H. Weede, Printer, High Road, Knightsbridge.

PATRIARCHAL LONGEVITY RE-ATTAINABLE.

"There is a word we hate to speak, a thought we dread to think, a thing at which we shudder. Our writers give it hard names, our painters sombre colours, and we reserve the saddest types and emblems to represent it. That word, and thought, and thing, is DEATH."—*Daily Telegraph.*

The sufferings of the human race through bodily disease and death are far greater than it is possible for language to express. So long and so terribly has humanity suffered that both religion and philosophy have come, almost universally, to hold and to teach that such appalling misery is the destined and inevitable consequence of man's physical existence. That it has long been the common lot, and must remain so to the end—until, indeed, the whole creation dissolve and pass away. The first declares that man was originally formed in the image and likeness of his Creator,—

" ———— To be immortal,
And the image of his own eternity,"

and that by the infraction of only one single command he became subject to pain and death. The second affirms that all nature is in a state of restlessness, that everything is changing,—passing from death to life and from life to death, and that one is but an effect of the other; that to die is the consequence of having already lived, to live is but a preparation to die; and that death is a natural necessity. Notwithstanding however that religion and philosophy so teach "how tenaciously do men in general cling to life! Even in circumstances where it appears almost a burden and a grief, when pain and sorrow seem as if they were the birthright of the sufferer, and but few intervals occur of cessation from disease, yet if the last mortal conflict seems approaching the spirit recoils from the struggle, and would fondly retain its grasp of life, even with all its attendant sufferings.... Man clings to the world as his home, and would fain live here for ever," "And can we see the newly turned earth of so many graves, hear the almost hourly sounding knoll that announces the departure of another soul from its bodily fabric, meet our associates clad in the garb of woe, hear of death after death among those whom we knew, perhaps respected, perhaps loved—without pausing to consider if we may not seek and haply find *more* than the mere causes—find the means of checking the premature dissolution that so painfully excites the deepest and most hidden sympathies of our nature?"

In the sacred records, which are held to be the foundation of religious faith and purpose, great encouragement is given to the hope and belief that even in *this material* world, such a state of suffering and wretchedness shall not always exist: and that philosophy is insensible and arrogant which declares that the instinctive and irrepressible yearnings for long-continued earthly life and happiness, which seem to be universal, are never to be satisfied. "They take very unprofitable pains who endeavour to persuade men that they are obliged wholly to despise this world and all that is in it even whilst they themselves live here: God hath not taken all that pains in forming, and framing, and furnishing, and decorating this world, that they who were made by Him to live in it should despise it; it will be well enough if they do not love it so immoderately as to prefer it before Him who made it."—*Clarendon*.

Although the desire for health and long life on earth is very natural and prevalent, there are and perhaps have ever been found many remarkable exceptions. But they are always the result of incidental impressions; and only show how far human nature is capable of responding to the action of external and educational influences. Some, through continual trial and disappointment, lose heart and hope; and, looking upon the world as little better than a dreary wilderness, pine in silence and suffering for deliverance from life and all its responsibilities

—too often risking all future consequences by determined self-destruction. Others, even those to whom a full proportion of the world's enjoyment has been allotted, having lost by death many or all of those most dear and cared for, long for the years to pass away, and for the time to arrive when they shall "be called to their account" in hope of rejoining the loved ones gone before; and many there are who find such consolation in religious devotion, and who so anxiously contemplate the higher and happier life which religion teaches to exist and promises as the reward of faith and righteousness, that all earthly joys and possessions appear insignificant and worthless. No desire exists and no value is attached to efforts tending to the prolongation of life beyond the ordinary period. Such efforts are not acceptable to all; and to many are utterly distasteful. The subject of this work is therefore only to be properly addressed to those who can conscientiously endorse such sentiments as the following:—" In this world there is, or might be, more sunshine than rain, more joy than sorrow, more love than hate, more smiles than tears. The good heart, the tender feeling, and the pleasant disposition make smiles, love, and sunshine everywhere.... A thousand gems make a milky way on earth more glorious than the starry clusters in the firmament."

It is recorded in Scripture that the Jewish Patriarchs lived to ages varying from upwards of one hundred to nearly a thousand years. Many have contended that their years were much shorter than those of the present time—not more, in fact, than one-fourth the period. If this were true, the days of Methusaleh, the oldest of the Patriarchs, would only have been 243 years instead of 969 as recorded. Terah the father of Abram would only have been 51 years of age instead of 205; and Abram himself only 44 instead of 175. The ages of Abram and the later Patriarchs generally, were, according to this supposition, considerably less, and therefore no contrast to the duration of life in our own day. By the same rule it would also follow that Enoch who "lived 65 years and begat Methusaleh" was then only between 16 and 17 years of age! Arphaxad the son of Shem, and Noah's grandson, "who lived

35 years and begat Salah," was then only 8 years and 9 months old! "Salah lived 30 years and begat Eber;" and "Nahor lived 29 years and begat Terah," so that Salah and Nahor were fathers when only just turned seven years of age! Going back to Adam himself, we find that he was 130 years of age when Seth his third son was born: and that before this period Cain had been married, and had a son Enoch, that "to Enoch was born Irad," and "Irad begat Mehujael," so that Adam was more than great grandfather when less than 33 years of age! From these considerations it is evident that the years of the Patriarchs were the same in length as ours, and that all ideas to the contrary are unwarranted by the evidence recorded.

Josephus, remarking upon the age of Noah as being 950 years, says, "But let no one, upon comparing the lives of the ancients with our lives, and with the few years we now live, think that what we have said of them is false; or make the shortness of our lives at present an argument that neither did they attain to so long a duration of life; for those ancients were beloved of God and (lately) made by God himself; and because their *food was then fitter for the prolongation* of life; and besides God afforded them a longer time of life on account of their virtue, and the good use they made of it in astronomical and geometrical discoveries. . . . Now I have for witnesses to what I have said all those who have written antiquities both among the Greeks and Barbarians."

Whatever has once occurred, is certainly again and for ever possible. The Jewish Patriarchs lived to extraordinary ages, some to nearly a thousand years; and therefore the re-attainment of such longevity is not an impossibility. Many object to all attempts of this character on what they consider scriptural authority. "The days of our years are three score years and ten," are words held to express the fiat of God as limiting human life to a few score years. This however is a great and injurious mistake; and could only have arisen from a fore-formed state of mind, or very careless reading of the context. The words are but the language of the Psalmist regreting that in his day the wrath of Heaven had been incurred, until, as a punishment, the days of the wicked were unusually shortened. "Thou carriest them away as with a flood.... they are like grass which groweth up. In the morning it flourisheth, and in the evening it is cut down and withereth. We are consumed by thine anger, and by thy wrath are we troubled. Thou hast set our iniquities before thee, our secret sins in the light of thy countenance. All our days are passed away in thy wrath, and we spend our years as a tale that is told." It is thus evident that "three score and ten" did not express the permitted term of man's existence, but simply the period to which it had been reduced by the most flagrant violation of God's commands, and doubtless of those natural laws which are essential to the preservation of health and the prolongation of active life. The subject, and the words quoted are found in the ninetieth Psalm, or in the "Prayer of Moses the man of God" who himself lived with "Eye not dimmed, nor his natural force abated" for nearly half a century longer than the period he was bewailing as the limit of human life. Indeed many hold that Moses really never died, in the common acceptation of the term death : but that he was taken to heaven in the body as were Enoch and others. It is not easy to avoid such a conclusion ; for as "Enoch walked with God, and *was not*, for God took him." "He was translated that he should not see death, and was not found, because God had translated him, for before his translation he had this testimony, that he pleased God." If translation was the reward of Enoch because "that he pleased God" it would be strange indeed if Moses by whom God had wrought so many wonderful events, and who could by "laying hands upon Joshua make him full of the Spirit of wisdom," whom "the Lord knew face to face," "whose sepulchre was never known," and whose natural powers when he disappeared were unabated, should receive any other reward than that accorded to Enoch and Elijah. It is even more strange, in a natural sense, that the most highly favoured leader of God's chosen people, the "man of God" in full health and strength of body and mind should go up "from the plains of Moab to the mountain of Nebo, to the top of Pisgah near Jericho," and there without sickness or infirmity of any kind suddenly die and disappear.

That Moses was rewarded by translation as were Enoch and

Elijah, would seem to be corroborated by the fact, that when Jesus "took Peter and John and James, and went up into a mountain to pray . . there talked with him two men, Moses and Elijah, who appeared in glory and spoke of his decease which he should accomplish at Jerusalem."

It is clear, from the evidence, that when Moses uttered the words " the days of our years are three score years and ten " he was not expressing the unconditional fiat of the Creator but was simply lamenting that the ignorance and wickedness of the people had so reduced the term of life that even if " by reason of strength they reached four score years, yet was their strength labour and sorrow, for it was soon cut off and they fled away." They only grew, and flourished and withered like grass.

The scriptures therefore do not discourage the desire and the effort to preserve existence on earth for the longest possible period. On the contrary, again and again are we enjoined " to righteousness that our days may be prolonged on earth."

" The fear of the Lord prolongeth days."

" What man is he that desireth life and loveth many days, that he may see good . . . . Depart from evil and do good: seek peace and pursue it." It is one of the most unequivocal promises of Scripture that he who seeks to do good, to promote the cause of truth and justice, to serve and honor his Creator, to obey His commands and fulfill the laws which He has impressed upon organic nature may hope to be rewarded with length of days the extent of which no man may predicate. " He shall call upon me and I will answer him; I will be with him in trouble; I will deliver and honour him; with LONG LIFE WILL I SATISFY HIM."

That practical science does not run counter to the encouragement afforded in the Sacred Scriptures may be gathered from the following quotations from eminent writers upon physiological and anatomical subjects:—

" The human body, as a machine, is perfect; it contains within itself, no marks by which we can possibly predict its decay; it is apparently intended to go on for ever."—*Anatomical Lectures by Dr. Monro of the University of Edinburgh.*

" Such a machine as the human frame, unless accidentally

depraved, or injured by some external cause, would seem formed for perpetuity."—*Medical Conspectus, by Dr. Gregory.*

"If a living organized being be examined at the epoch of its greatest perfection, when the structure is sufficient to perform its functions, and the functions are adequate to maintain the organization, a mutuality of cause and effect is perceived which almost promises immortality."—*Sketches of the Philosophy of Life, by Sir T. C. Morgan.*

"We have seen that there is within the animal frame a system of operations by which a constant supply of nourishment is afforded to make up for the daily waste and decay; and that every part is undergoing a renewal. To view a man then in the full vigour of life, we might suppose that, excepting accidents, he was calculated to go on in the course of existence for an indefinite period."—"*The Human Body,*" *a Pamphlet by the Messrs. Chambers of Edinburgh.*

"There certainly appears no reason why an object once endowed with life should not live for ever; for the state of maturity might be prolonged for ever as it is: there is nothing impossible in such a state.... If we could imagine a physiologist seeing for the first time an organized structure, such as the human frame, in a state of perfection, however closely he might examine it, and however intimately he might know the structure, he could not, without the knowledge of experience, pretend to say there appeared any reason why death should occur; he could not indeed conceive such a thought as death."—"*Body and Soul,*" *by Dr. Redford.*

"The head acts because the heart acts, and the heart acts because the marrow of the brain and spine acts, a seemingly perpetual motion, for the death of which there seems no natural necessity, except accidental obstructions, or that habit of body which tends to hardness, and is what is called old age. This hardness interrupts motion, and ultimately causes death: but were it not for this growing hardness, or the obstruction caused by disease, there seems nothing to prevent the mutual action of head and heart from being everlasting."—*Family Herald.*

"At some future day there can be little doubt that the value and duration of life will be extended greatly beyond what it is at present—greatly beyond, perhaps, what we at present can imagine."—*Dr. Thomson's Medical Dictionary.*

The Registrar-General of England, in one of his valuable Reports observes, that, "The prolongation of the life of the people must become an essential part of family, municipal, and national policy. Although it is right and glorious to incur risks, and to sacrifice life for public objects, it has always been felt that length of days is the measure and that the completion by the people of the full term of natural existence is the groundwork of their felicity. For untimely death is a great evil. What is so bitter as the premature death of a wife, a child, a father? What dashes to the earth so many hopes, breaks so many alliances, blasts so many auspicious enterprises as the unnatural death? The poets as faithful interpreters of our aspirations, have always sung, that in the happier ages of the world this source of tears shall be dried up."

In the "Golden Legend," by Professor Longfellow, one of the characters is made to utter the following sentence, in reply to the question " can you bring the dead to life ?"

" ——yes, very nearly :
And what is a wiser and better thing,
Can keep the living from ever needing
Such unnatural strange proceeding:
By showing conclusively and clearly,
That death is a stupid blunder merely,
And not a necessity of our lives."

In the Poem of "Queen Mab," by Shelley, the following passage occurs :—

" Man, once flitting o'er the transient scene,
Swift as an unremembered vision,
Shall stand immortal upon earth."

Sir Walter Scott, in his "Kenilworth," speaking of the future says—"The happy period is brought nearer to us, in which all that is good shall be attained by wishing its presence, all that is evil escaped by desiring its absence; in which sickness, and pain, and sorrow shall be the obedient servants of human wisdom, and made to fly at the slightest signal of a sage. . . . . When sages shall become monarchs of the earth ; and death itself retreat from their frown."

This language of the poets and the unavoidable yearnings of the human heart and soul are even surpassed by the prophetic teachings and promises of the sacred writings both Canonical and Apocryphal.

"The face of the covering cast over all people, and the veil that is spread over all nations shall be destroyed, and death swallowed up in victory."

"I will ransom them from the power of the grave; I will redeem them from death; O Death, I will be thy plagues; O Grave, I will be thy destruction."

"The righteous shall never be removed, but the wicked shall not inhabit the earth."

"As a whirlwind passeth, so are the wicked no more; but the righteous are an everlasting foundation."

"In the way of righteousness is life; and in the pathway thereof there is no death."

"Behold, the righteous shall be recompensed in the earth."

"Verily, verily I say unto you, if a man keep my saying he shall never see death."

"And God shall wipe away all tears from their eyes; and there shall be no more death; neither sorrow, nor crying, neither shall there be any more pain, for the former things are passed away."

"The last enemy death shall be destroyed."

"Seek not death in the error of your life, and pull not upon yourselves destruction with the works of your own hands."

"God made not death, neither hath He pleasure in the destruction of the living."

"For God created all things that they might have their being; and the generations of the world were healthful, and there is no poison of destruction in them, nor the kingdom of death upon the earth; for righteousness is immortality: and ungodly men with their works and words called death to them, for when they thought to have it their friend they consumed to nought."

"God created man to be immortal, and made him to be an image of his own eternity."

"I have no pleasure in the death of him that dieth saith the Lord God, wherefore turn yourselves and live ye."

"Keep my commandments and live."

It may not be denied that this language of Scripture has a spiritual application; but it must not be claimed that it is exclusively spiritual. Whatever is true is true universally; is true in all respects; is true not alone spiritually; not alone physically; but is true both materially and spiritually. The Scriptures, if true at all are true entirely. They speak of spiritual progress and immortality, and also of bodily progress and immortality. They speak of the preservation and perfection of the entire man; of the soul, of the spirit, of the mind, and of the body; and of the possibility of the whole together avoiding natural death and passing away from the earth, into immediate heavenly existence. They even speak of perfection and existence upon earth until the second coming of the Messiah, and the dissolution of the material world.

"I pray that the whole spirit, soul, and body be preserved blameless until the coming of the Lord Jesus Christ."

"For He shall descend from heaven with a shout, with the voice of an Archangel, and with the trump of God. And the dead in Christ shall rise first; then we which are alive and remain shall be caught up, together with them, in the clouds to meet the Lord in the air; and so shall we ever be with the Lord."

"Then the heavens shall pass away with a great noise, the elements melt with fervent heat, and the earth also and the works that are therein shall be burned up, and all things shall be dissolved."

Thus we see that the Creator of the world has promised that a time shall arrive when man may become both spiritually and bodily perfect, when he shall be rewarded with unbroken earthly existence, until all shall dissolve and pass away; and being then blameless and deathless shall be translated from earth to heaven, there to live for ever in the immediate presence and influence of the "Heir and final Judge of all things."

We have seen also that no human skill or ingenuity can discover an imperfection in the human structure. The "work of God was perfect." "He saw everything that he had made; and, behold, it was very good." Anatomy and physiology entirely fail to prove that the human body is necessarily incapable of unending existence. They are completely powerless in proving that life may not be continued indefinitely. Science then with all its most recent and cogent powers affords not a particle of evidence against, nor do the Sacred records anywhere forbid, the desire and the effort to realize a term of active life upon earth at least equal to that enjoyed by the most favoured Jewish Patriarchs. The Scriptures teach that man when first created received as a special gift from his Creator the "breath of life," when he "became a living soul." That the maintainance of the living state was made to depend upon obedience to certain commands. That the breaking of these commands created a constant liability to suffering and death. That the period of life was made immediately dependant upon the will and favour of God. And he would be an injudicious if not a reckless philosopher who should deny that both then and always, and now as much as ever in the past, the degree and the duration of life are in the hands of Him who first brought man into the condition of a "living soul."

It therefore follows that however we may investigate, experiment, hope or struggle to realize one of the strongest desires of our nature,—health and long life, all will be useless unless we are in all things worthy and Heaven is at all times willing. However anxiously our instincts may lead us to desire vigorous and long continued earthly existence, all our efforts will be in vain unless we are specially permitted to possess the necessary understanding. A "veil has been spread over all nations" through spiritual and physical unrighteousness, and it will not be lifted until we are able and resolved to "cast away from us all the transgressions by which we have transgressed and make ourselves a new heart and a new spirit." If there is a God at all, He is master of all. "All our times are in His hands, our times of health and sickness, of life and of death." The physician of long experience must have been a careless observer if he has not been forced to conclude that his efforts have often been unsuccesful from causes, to him, invisible and incomprehensible. That often, cases have occurred in his practice, when the patient was not, at first, so great a sufferer as many who had quickly recovered under his care and treatment; and yet all his remedies, and all that could be suggested on consultations with his medical brethren have failed to arrest the progress of disease, and the gradual all-conquering approach of death. Many physicians and surgeons as well as nurses and ministers of religion have known cases where the patient at an early stage of his sickness has felt and expressed a foreboding that he should not recover. Some have been able to state the very hour of their departure; and even when all around have thought they were improving, a sudden relapse and death have occurred within or at the very time which had been predicted. When, in the pride of curative skill and experience, it was thought that every influence and affinity of the morbid and counteracting elements were traceable and well understood, every effort has been set at nought, and medical wisdom and devotion proved to be vain and useless.

With such cases as these before us can we be so blind as not to admit that a Higher Power had willed the giving up of the life which had been only conditionally granted? The sentence so often heard from the sick and dying "my time is come," or "my race is over," is therefore something more than simply a death-bed form of expression. It is the utterance of the soul after a mysterious warning to prepare for a coming and final change.

If we now attend to certain practical evidences we shall see that there is no definite period or number of years beyond which it is impossible for life to be maintained; and within which death must of necessity occur. This will be obvious both from the investigations of anatomists as recorded in the several statements already quoted, and from the following instances of longevity:—

"The Ancient Britons only *began to grow old* at one hundred and twenty years of age." *Plutarch, de Placitis Philosophia.*

In Pinnock's Edition of *Goldsmith's History of England*, the following note occurs:—

"It is stated by Plutarch that the Ancient Britons only

began to grow old when a hundred and twenty years of age! Their arms, legs, and thighs, were always left naked, and for the most part, were painted blue. Their food consisted almost exclusively of acorns, berries, and water."

Dr. Henry in his *History of England* states that they were remarkable for their "fine athletic form, for the great strength of their body, and for being swift of foot. They excelled in running, swiming, wrestling, climbing and all kinds of bodily exercise; they were patient of pain, toil, and sufferings of various kinds; were accustomed to fatigues, to bear hunger, cold, and all manner of hardships. They could run into morasses, up to their necks, and live there for days without eating."

"Boadicea, Queen of the Ancient Britons, when about to engage the degenerate Romans, encouraged her army by a fervent and eloquent speech; and amongst other reasons why they should conquer she says 'The great advantage we have over them is, that they cannot like us bear hunger, thirst, heat or cold; they must have fine bread, wine, and warm houses; to us every herb and root are food; every juice is our oil; and every stream of water our wine."

The aborginal inhabitants of New Zealand and of other islands of the southern region are known to be remarkably healthy and long-lived. A gentleman who has lived among them for upwards of seven years, says, that he has known many of them who could not remember their ages to within ten to twenty years. They are said to be able to go to war, to follow the chase, and to obtain a full supply of their wants by hunting, fishing and roaming the forests and plains; and to be equal, in many respects, to the finest young men of Europe long after they have reached a hundred years of age!

"The Macrobians lived to a hundred and twenty years old, and some to a much longer period."—*Herodotus*.

The Gymnosophists of India were never afflicted with disease, and lived to ages ranging from one hundred and fifty to two hundred years.

Those primitive Christains who through persecution fled to the deserts lived, many of them, to upwards of a hundred and some to a hundred and fifty years old.

"Pliny records that in the year A.D. 76, in the reign of the Emperor Vespasian a census was taken, and there were living in that part of Italy which lies between the Apennines and the Po only, 124 men who had reached a hundred years and upwards, viz. 54 of a hundred; 57 of a hundred and ten; 2 of a hundred and twenty-five; 4 of a hundred and thirty; 4 of from a hundred and thirty-two to a hundred and thirty-seven; and 3 of a hundred and forty.... Several Roman actresses lived to a great old age. One Luceja, who came on the stage very young, performed a whole century; and even made her appearance publicly when in her hundred and-twelfth year."—*Hufeland*, p. 70.

"A Dane named Draakenberg born in 1626, served as a seaman in the Royal Navy till the ninety-first year of his age: and spent fifteen years of his life as a slave in Turkey. When he was a hundred and eleven and had settled to enjoy tranquility he resolved to marry." He did so and outlived his wife a long time. "He died in the year 1772 in the one hundred and forty-sixth year of his age."—*Ibid*, p. 79.

When the Brazilians were first discovered "it was not uncommon to see men one hundred and twenty-five years of age, and some a hundred and forty."

In Cottle's "Alfred" several Monks are named as having lived in the Monastery of Croyland to great ages. Father Clarenbald died A.D. 973 at the age of one hundred and sixty-eight years. In the same year Father Swarling died aged a hundred and forty-two; and Father Turgan died in the following year at a hundred and fifteen. About the same time also a Bishop of St. David's died aged one hundred and sixty years.

St. Patrick, the patron saint of Ireland, died about A.D. 460, at the age of one hundred and twenty years.

Thomas Parr, a native of Shropshire, who died in 1635, although greatly afflicted in his younger days, lived to the age of a hundred and fifty-two years. He married at the age of eighty-eight, "seeming no older than many at forty." He married a second time at the age of a hundred and twenty-one; and when a hundred and forty-five years old he could run in foot races; thrash corn, and perform the ordinary work of an agricultural labourer.

Henry Jenkins, a native of Ellerton in Yorkshire, died in 1670, at the age of one hundred and sixty-nine. A child was born to him when ninety years of age; and when he was a hundred and sixty years old he walked to London, a distance of 200 miles, to have an audience with the King, Charles II.

Spottiswood records that Kentigern, afterwards called St. Mougah, or Mungo, lived to the age of one hundred and eighty-five years.

"Joseph Creole died in Caledonia, a little town of Wisconsin, on the 27th of January, 1866, at the age of one hundred and forty-one years. He was an inveterate smoker. He was twenty years older than Jean Claude Jacob, a member of the French National Assembly, who was called the 'Dean of the Human species and who died at the age of one hundred and twenty-one. . . . Of late years a sense of loneliness seemed to sadden him, and he frequently remarked, with a startling air of sadness, that he feared that perhaps 'death had forgotten him; but he would always add with more cheerfulness that he 'felt sure that God had not."—*Liverpool Courier, March* 16, 1866.

"Aunt Milly, a colored woman, died at the house of her former master, Captain Harris, on January 7th, in the hundred and thirty-sixth year of her age.

"Another colored woman named Caroline James, the mother of thirty-five children, has just died in Richmond, at the age of one hundred and thirty years."—*Ibid, March 9th*, 1867.

"LONGEVITY.—Springhead, nestling in a lovely valley of flowers and blushing fruit sinuous with acres of watercress, has long been a popular resort of Londoners; for apart from its natural attractions there was an aged female, Mrs. Clayton, mother of the proprietress on the north side of the stream, that every visitor desired to see. She was born in January, 1760, and, until lately, assisted her daughter, Mrs. Arthur. Her health was uniformly good; she generally rose at six in the morning, and retired at nine in the evening, and walked often to Gravesend, a distance of three miles, without apparent fatigue. This she did within two months of her death. On the 3rd ult. whilst engaged in the cress-house, she was seized with a trembling fit, the precursor of dissolution, from which time she gradually sunk, until Sunday the 14th when after taking an affectionate leave of her family she closed her eyes as if in sleep, and gently passed away, aged 107 years and seven months."— *City (London) Press, August 3rd, 1867.*

"In the year 1566 a native of Bengal, named Numa de Cugna, died at the age of three hundred and seventy years. He was a person of great simplicity, and quite illiterate, but of so extensive a memory that he was a kind of living chronicle, relating distinctly what had happened within his knowledge in the compass of his very long life, together with all the circumstances attending it. * * He asserted that in the course of his life he had had seven hundred wives. The first century of his life passed in idolatry, from which he was converted to Mahometanism, which he continued to profess to his death."— *Maffeus' History of the Indies;* and confirmed by Ferdinand Costequedo, Historiographer Royal of Portugal.

The Egyptians arrive at a great age. "Dr. Clott speaks of a man whom he had seen, one hundred and thirty years old, without any other infirmity than cataract in one eye: and he knows another now living at one hundred and twenty three years of age; who enjoys a sound state of health; and has several children, the eldest of whom is eighty-two, the second seventy-four, the third three years old, and the fourth only a few months."—*Foreign Quarterly.*

Amyntas and Amaryllis, King and Queen of Arcadia, during the latter part of the "Golden Age" it is said "lived a long and happy life. . . . their generation was very long-lived, there having been but four descents in above two thousand years. His heir was called Theocritus, who left his dominions to Virgil. Virgil left his to his son Spenser, and Spenser was succeeded by his eldest born Philips."—*Note to "Arcadia," by Sir Wm. Jones.*

Many instances of great but variable longevity may be found among the lower animals. The wild hog is said to live in its native state, free from disease, to the age of three hundred years. The elephant has been known to live to a great age, sometimes to three and four hundred years. When Alexander the Great had conquered Porus, King of India, he took a great elephant which had fought very valiantly for the King, and named him Ajax, dedicated him to the Sun and let him go with this inscription—" Alexander, the son of Jupiter, hath dedicated Ajax to the Sun." The elephant was found, with this inscription, three hundred and fifty years afterwards.

In *Reynold's Miscellany* for Feb. 26th, 1859, an account is given of an elephant called Hannibal which had then recently died in a travelling circus in America. It is said " He was extremely old. We have heard his age stated variously at from five hundred to one thousand years."

Thomson thus speaks of the longevity of the Elephant,

" With gentle might endued,
Though powerful, yet not destructive; here he sees
Revolving ages sweep the changeful earth,
And Empires rise and fall;—regardless he
Of what the never-resting race of man project."

The swan is said to live to the age of three hundred years.

" Some time ago, a male swan, which had seen many generations come and go, and witnessed the other mutations incidental to the lapse of two hundred years, died at Rosemount. He was brought to Dunn when the late John Erskine, Esq., was in his infancy; and was then said to be one hundred years old. About two years ago, he was purchased by the late David Duncan, Esq., of Rosemount; and within that period his mate brought him forth four young ones, which he destroyed as soon as they took the water. Mr. Mallison Bridget (in whose museum the bird is now to be seen) thinks it might have lived much longer but for a lump or excrescence at the top of the windpipe, which, on dissecting him, he found to be composed of grass and tow. This is the same bird that was known and recognised in the early years of octogenarians in this and the neighbouring parishes by the name of the ' Old Swan of Dunn.' "—*Medical Gazette.*

The eagle is known to live to a great age. *Tacitus* says it attains to five hundred years.

Some of the parrot species are affirmed to live in their wild or natural state to ages ranging from five to seven hundred years.

The rook, raven, crow, hawk, goose, pelican, heron, crane, sea-gull, and other birds of like nature are believed to live to a great age, to more than a hundred years.

Some kinds of fish are very long lived.

"Gesner says, that the longevity of the pike is almost incredible; he mentions as an instance, one that was taken in Hailborn, in Swabia, in the year 1497, with these words engraven on a ring—'I am the fish that was first of all put into this lake, by Frederick Second, Oct. 5th, 1230.' This gave it the age of two hundred and sixty-seven years."—*Rhind's Six Days of Creation.*

Some writers affirm that the whale, shark, and other marine animals live to ages of a thousand years and upwards.

Serpents, it is thought by many, never die of "old age," or "natural decay" but are capable of endless existence. Hence when formed into a ring or circle by bringing its extremities together, the serpent has been, from the earliest ages, an emblem of immortality and eternity.

This immortality of the serpent is thought by some to be confirmed by the teaching of Scripture. If it be true, as some believe, that the serpent which at the beginning of creation beguiled Eve, and which was, as a part of the curse pronounced against it, destined to have its head bruised by the seed of the woman, is the same creature as that spoken of in Revelations, as, in the end, to be overcome and "cast into the lake of fire and brimstone,".... to "be tormented, day and night, for ever and ever," this belief would appear to be reasonable.

"The wisdom and subtilty of the serpent are frequently mentioned in Scripture, as qualities which distinguish it from other animals; and several are the instances wherein it is said to discover its cunning: 1, when it is old, by squeezing itself between two rocks, it can strip off its old skin, and so grows young again—2, as it grows blind, it has a secret to recover its sight by the juices of fennel—3, when it is assaulted, its chief care is to secure its head, because its heart lies under its throat, and very near its head;—and 4, when it goes to drink at a fountain, it first vomits up all its poison, for fear of poisoning itself as it is drinking."—*Calmet's Dictionary.*

"The serpent was more subtle than any beast of the field, which the Lord God had made."—*Genesis* iii. 1.

"Be ye therefore wise as serpents and harmless as doves."—*Matthew* x. 16.

"Some species of fish and certain snakes are said to live till some accident puts an end to their *indefinite term of life.*"—*Southey.*

The instances above given are fully sufficient to demonstrate that NO DEFINITE PERIOD HAS BEEN ASSIGNED AS A LIMIT TO THE DURATION OF LIFE.

Secondly. That there exists in nature an analeptic or restorative principle and action is proved by the following cases:— Numa de Cugna, the native of Bengal whose long life is referred to at page 18,- "Had four new sets of teeth; and the colour of his hair and beard had been very frequently changed from black to gray, and from gray to black."

In the twenty-third volume of the *Philosophical Transactions,* Dr. Stare states that his grandfather, a native of Bedfordshire, died in his one hundredth year "of a plethora for want of bleeding." He had remarkable health and vigour; and at the age of eighty-five had a complete set of new teeth; and his hair, from being of a snowy white, gradually became darker."

One of the Egyptians whom Dr. Clott saw at a hundred and twenty-three years old "at the age of eighty-two cut six new teeth; which he was obliged to have extracted, on account of the pain and inconvenience they occasioned him."

"Philip Laroque, of Frié in Gascony, a butcher, died at a hundred and two. At the age of ninety-two he cut four new teeth."

"Helen Gray died at the age of a hundred and five, she was of small stature, lively, peaceable and good tempered, and a few years before her death acquired new teeth."—*Hufeland, p* 84.

"A Magistrate named Bauborg, who lived at Rechingen in the Palatinate, and who died in 1791 in the hundred and twentieth year of his age. In 1787 long after he had lost all his teeth eight new ones grew up. At the end of six months they again dropped out, but their place was supplied by other new ones, both in the upper and the lower jaw: and nature, unwearied, continued this labour four years, and even till within a month of his death. After he had employed his new teeth for some time with great convenience in chewing his food, they took their leave, and new ones immediately sprung up in some of their sockets. All these teeth he acquired and lost without any pain: and the whole number of them amounted at least to fifty."—*Ibid*, p. 92.

"A short time ago the *Times* newspaper gave an account of an old lady more than eighty years of age, who had cut her third set of teeth; and her features, it is said, have now the juvenescence of thirty years. Many such facts could be collected. We are therefore bound, perhaps, to give credence to certain good authorities when they assert that such natural changes have occurred in the entire body, that the powers of youth have been restored to persons with whom they have been familiar.

"*Valescus de Taranta* relates that there was an abbess in the nunnery at Monviedra who reached the age of one hundred years, and was then very infirm: but the lost powers of nature unexpectedly came back to her. Black hairs sprouted from her head, and the white hairs were thrown off; all the teeth returned into her mouth; wrinkles were lost from her face; her bosom swelled, and she became at last as fresh and lovely as she had been at the age of thirty.

"Several well-authenticated instances are likewise recorded of rapid change in the colour of the hair. By an inscription on a tomb stone at Breslau, it appears that one John Montanus, who was a Dean there, recovered three times the colour of his hair. . . . Does it therefore appear incredible or impossible that man may occasionally after his "three score years and ten," again exhibit the powers and physical qualities of youth?"—*Family Herald*, July 25th, 1857.

"The *Auxilia Breton* mentions a curious circumstance. It states that a gendarme named Labo of the Department of the Ilouet Valaire, who had a gray beard and hair presented himself a few days ago perfectly black! He said that he had had a determination of blood to the head, which caused his head to swell and become black, as did also his beard and hair and part of his body. He had felt great pain for a time, but that afterwards he found himself much better; that then his skin resumed its natural color but that the hair and beard remained black. Two comrades of the gendarme, one of them a corporal, confirmed his statements."—*Morning Advertiser*, *April* 12 (?) 1855.

"RENOVATION IN OLD AGE—I lately met a gentleman who

mentioned to me the following particulars in respect to himself, one or two of which may be worth noting as rather remarkable in the history of our species. He was born in the year 1781, and is as hale and active as at any previous time of life; sleeps well, eats well, and is in full possession of all his mental faculties; the eyesight good, but obliged for close reading to use spectacles. His hair *white*, is now returning to its former colour, *black*, and he is in process of getting a *new under tooth*, about half way (as I saw it) shooting through the gum. He never wore flannel next his skin, or otherwise on his person; takes the cold bath regularly, with a cheerful good complexion, and I believe occupies much of his time in intellectual studies, and in official duties as a respectable elder of the Church of Scotland."—*Notes and Queries.*

What nature has done repeatedly, although apparently by accident only, is evidently a natural possibility; and we may reasonably hope at some future day to discover the laws and principles which operate in such cases; and also to be able at our will and pleasure, and for special purposes, to induce and regulate their action. Such a purpose is not contrary to the spirit and letter of the sacred writings; and certainly is in accordance with the promptings of our nature. "May good health and long life attend you" is one of the commonest forms of utterance among friends and relatives; and the highest expression of loyalty by the Mussulman devotee is "May our Sultan live a thousand years!"

"The ordinary workings of Providence are according to certain fixed laws, regard and obedience to which meet with reward; while neglect and infraction are deservedly punished. The study of these laws, and their application, is the part of wisdom and prudence, as much as the dependence on Divine Power and Goodness is the part of true piety."

"Like the pious pilgrim to the Holy Land, toil on in search of the sacred shrine, in search of truth—God's truth—God's laws—as manifested in His works, in His creation."—*Prince Albert.*

"Man has been made susceptible of experience; and consequently more and more perfectible; it is absurd then to wish to arrest him in his course, in spite of the eternal law which impels him forward."—*De Marsais.*

"Whoever has attentively meditated on the progress of the human race cannot fail to discover that there is now a spirit of inquiry amongst men which nothing can for any lengthened period control. Reproach, and threats, and persecution will be in vain. They may embitter opposition, and engender violence, but they cannot abate the keenness of research. There is a silent march of thought which no power can arrest, and which it is not difficult to foresee will be marked by important events."—*London Journal.*

"Philosophers tell us that the effect of a blow with the hand on the thin air is felt for ever throughout the vast space which the atmosphere occupies; and keen observers assure us that a truth once uttered abides for an eternity in the public mind,—that apparently it may at first be unheeded, and much time elapse before it is fully manifested; but that there is a Divinity in it that ultimately shapes its end. Great facts, rational proposals, useful designs, have been for a time despised, neglected, or ridiculed; but one after the other they turn up in due season to reproach ignorance, and benefit mankind."—*Liverpool Journal.*

"We touch not a wire but it vibrates in eternity: and there is not a voice that reports not at the throne of heaven."

"The effort to extend the dominion of man over nature is the most healthy and most noble of all ambitions."—*Lord Bacon.*

"It is perfectly vain to attempt to stop investigation...... Depend upon it, if a chemist, by bringing the proper materials together could produce a human body he would do it: and why not? There is no command forbidding him to do it—his inquiries are limited solely by his own capacity."—*Professor Tyndall, in lecture before the British Association, Dundee, Sept. 5th,* 1867.

"Humanity is yet underground; so much matter envelopes and crushes it; so many superstitions, prejudices, and tyrannies form a thick vault around it, and so much darkness is above it.. .. yonder, far in the distance, a luminous point appears. It increases—it increases every moment; it is the future—it is realization—it is the end of woe, the dawn of joy—it is Canaan, the future land where we shall only have around us brethren, and above us heaven."—*Victor Hugo.*

Having shown that neither practical science, nor the sacred writings assign an impassible limit to the duration of life; that there exists in nature a restorative or analeptic power and tendency; and that among the varied objects of human research and progress that of preserving life is one of the most important, we may now enquire into the causes which operate in checking and ultimately completely arresting the powers and functions of living structures. The first step in the inquiry is to ascertain the differences which exist between a young and vigorous animal, and one which has passed through the various stages of life to the end, when it is said to die of "old age" or "natural decay."

First, mechanically, in animals which are killed for food it is found that the flesh, liver, cartilage and other eatable parts of the oldest are much more solid and dry than the corresponding parts in the young.

The bones are more dense and brittle. Any one can readily distinguish the bones of a lamb or calf, for instance, from those of a sheep or an ox, not only by the size but by the difference in weight, texture, porosity, and form. The bones of any young animal are light, spongy, elastic, and saturated with semi-fluid marrow: while those of the aged are heavy, dense, rigid and nearly marrowless.

The substance of the brain, spinal column, and nerves is more solid and resisting in old than in young animals.

The substance of the eye presents a remarkable difference in these respects. In youth the eyes are bright, clear, sparkling, and crystalline, and the sight quick and powerful: in the aged they are dull, muddy, glazed, without expression, and lifeless, and the power of vision faint and indistinct.

The whole nervous system of the aged animal is less delicate and susceptible than that of the young.

The whole body, as well as all the parts separately, are specifically heavier in old than in young animals of every kind.

"The most considerable differences that are found in one and the same person, during his whole life, are in his infancy, in his maturity, and in his old age. The fibres in the brain, in a man's childhood, are soft, flexible, and delicate; a riper age dries, hardens, and corroborates them; but in old age they grow altogether inflexible, gross, and intermixed with superfluous humours, which the faint and languishing heat of that age is no longer able to disperse; for as we see that the fibres which compose the flesh harden by time, and that the flesh of a young partridge is, without dispute, more tender than that of an old one, so the fibres of the brain of a child, or a young person, must be more soft and delicate than those of persons more advanced in years."—*Malebranche.*

SECOND, Microscopically, great differences are found. The one is highly vascular, arterial, membranous, glandulous, porous, filled with animal juices and fluids of every kind, and all in a state of high activity, and change. The other is much less vascular, scarcely at all arterial, but greatly venous, the membranes, glands, and cells or pores, almost obliterated, and the fluids thick—tending to set, and nearly motionless.

THIRD, Chemically, great differences in the temperature, electrical condition, and composition of the whole system, and of all its parts. If the blood, milk, and other fluids and juices of the aged animal be analysed, they are found to contain a much *larger amount* of *solid matter* than is found in the same portions of young animals. If the flesh and solid parts generally are examined they are also found to contain a much larger proportion of solid matter. This solid matter is chiefly albumen, fibrin, gelatine, and compounds of lime and magnesia; but as age advances the albumen diminishes, and the fibrin, gelatine, and earthy compounds increase.

"There is much more albumen in the flesh of young animals than that of old ones; but more fibrin in the latter than the former: from the flesh of an old horse, for example, there was not found the tenth part of the quantity of albumen which was furnished by an equal weight of ox flesh."—*Liebig.*

If the bones are analysed they will be found on the average to consist of phosphate of lime, 50 parts; carbonate of lime, 10; sulphate of lime (with sometimes traces of magnesia and other earths), 10; and gelatine, 30 parts; making together 100.

"PARALLAX" ON ZETETIC ASTRONOMY.—"The gentleman who has adopted this *nom de plume* delivered his first lecture at the Public Hall on Monday Evening last. There was a large and highly respectable audience, the room being crowded. The lecture, which was a clear and elaborate exposition of the extraordinary science of Zetetic Astronomy, was listened to with the greatest attention. He contended that &c. (details follow). If we may judge by the applause by which some of the Lecturer's arguments were confirmed, we should say that many of those present were ready to exclaim—'Behold a greater than Newton is here!' A hot discussion followed, in which the Rev. J. Nixon Porter and other gentlemen took part, but 'Parallax' maintained his ground."—*Warrington Guardian*, March 24, 1866.

"EARTH NOT A GLOBE—On Monday last a gentleman adopting the *nom de plume* of 'Parallax,' a very appropriate name, seeing that the basis of his arguments is the relation to each other of parallel lines—commenced a series of lectures at the Public Hall on 'Zetetic Astronomy,' a system directly opposed to the great Newtonian theory, which has obtained amongst us for so many years.... That he is a clever man, and has studied the matter deeply, and that he is master of his subject and thoroughly convinced of its truth, is apparent; and his arguments are certainly very plausible. The lectures drew large audiences, and among those present we noticed the Rev. W. Hamilton; the Rev. J. E. Weddell; the Rev. J. Nixon Porter; Alderman Dr. Smith; Councillor Neild and family; Dr. T. S. Smith; Messrs. H. and E. Rylands; Mr. G. Webster and the Misses Webster; Mr. Cooke; Mr. Stewart; the Misses France; Mr. Greening; Mr. Potter; Lieutenant Bolton; Miss Bolton; Mr. G. H. Bolton and family; Mr. H. White; Mr. C. Barlow; Ensign Cartwright; Councillor Silcock, Mr. L. Cartwright, &c., &c. 'Parallax' commenced by explaining the word 'Zetetic,' which had been adopted because they did not sit in their closets and endeavour to frame a theory to explain certain phenomena, but went abroad into the world and thoroughly investigated the subject. (A long report of the three lectures here follows). Lengthy and animated discussions ensued; votes of thanks were passed to the Lecturer and the Chairman, the Rev. Nixon Porter, who declared that he was much struck with the simplicity and candour with which the lecturer had stated his views; and after a promise by 'Parallax' that he would pay another visit to Warrington in a few weeks, the audience dispersed."—*Warrington Advertiser*, March 24, 1866.

"THE EARTH NOT A GLOBE—Lectures on the above subject were delivered this week in the Royal Assembly Room, Great George Street, Liverpool, by 'Parallax' (a gentleman known to the Literary World by a work on 'Zetetic Astronomy,' and who came somewhat prominently before the Liverpool public 14 or 15 years ago through the columns of the *Mercury*). The Hall was well filled by respectable and critical audiences. He commenced his first lecture by comparing the Newtonian principle of Astronomy with the Zetetic (which must prove all and take nothing for granted); and endeavoured to demonstrate in a comprehensive and logical manner that the Earth is not a Globe but a plane, that in fact all theories of the earth's rotundity are fallacious, and that the followers of Newton and other philosophers had been adopting and believing a 'cunningly devised fable.' The lectures were illustrated by numerous diagrams and experiments, and were listened to with the greatest attention by all present. 'Parallax' appears to have studied the peculiarities of his subject thoroughly, and was frequently warmly applauded during the delivery of his lectures."—*Liverpool Mercury*, Oct. 3, 1866.

Selections from The Earth

EXPLANATION OF MIDDLETON'S ATTEMPTED DIMENSIONS of THE EARTH.

The distances on this chart are those found by the modern steamship.

Dimensions mean the compass within which the whole Earth lies. This dimension is much smaller than one would suppose.

The latitudes are out as much as 30 degrees on the China side, and countries thought tropical are really Arctic in consequence. This has been explained in The Earth Magazine for the months of March and May, 1902.

The Great Secret of the Earth lies in the Gulf of Pichili, which is quite 30 degrees out of the Globe's latitude. Rivers in the Gulf of Pichili freeze SOLID from November to March. This severity of climate upsets the Globe's latitudes, and allows of a reasonable Ground Plan of the Earth.

The Longitudes are fairly representative.

MIDDLETON'S ATTEMPTED
DIMENSIONS OF THE EARTH

By Lady Blount

CELESTIAL PHENOMENA.

The following article will form a reply to several enquirers.

"The heavens declare the glory of God." In spite of "the fool" having "said in his heart, there is no God," the above statement, which forms the opening words of the 19th Psalm, is an admirable fact, which can be grasped and appreciated by all classes and kinds of God-fearing men and women.

"And the firmament showeth His handiwork." These words are as true now as they were when they were written by the Sweet Psalmist of Israel hundreds of years ago.

The Psalmist goes on to say: "Day unto day uttereth speech, and night unto night showeth knowledge." Let us listen to the "speech' uttered, and profit by the "knowledge" shown daily.

"There is no speech nor language where their voice is not heard," yet the heavens speak not in an unknown tongue, nor to any one nation, but the significance of what they declare may be understood by enlightened men of various nationalities; and without the gift or the cultivation of tongues.

"Their line (or rule) is gone out through all the earth, and their words to the end of the world." This shows that the heavenly bodies have influence, power or "rule:" all through the earth.

"In them hath he set a tabernacle for the sun."

These statements harmonize with the statements made in the beginning of Inspired Writ, viz.: that the sun and moon are "two great lights" which were made by the Creator to give light upon the earth—and to rule over the day and over the night, and divide the light from the darkness—and also to be for signs and for seasons and for days and years."

The Book of Job is supposed by many students not only to be the oldest book in the Bible, but the oldest book in existence.

Different stars, and constellations are referred to and named in it by names which are familiar to us, and it is evident that the knowledge of pure astrology (which was originally one and the same thing as astronomy) has been handed down to us from the Creator, through Adam and Seth. Josephus informs us that to the antediluvians we are indebted for very much that is known on this subject.

He states that Adam was instructed by the Creator Himself who ordered Seth to write the rudiments of the knowledge regarding the heavenly bodies which He had imparted to Adam upon permanent tables of stone, which Josephus says he had himself seen.

And these tables included a tabulation of eclipses.

The path of the moon is like the path of the sun—a spiral—and when the moon's path crosses that of the sun, there is an eclipse, if both bodies are in conjunction.

We believe that knowledge relating to the stars as "signs" has to some extent ceased, or been perverted.

The sign of the "Star in the East" was understood by wise men, or Magi, when our Lord was born on earth (*Matt.* ii. 9). Since then we have no inspired record of men being guided by the stars; but it is quite possible that a deeper knowledge may be imparted to faithful followers of the Truth respecting God's Works in Creation.

When we study the Word of God, and consider the perfect order of the universe, we cannot help perceiving a perfect and divine precision underlying all visible things created; and an invisible power *behind* the scenes, directing and governing the whole. Even the weakest of sin-crippled human intellect can scarcely fail to see the Creator's care of, and provision for, His creatures, if they only study His Works. Our Lord told us that the hairs of our heads are numbered, as also are the stars in heaven above.

But in addition to this we have the sure testimony of Inspired Writ in *Isa.* xl. 26, wherein the prophet by power of the Holy Spirit says :

"Lift up your eyes on high, and see who hath created these, that bringeth out their host by number. He calleth them all by name; by the greatness of His might, and for that He is strong in power, not one is lacking. (R.V.)

The psalmist also states that "He telleth the number of the stars; He giveth them all their names."—*Ps.*cxlvii.4 (R.V.)

The heavens are described in Scripture as "spread out as a canopy or tent"—and so do they appear.

> The outstretched heavens above appear a "dome,"
> To everyone on earth, where'er he roam.

In the beginning it appears that Draco was the Polar star, the change having taken place "owing to the slow recession of what is called the pole of the heavens. The same movement which has changed the relative positions of these two stars has also caused the constellation of the Southern Cross to become invisible in northern latitudes. The Southern Cross was just visible in the latitude of Jerusalem at the time of the first coming of our Lord. Since then, through the gradual recession of the Polar star, it has not been seen in Northern latitudes." See *The Witness of the Stars*, by Dr. E. W. Bullinger.*

The sun takes his course through the Heavens, passing through the twelve signs of the Zodiac, or about one sign for each month.

The sun's path through the constellations is called the ecliptic. But there is also an annual difference, because the sun does not come back exactly to the same spot in the sign when he commences the year, but a little behind. If the fixed stars daily revolved around the earth at exactly the same rate of speed as the sun (which they do not), and if the sun started at exactly the same place each succeeding year, the signs would correspond with the months, and calculations regarding star motions would be simplified.

The sun passes through the twelve signs every year, but in consequence of the slow precession of the equinoxes the sun commences the year in the same sign for centuries.

The celestial equator is the sun's path around the heavens, at the vernal or the autumnal equinox, and if it were possible to stand in the centre of this great circle, the stars and sun and moon would appear to move around without

---

* Published by Messrs. Eyre & Spottiswoode, Great New Street.

ever rising or setting. But both north and South of the equator the stars rise and set obliquely. On the equator they appear to rise and set at right angles to the spectator.

The points where the two great circles cross each other, or intersect, are called the equinoctial points and the slow and gradual movement of these points is termed the "precession of the equinoxes."

The relative speeds and various motions of the heavenly bodies are governed by their various heights and declinations. "The more rapidly a star or planet revolves, and the higher its distance from the earth, and the greater its distance from the centre of revolution, and vice versa, the more slowly a star goes around and above the earth, and the less is its height above the earth and the less its distance from the centre of revolution. Thus their velocities are proportionate to their heights, and to their distances from the North Centre." But to make their daily circles "Zetetes" states that: the whole of the "ether," or whatever other name we like to give to the subtle matter above our atmosphere which fills all the space between the earth's plane surface and the firmamental vault of heaven—and in which all the heavenly bodies are contained and move, the whole substance of this subtle fluid is in a state of flux, like a great stream continually going around the polar centres, and carrying all the heavenly bodies with it at various heights, according to their different densities, all being light and comparatively small bodies. But this flux is not like an ordinary stream where all the currents flow at the same general level and rate, nor is it like those streams which flow quicker in the middle and slower at the sides, but rather, in shape at least, it is something like the great Maelstrom, or whirlpool, off the coast of Norway; that is, like a great funnel with the tube or hole pointing downwards through which the mighty current flows into unknown subterranean, or submarine regions.

This explains also the action of the dipping needle.

A correspondent owns that "the sun *appears* to travel in a daily circle around the heavens, but this is only an optical delusion and that it is really the earth that is moving."

Now with all due respect to the one who makes the above statement; in reply to his query as to how we can disprove this. I should like to ask him and others of the same mind if they can prove that the apparent motion of the heavenly bodies is not real. Experimental tests have been made to see if the earth has any motion, and no motion such as is assumed by the astronomers has ever been discovered. Why then should we not believe that the "apparent" motion of the heavenly bodies is real? We want something better than the unfounded assumptions of modern astronomers. We can see the heavenly bodies move, and being only comparatively small bodies of light revolving round and above the earth in the great ethereal stream, it is to us much more reasonable to believe that these 'lights" are in motion circling over the earth, as they appear to do, than to believe that a ponderous body, like the earth, weighing millions upon millions of tons should be suspended in space like a feather; rushing away through "space" in its so-called orbit, and tumbling topsy turvey, carrying us all head over heels in its regular and periodical revolutions while tearing away forever round the sun! It is absurd and we cannot believe it.

This then is our reply for the present to correspondents who merely affirm that "although the sun appears to travel a daily circle it is only an optical delusion."

We have given proof that the delusion is on the other side. It has been proved in *The Earth*, by mathematical calculations as well as by the evidence of our senses, that the sun is neither large enough nor high enough to light all the earth at one time but only about one half of it. And this is not only owing to the comparatively small size of the sun, but as has been shown by "Zetetes," it is also owing to the fact that when the rays of the sun strike our atmosphere they are refracted, or bent out of a straight path; and when by perspective the sun goes sufficiently far off from any particular locality its rays are deflected by the atmosphere so that they do not reach the earth at all, and darkness ensues. This is a wise provision of the Creator that when darkness comes over the earth, the creatures of His power being only of limited strength may take their nightly repose and be refreshed with sleep. The sun still going his daily round lights up other parts of the earth far distant from us and comes round next morning to wake us with his bright and cheering rays. Thus again do the "heavens declare the glory of God and the firmament showeth His handiwork."

STELLAR MOTION.

Of all persons in the world Zetetics must not form *hypotheses*; or draw hasty conclusions; or build up any theories on what they observe. They are emphatically *seekers* while others are *reasoners*.

The primer of astronomy prepared by the late Richard Proctor for the London School Board commences with the words:

"*Astronomy is a science whose facts are based upon reasoning.*"

Never was a truer word written as to what is called "modern astronomy."

Never was a falser statement recorded if it refers to any true "science."

Never was a sadder confession made by men who arrogate to themselves the title of *scientists*.

"Facts" are facts, and "reasoning" is reasoning. A "fact" is *a thing done*, which no reasoning in the world can ever affect or alter. "Reasoning" can be based on a fact; but a "fact" can never be based on reasoning.

Science is the Latin *Scientia* which means knowledge. It is *what we know*, apart from all reasoning. Hypothesis is the opposite of this, and is only *what we think*.

Zetetics must be careful to remember these elementary truths; and while they *seek* out facts, and observe them, and collect their *data*, they must not fall into the mistake of "astronomers" and substitute their reasoning about these facts for the facts themselves.

Euclid furnishes us with a useful guide as to the observation of facts, or rather as how we should deal with them and reason about them.

When it cannot be proved that one thing is equal to another, it has to be proved that the one cannot be greater, and that it cannot be less than the other. Then the conclusion is that it must be the same in size.

So with an observed fact; say the disappearance of ships at sea. We must not say that it is caused by the globular shape of the earth, because that is only assuming the very thing that we are *seeking* to find out. It must be shown to us that such disappearance could not be caused in any other way. It has not been so shown, and therefore it proves nothing; and nothing is proved beyond the *fact* that ships do disappear. That is a fact.

Now we, who are *seeking* for truth, contend that the disappearance of ships at sea may be caused by other phenomena. It may be due to the laws of perspective. Or, there may be other causes of which we are ignorant.

One thing is clear, and that is, that such disappearance does not prove the rotundity of the earth, because it has not been proved that there is no other explanation to be given.

Or, take the tides. The fact as to the tides is observed by all. But that they are caused by gravitation is not proved; unless and until it is proved that nothing else could cause them.

Do not let us be drawn into the snare which besets astronomers. They are tethered by their hypotheses, and can never really discover anything, because whatever they observe it must be explained by their hypotheses. It is a case of *chose jugée*, with them; but that is contrary to the very foundation of real science.

Or, take *Eclipses*. We know *what we see* and nothing more. By a careful *observation* their cycles were known ages ago, many centuries before modern astronomy was dreamed of.

We know also, the popular explanation, that eclipses are caused by the motions of the earth and its shape; and therefore the conclusion is drawn that the earth is a globe. But this is the very thing which we are seeking to prove. It is quite true if that be the *only* explanation that can be given. But this is the very point in question.

Our position is that it has not yet been shown that this is the *only* reasoning that explains the phenomenon.

We should be just as illogical on our side if we said that eclipses *must* be caused by the revolution of a dark satellite.

That would, of course, cause the phenomenon, therefore we say this *may* be the cause. But, nevertheless, we admit that it is not proved. Eclipses do not *prove* that the earth has a dark satellite ; but a dark satellite would cause eclipses ; but this is no proof that such a satellite exists, or that eclipses are caused by it.

We are *Zetetics*, or *seekers*, and of all people in the world it is not for us to base our facts upon reasoning ; but to observe our facts and collect our *data*, drawing no conclusions until we have them all. We must neither accept as proof from others, or state it as from ourselves that such and such must be the explanation of a fact, until all other explanations are exhausted and can be ruled out. Then we shall all be shut up to the only explanation that is left.

As Zetetics we cannot be called upon to *explain* anything, we are *seekers*, and our very position protects us from any demand that we shall or must explain any phenomena.

About the earth we do know many facts. Its motion has never yet been proved ! But proofs innumerable, from Geodesy to Ballooning, prove that it is so founded and established that it cannot be moved.

But when we " consider the heavens ·' the case is different. There we observe that everything is in motion. We observe that there is a great variety in these motions. There is the motion of the sun, and the motion of the moon ; the motion of the planets. These latter are called *asteres planates*, or wandering stars, because they have a peculiar motion of their own. Astronomers speak of "apparent motion" and "real motion," but this is assuming the very thing that has to be proved. We say that we believe the "apparent" motion *is* the "real" motion, and we must believe this until the contrary be proved.

Then there is the motion of *Comets ;* the motion of the fixed stars, rising in the east and setting in the west ; others never rising and setting but revolving round a central star.

Do we know all about stellar motion yet ? Have we all the *data ?* Is there nothing more to be learned ? May we say we have all the facts and may sit down and reason about them ?

Take for instance the motion of the stars in what is called the South Pole. Our grievance is that we cannot get at the facts. One writer says that the stars there do not rise and set, but that they revolve round a central point, as the stars revolve at the North Pole. Others, who are travellers, speak of the Southern Cross rising and setting.

What are the facts? This is the simple question that we as Zetetics, seek to find out.

The writer once enquired of a friend in the South, and begged him to watch the motions of the southern stars. He put our letter in the newspaper there, instead of observing for himself and for us ; and the following answer was given :

"Of course, it cannot rise or set, because the earth is a globe, having a southern Pole."

But this is the very thing we were seeking to find out, which is thus quietly assumed ! And thus we are at present shut up to and shut out from that which we are *seeking* to find out.

Now, suppose, it shall be clearly shown that the stars in the extreme south do have a different motion, and do revolve round a common centre, as in the North. That would not prove either that the earth is a globe, or that it moves. It would only prove that certain stars have that peculiar motion, different from other stars. It would not prove anything as to the shape of the earth, unless and until it should be also shown that the stellar motion in question could not arise from any other cause.

This, of course, could not be shown.

We are anxious in this article to warn Zetetics not to fall into the snare of astronomers, and be so ready to come forward with their explanations. It is not for us to explain but to *seek*.

Our opponents are only too eager to get us to give our explanations of the phenomena, in order that they may deal with ours as easily as we deal with and expose the falsity of theirs.

Do not let us come down from our high platform.

Do not let us leave our impregnable position. Do not let us take off our strong armour and lay ourselves open to certain defeat.

Let us be content to examine their observations, and expose their false reasonings, while we continue our special work and *seek* after the truth.

THE MIDNIGHT SUN—N. AND S.

"The Sun and Moon Stood Still."—*Hab*. iii. 11.

The sun's motion is not only a fact according to the Holy Scriptures, but also a fact which is self-evident to our senses without any artificial aid. Tully says that the sun is called *Sol* because it is the " only " heavenly body of that magnitude, and because when it rises it puts out all the other orbs and " only " appears itself. Both common sense and experience verify this statement, which is in accord with the first chapters of Genesis, wherein we read : " God made two great lights, the greater light to rule the day."

It seems an undeniable fact that the midnight sun is seen in the southern hemisphere (so-called), for we have it on no mean authority, viz. : the Perth, W.A., Astronomer Royal, the testimony of the crew of the *Belgica*, and the whole exploring party on board the *Discovery*, that it is thus seen in the South as it is in the North.

Mr. Ernest Cook states in the last letter which I received from him, that, " the midnight sun ought to be seen in the middle of summer (December)......at all places south of latitude 66½ S. As a matter of fact there is no known habitable land in the Antarctic regions, so we hear nothing about it, but I read only a few days ago an account of the exploring vessel, the *Discovery*, I think it is called, wherein it was stated that the crew played cards on deck at midnight with the sun shining down upon them on Christmas Eve."

This corresponds precisely with the accounts which we received here in England. In more than one report it was stated that at the termination of a long night of four months' duration, they had about two minutes' daylight, then after twenty-four hours' night they had a day that lasted ten minutes, and after this the days increased in length, becoming longer and longer until the sun remained above their horizon for several months as it is observed during our Summer at the North "Pole" regions. I have also received confirmation from the Perth Astronomer Royal that there is a star which is practically a south "polar star," and that the southern constellations revolve round a central point as the stars revolve at the North Centre. As to the motions of the southern stars, it has been asserted that they cannot rise or set "because the earth is a globe." But this is only assertion. The *fact* which we learn however is, that "certain stars in the extreme south have a different motion from those in the North."

By modern scientists it is deemed a legitimate deduction to conclude that the belief that the earth is a revolving globe moving rapidly in an orbit round the sun, with its axis of revolution inclined to the plane of the ecliptic, is verified by the fact that its "poles" are alternately illuminated by a long continued day lasting for a period exceeding one hundred ordinary days and nights.

But let us examine whether this phenomenon will afford any proof in support of the globular theory. In *South Sea Voyages*, by Sir James Clarke Ross, it is stated: "In lat. 65° 22' S. long. 172° 42' E., on Jan 4th, at 9 p.m., the sun's altitude was 4°. The setting sun was a very remarkable object, being streaked across by five dark horizontal bands of nearly equal breadth, and was flattened in a most irregular form, by the greater refraction of its lower limb, as it touched the horizon at 11° 56' 51". Skimming along to the eastward, it almost imperceptibly descended until the lower limb disappeared exactly 17 minutes and 30 seconds afterwards. The difference in the horizontal and vertical diameter was found by several measurements to amount to only 5' 21" the horizontal being 32' 31", and the vertical diameter 27' 10", that given in the Nautical Almanac being 32' 34"." At p. 207, vol. 1, it is said: "In lat. 74° S, long. 171° E, on Jan. 22nd, 1841, it was the most beautiful night we had seen in these latitudes. The sky was perfectly clear and serene. At midnight, 12 o'clock, when the sun was skimming along the southern horizon, at an altitude of 2º, the sky overhead was remarked to be of a most intense indigo-blue, becoming paler in proportion to the distance fron the Zenith."

From these quotations it appears that Lieutenant Wilkes saw the sun set at a few minutes before 10 o'clock. Captain Ross, a few days before, said that the sun did not entirely set or disappear until 14 minutes past 12 o'clock—the sun remaining above the horizon two hours longer than it did to Lieutenant Wilkes a few days later in consequence of "unusual refraction." It is not stated whether the sun was seen in the northern or southern horizon; but as the earth is a plane, and the sun's path is concentric with the northern centre, it was skimming along to the eastward, beyond, or on the other side of the northern centre.

The sun rising at E (the east) would, during the day, move from east to west (from E to W); but during the night it would be seen by the operation of great refraction " skimming along to the eastward from W to S and E. Captain Ross saw this phenomenon, but not Lieut. Wilkes, who reports that the sun set a little before 10 and rose about 4 o'clock. Captain Weddle was in lat. 74º 15′ S. on Feb. 20th, 1822, and he stated that "the sun was beneath the horizon for more than six hours." Captain Ross in his record states: "at midnight, in lat. 74º S, the altitude being only 2º, the sun was skimming along the southern horizon."

In M. Chaillu's book, *The Land of The Midnight Sun*, he says that between June 13th and 16th he sailed towards the Midnight Sun in a steamer leaving Stockholm for Haparanda, the most northern town in Sweden, 65º 51″ N lat., 41 miles S of the Arctic Circle as marked on maps. It is in the same latitude as the most northerly part of Iceland.

At the North Centre "the sun is to be seen for six months." These quoted words are followed by the statement that the sun is seen at the Arctic Circle for one whole day, at the base of the North Cape from May 15th to August 1st. At the Pole the observer seems to be in the centre of a grand spiral movement of the sun, which further south takes place north of him.

The known location of the Midnight Sun is in keeping with the statement that the earth and sea together form a vast plane—for, water being level is the best proof that the earth must be a plane, because if water is level, the land about it is level, and this M. Chaillu unconsciously proves in his book.

If the earth were a globe God would not have given us the wrong order of Creation in the opening of the Holy Decalogue, wherein He distinctly states, that heaven is "above, the earth beneath, and water under the earth." The laws of Nature (which were created by God) cannot be violated, and that it is impossible to look round a globe the annexed diagram will clearly illustrate:

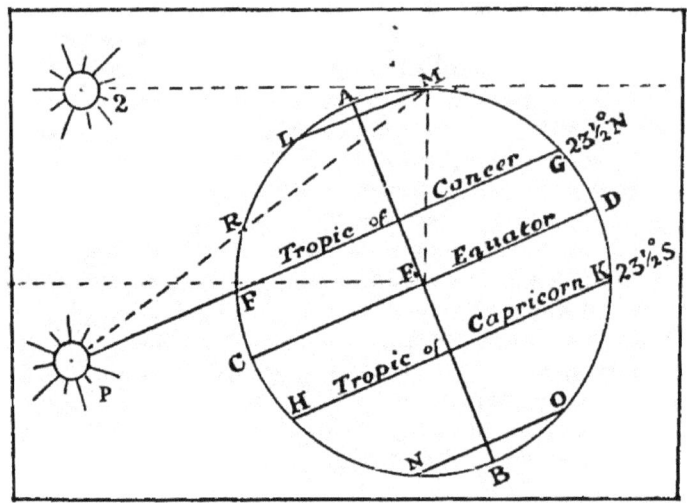

An observer at M can see the sun at midnight above the horizon as shown by the dotted lines, that is, looking directly over the North Pole (M Q); the horizon is a straight line tangential to the surface of the assumed globe at the point of observation. It must therefore be placed at right angles to the dotted line E M; but we will leave the reader to examine for himself the bearing of this case in an unprejudiced attitude. Then it will be clearly shown to be such an one, that for the observer at M to see the sun at midnight at P, whether it be from north or south of the globe, he would have to look downwards and his vision pierce through the globe for over 5,000 miles. The sun being seen *there* leaves the Newtonian philosopher with "a hard nut to crack." No wonder that Haeckel entitled his book: *The Riddle of the Universe*. To get over the difficulties terrific distances have been assumed, and the sun has been asserted to be a million times larger than this earth, whereas there is nothing to prove that it is more than the reasonable diameter—about 30 miles—especially seeing that the sun cannot illuminate more than 90 degrees of the semi-surface of the earth, representing an arc of 5,400 miles with a diameter of 1,800 miles.

It seems evident while we cannot account for what Commanders Wilkes, Bison, Sir J. Clark Ross, and other navigators in high southern latitudes saw, namely, from January 10th to February 19th, that there was night, and also on the 2nd, 6th, 7th, 8th, 11th, 12th, 14th, 17th, and 19th of February—but no mention being made of the missing dates—we have concluded that the sun had been "up" on those dates, and that mention would have been made of so wonderful a phenomenon.

I say that while we cannot account for everything, nevertheless, it is evident that, according to the statements of recent navigators and explorers in high southern latitudes, similar phenomena, relative to the midnight sun in southern latitudes, is annually seen, as witnessed in the North. But while the sun can be seen at midnight far South on a Plane Earth, on a globe it would be an impossibility, because the observer would have to either look round the corner or to see through a solid globe.

It is self-evident that an orb, as the sun evidently is, would be more likely to give its light simultaneously in northerly and southerly directions over a Plane Earth, than if it were a globe; in fact, the globular theory of the earth is proved to be an impossibility through the evidence of the Midnight Sun in northern and southern regions.

We ask: Where and when, and by whom was ever a degree of longitude measured south of the Equator? We think that it never has been done. But if it could be done it would not settle the question of distance in the South, and the true shape of the earth.

When we speak of distances then we have done with "degrees" so far as longitude is concerned. We have no evidence of certain facts, special observations, and scientific experiments; therefore, if we use the evidence of our senses in harmony with the Revealed Word, which is the Truth, we shall prove the truth of the words, that the sun's "going forth is from the ends of the heaven, and his circuit to the end of it."

It is a fact that the Midnight Sun at or near the southern "pole" (in the heavens) is seen at the same time in northern regions. Also, I believe that the Midnight Sun at the North "Pole" is visible at the same time in New Zealand. But it is near mid-day there. Therefore, we know the extent of "his circuit" in both the northern and southern heavens.

The sun is never seen above the horizon on the 21st December further north than the Arctic circle, $23\frac{1}{2}°$ from the northern centre, or $66\frac{1}{2}°$ north of the Equator. This is the hardest fact for the Newtonians to face that ever was put before them if they would face it fairly, instead of assuming to be able only to see the sun as far as New Zealand when it is known to be shining over the North Pole (so-called).

Now I will repeat another fact, often mentioned, namely, that the shape of the heavenly bodies and their motions have nothing necessarily to do with the shape of the earth, or with the fact that God "hath founded the earth upon her basis that it should not be removed for ever."

It seems evident that the sun's path, in his revolution around and above the earth, expands and contracts alternately. The movements in some respects may be compared to the mainspring of a watch, though the body of the sun moves per se, and when we speak of the northern and southern "declination," it is only another form of saying that the sun's path is nearest the polar centres at one period, and farthest away from them at another. Whether the sun's path was once very near to the earth's arctic (or polar) centre is not now under consideration, though, en passant, there are apparent evidences that the conditions and productions found within the tropics once existed in northern regions.

According to the testimony of different persons, equal days and nights occur when the sun is on the Equator, and long days and short nights occur at the extremities or when the sun is in the tropics. The longest days come in the northern parts when the sun is on the tropic of Cancer, and the shortest days in these parts when the sun is on the tropic of Capricorn, and vice versa for the southern parts.

The following diagram will show the impossibility of the North Pole, or the South Pole, ever receiving the sun's light if the earth were a globe.

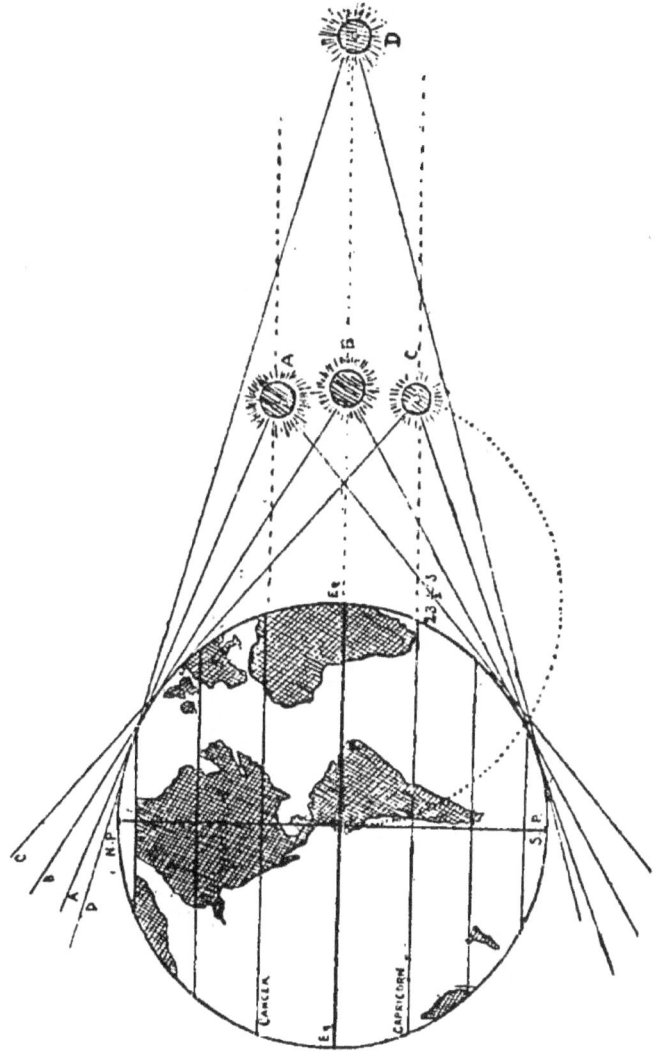

Let this diagram represent the globe, with the tropic of Cancer 23½° N, and the tropic of Capricorn 23½° S of the Equator. Eq representing the Equator. Let A represent the position of the sun from the earth, when in the tropic of Cancer. It is placed at a distance from the earth about equal to the earth's semi-diameter. This is more than its real distance from the earth by at least a thousand miles. But we have placed it further off than we need to in order to show the impossibility of the sun's ever shining on the North Pole, so-called. From the point A the centre of the sun when on the tropic of Cancer draw a line as a tangent to the sphere towards the North Pole, and produce it to A. It will be seen that this line does not touch the North Pole at all, so that according to the astronomers' theory the sun's direct rays should never be seen at the North Pole, not even when the sun has crossed to north declination and reached $23\frac{1}{2}$ N. Much less would the sun's rays reach the North Pole when the sun is on or over the Equator at B, as the line B b shows drawn at a tangent. This line would touch the globe further from the North than the line A a. And of course it is worse still when the sun is in the tropic of Capricorn, as may be seen by the line C c.

And the same line of reasoning shows that on the globular theory the sun never could shine on the South Pole, much less the midnight sun ever be seen there! If readers will draw their own diagrams carefully, and make them much larger these points will come out more clearly.

When the sun is over the equator, and the days and nights are equal all over the world, the sun could not be seen at either of the poles according to the spherical hypothesis; not even if we place it right away from the earth at D. But according to reports the sun can be actually seen at these places at the times indicated, therefore these facts prove clearly that there is something wrong with the globular theory.

That the sun revolves in a spiral orbit over and round the earth, is evident from observation; but some globularists say: "If the earth were flat we should always see the same stars wherever we might be—whether in London or Cape Town." Also: "If the earth were flat the sun would rise and set upon all the countries of the world at the same time." But this is not true.

SOUTH AND NORTH:

Their respective Stars and Motions, etc.

It has been asserted by upholders of the globular theory that the earth has been proved to be a rotative and revolving globe because the stars in the southern " hemisphere" move round a south polar star in the same way that those of the North revolve round Polaris.

Therefore, in consequence of the foregoing assertion and the mythical conclusions and deductions derived therefrom, the Ed. wrote to Mr. Ernest W. Cook, Government Astronomer, Perth Observatory, West Australia, relative to celestial phenomena, star motions, magnetism, etc. That gentleman has kindly replied to many of the queries I put to him.

He says: "There is a point in our sky round which all stars appear to revolve......The axis of our largest telescope is directed precisely to that spot......It is not an absolutely fixed spot in the sky." He goes on to say that the position varies from day to day very slightly. This variation is (he says) in our accepted theory caused by the attraction of sun, moon, and planets. There is not any star in this exact spot; in fact the spot itself varies slightly. There is, however, a small star called Sigma Octantis, and in reply to my question Mr. Cook says: "The sun and moon always appear to revolve round this point the whole year through." Making some allowance for their gradual change in declination, *i.e.*, their motion is more in the form of a spiral.

I am grateful to our esteemed friend, the noted Perth astronomer, for the valuable information he has kindly forwarded. For we are assured that no results of solid facts and true experiments regarding either northern or southern constellations could possibly reveal any proof of either the earth's mobility, or rotundity. In studying the laws of the universe the minds of atheistic astronomers are handicapped with the belief that because the sun and moon and the planets are globular, therefore the earth must necessarily be a globe, and they start off with the idea that the earth is a *heavenly* body—quite forgetting, or ignoring, the teaching of the Holy Scriptures. The earth is God's footstool, and it is " founded " that " it shall not be removed for ever."

"Heaven is *above*, the earth *beneath*, and water *under* the earth; this is the true order of the universe as set forth in the second Commandment. And regarding the heavenly bodies, it is written, in the true account of Creation which comes from the Creator Himself: "God made two great lights; the greater light to rule the day, and the lesser light to rule the night: He made the stars also. And God set them in the firmament of the heaven to give light upon the earth."

So far as any man does not believe these God-given statements his mind is warped; and to say the least he stands at a great disadvantage, of which we are assured.

We are stating a palpable fact when we say that "the man in the street" may see that the North Pole star is the centre of a number of constellations which move over the earth, in a circular direction, and those nearest to it, viz.: the Great Bear, etc., are always visible in England during the whole of their 24 hours' revotion.

Those further away southwards rise NNE, and set SSE; still further south they rise East by North, and set West by North. The farthest South visible from England—the rising is more to the East and South-East, the setting being to the West and South-West. As a matter of fact all the stars visible from London rise and set in a way which is incompatible with the doctrine of rotundity; *e.g.*, if we remain all night on Hampstead Heath, standing with our backs to the North, and note the stars in the zenith of our position, we shall see that the Zenith stars will gradually recede to the North-West; the same stars rise towards our position from NE, demonstrating that the path of all the stars between ourselves and the Northern Centre move round the North Pole star ("Polaris") as a common centre of revolution.

This is just what they must do over a plane such as the earth is proved to be by Zetetics. Upon a globe, zenith stars would rise, pass over head, and set in the plane of the observer's position. Now if we watch in the same way the zenith stars from Sydney, Melbourne (Australia), New Zealand, Rio Janeiro, and other places in the South, the same phenomenon is observed—and we know (from special observations made) that the zenith stars rise from the morning horizon to the zenith of the observer, and descend to the evening horizon—and we are informed that from and within the equator, the North Pole star (Polaris) and the constellations Ursa Major, and some others, can be seen from every meridian simultaneously. On the other hand: in the South, the whole of the remarkable constellation of the Southern Cross cannot always be seen even as far north as Perth. Yet it appears that all the constellations of the South revolve around a southern centre, or pole. But, nevertheless, *the earth has no such* "pole" *or centre*, such as is maintained by globularists, and described in the extreme opposite point to N on paste-board globes and maps.

The Cross, which is to navigators a token of peace, and according to its position, indicates the hours of the night, is not always visible—nor always seen far above the horizon just as the Great Bear is at all times visible upon, and north of the Equator. Humboldt states, when he saw it, that it was strongly inclined, showing that it was rising in the East, and his account leads us to regard it as sharing in the general sweep of the stars from east to west in common with the whole firmament of stars—but in any case giving evidence that the earth is a plane.

Mr. Cook states ; "I do not know where the South Magnetic Pole is situated. We hope to find this out on the return of the Antarctic Expedition." We shall be interested in studying this portion of the recorded results of the Expedition ; and we trust that much helpful information will be the result.

But the existence of a South Centre in the heavens, commonly called a South Pole, around which the sun may turn in his appointed course in the heavens, disproves *not* that the earth is a plane, and immoveable. Nor would it disprove an extended Southern Circumference *beyond* which God has not yet permitted men to penetrate.

Note from Lady Blount

A short time ago I published a pamphlet under the above title, with a diagram on page 17, which also appeared in *The Earth* for October and November, 1903, Nos. 39 & 40, p. 275. The diagram was a representation of the globe with the equator as a straight line, and the tropics of Cancer and Capricorn also were shown as straight lines at a distance of 23½ degrees from the equator; the lines produced beyond and outside the globe were to show the sun's relative position when in the tropics, and also to show that when in these positions the direct rays of the sun cannot reach the two poles. Readers should refer to that diagram, and compare it with the diagrams which follow.

Two or three correspondents have been pleased to criticise the diagram above referred to, as not exactly representing the globular theory. So it is necessary to write this article, and to give a few further thoughts upon this subject.

Many of our readers have, perhaps, never realized how very difficult it would be to represent the globular theory exactly. It would be impossible for us to do so. The astronomers themselves never do so. High-class works on geography and map projection generally have the same defect. Why therefore should I be required to give what is not found, either in works on astronomy or in recognized standard atlases? However, I gave some approach to the theory; something which I think fairly represents the theory, while at the same time comparing that theory with some known facts. I cannot yield to the globular theory, nor accept all its wild hypotheses.

Now it has been thought by the correspondents above referred to, that I ought not to have made the lines in my diagram, representing the tropics of Capricorn, Cancer, and the equator, parallel straight lines, nor have produced the line say representing the tropic of Cancer to A (see diagram referred to). And it was thought that the diagram in *Celestial Phenomena* does not give the sun in its true position on the globular theory.

It has been said that the sun should be placed on a line drawn from the centre of the globe through the end of the line representing the tropic of Cancer as at E, and beyond in the following diagram I. This diagram I shall refer to later on.

Even then we shall find this would not be in exact accordance with the globular theory, as I will show later on. But it is thought that the line should be produced from centre E through E^1, and beyond, so that the observer at E^1 would see the sun vertical at noon. And vertical to a globularist means that an imaginary line should pass from the centre of the earth into " space," through the point where the observer is said to stand.

This then fairly represents the globularist's objection, with which I shall proceed to deal. But I have some remarks to make first, under heading of my new diagram I.

Map Projection.

Diagram I

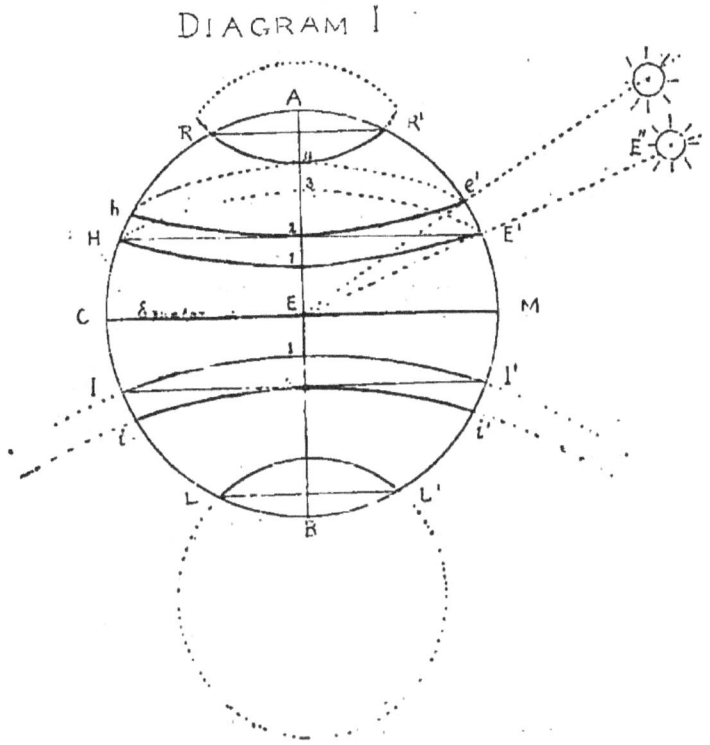

The above diagram represents the general projection given with the Map of the world, that is with one so-called hemisphere. The equator C E M is given as a straight line; the tropic of Cancer—H I E¹—as a curved line, curving towards the North; and R R1 as another curved line—the Arctic Circle—also curving towards and around the North "Pole"—A. South of the equator we have the line I1 I¹, as the tropic of Capricorn, curved inwards towards the South, that is in a direction opposite to the northern tropic; and lastly, the curved line—L L¹—round the so-called South "Pole." And we have been taught to receive this account of globular projection without any questioning. But let us examine it a little.

First let us ask what determines the points I1 and E_1 for the tropic of Cancer?

It will I suppose be replied that they are $23\frac{1}{2}$ degrees from the points C and M on the equator, measured along the curve towards the North Pole. Then if the point E^1 be $23\frac{1}{2}$ degrees from the equator, measure along the curved line M E^1 A will the point (1) also be the same number of degrees from the point E taken as being on the equator? If not, why not? If it be the same, then we have the fact cropping out, that on all maps of the world the degrees measured along a straight meridian from E to A are not as large as those measured along the curved meridian M E^1 A. And if each degree measures, as we are told it does, 60 geographical miles, then the distance in such miles from M to A, along the curve, would be 5,400 geographical miles; while from E to A the line would be only about 3,436 such miles, for anyone can see that the distance from E to A is considerably less than the distance along M E1 to A.

So that all our maps of the world are out of the truth, with respect to the size of countries measured from the equator, either towards the North or towards the South, even on the globular assumption. And the scale of miles is also wrong in this direction, as given with such maps. Also as the meridians recede from the centre to either side the scale is always altering until we reach the outside circle. But if we were to take E^{11} as the true scale for the $23\frac{1}{2}$ degrees, then $h^2 e^1$ would represent the tropic of Cancer: that is the upper curve of the two. And the same may be said of the two lower curves—I1 I^1 and i2 i^1. Which of these represents the true tropic? I leave readers to take their choice.

But notice what a difference it would make to the sun's position North. In one case the globularist would contend that the sun should be seen along the line E E^1, somewhere in the direction of E^{11}; and in the other case somewhere along the line E e1, or about e^{11}.

Readers may take their choice; for both positions are founded on globular assumptions! And both tropics, whichever we take North and South, are untrue to the lines of perspective. In the North, the Arctic Circle R R^1 would shoot off northwards into space; and in the South the Antarctic Circle, L L^1, would also shoot off into space *in an opposite direction*. But I will leave for the present globular map projection, and ask my readers to notice diagram II.

"PARALLELS OF LATITUDE."

DIAGRAM II

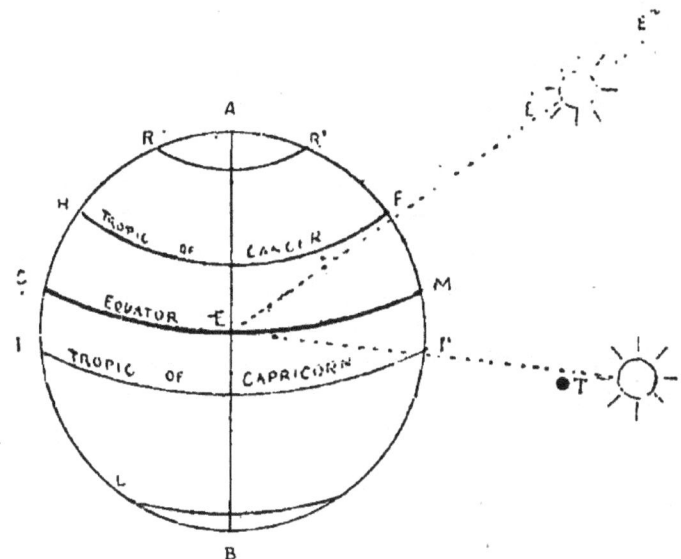

We now have briefly to consider diagram II., which is based on a more natural projection.

If the spectator be supposed to be in such a position that he can see the Arctic Circle as a curve, and not a straight line, then the other great circles should be shown in a similar position as regards their curvature. In other words, the tropic of Cancer—H E^1—should curvate towards the North; the equator—C E M—should do the same; the tropic of Capricorn—I I^1—should also curve in the same general direction; and the Antarctic Circle—L L^1—the same; all of them traversing the earth in the same general direction as the rest of the parallels of latitude.

These circles are known as "parallels of latitude," and therefore they should *all* be PARALLEL! But this would expose the position of geographers and astronomers in making the *parallels* north of the equator curvate in one direction, while the so-called "parallels" south curvate in another and opposite direction! I fear there is more trickery about the globe and its delineations than most of our readers are yet aware!

Let us now notice the relative position of the sun in the tropics. We will draw a line from E as the centre of the supposed globe, and pass it through E^1, towards the sun at E^{11}, for the tropic of Cancer. Similarly we will draw a straight line from E through I^1 towards T, for the position of the sun when in the tropic of Capricorn. How does that suit our opponents?

If someone should suggest that the diagram of the globe should be tilted, and that the "axis"—A B—should be inclined 23½ degrees from the vertical, all they need do is to tilt the paper just so much—or as much more as they like! It is more convenient for printing as we have placed it.

But we should like to know why the globe should be so tilted; and whether it is deemed more proper to tilt the "axis" 23½ degrees to the right, or to the left? Perhaps some astronomer might be able to enlighten us on this point, and give us reasons for his hypothesis. But I must pass on: these two diagrams are merely preliminary to what I have to say in connection with diagram III.

DIAGRAM III

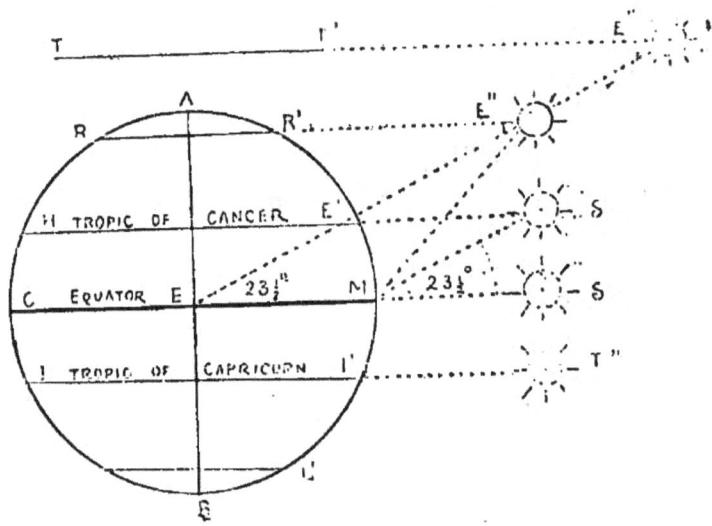

True Parallels of Latitude.

If we want one general view of the so-called "globe," with the Equator as a straight line, we must make all the lines denoting latitude, both north and south, parallel to the equator. I have so placed the leading parallels of latitude in diagram III. The central line C E M represents the equator; H E^1 the tropic of Cancer; and R R_1 the Arctic Circle. South of the equator I I^1 would represent the tropic of Capricorn; and L L^1 the Antarctic Circle. The line A B would represent the supposed "axis" of the globe, as it passes through the centre of the earth at E.

In a former article and diagram the sun was placed on a continuation of the equatorial line as at S, so that a spectator at M would see the sun on the 21st of March, directly over his head in the direction of S. But when the sun arrives at the tropic of Cancer, in the northern midsummer, it is said to be $23\frac{1}{2}$ degrees north of the equator. In other words, the same spectator at M, on the equator, would see the sun at S^1 $23\frac{1}{2}$ degrees from his former vertical position at M S. Therefore, to place the midsummer sun there corresponds with fact; but it does not correspond with the astronomical theory, so the objector says that the midsummer sun should be placed in the line E E^1 E^{11}.

That is the $23\frac{1}{2}$ degrees, they say, should be measured from the centre of the globe! Yet no one in this world ever saw the sun from that position; so that I am required to sacrifice fact to fancy; and instead of putting the sun at S^1, where it is actually seen in summer, I am asked to place it at E^{11}, as though it were seen from the centre of the earth!

To please the objector I will place the sun there for arguments' sake, and then let us notice what follows. When the sun is at E^{11}, the spectator on the equator, at M, would see it at some angle nearer to 40 degrees from the vertical than $23\frac{1}{2}$ degrees. This angle would be greater or less great according to the various distances at which the sun might be placed, but it would never some down to the required $23\frac{1}{2}$ degrees. Besides Zetetics have on several occasions given proof that the sun is not at such a great distance from the earth. But we have have placed it as far off as it was in the former diagrams, and no objection has been raised to the distance of the sun from the earth, but only to the angular position given.

Now, as a matter of fact, a spectator at the equator sees the sun at $23\frac{1}{2}$ degrees from the vertical; therefore, the sun's position at E^{11} *is not its true position.* This may be seen by making at S M S^1 an angle of $23\frac{1}{2}$ degrees; and afterwards drawing a line from M to E^{11}, making, with S M, an angle nearly twice as great!

Again, if the sun be placed at E^{11}, and we draw a line parallel to the equator across the so-called "globe," it would about coincide with the line R R^1, and so the tropic of Cancer would be super-imposed on the Arctic Circle! Would this suit our globular friends?

But why should the objector stop at E^{11}? Why not go on to E^{111}? In this case we should have the tropic of Cancer, if represented at all, outside the globe, a long way north of the North Pole itself—say at T T^1! If we must take the globular theory for our standard, we should find it impossible to properly represent it on paper. We should have to continue the line from the centre of the globe, at E through E^1, on to E^2, on to E^3, on and on for 92 millions of miles! This would be the globular theory with a vengeance.

But who could represent it? And yet some have objected because I have not been true to the theory in every detail. It is impossible to be true to it. The astronomers themselves are never true to it; nor are the geographers and those who bring out map projections. Some of the diagrams in the best astronomical works outrageously misrepresent their own theories, and the reader is thus deceived. I could give instances, but it would make my article too long, and require too many diagrams.

What I have already shown ought to be sufficient. But I will point out another fact. If the sun were a million times larger than the globe, the globe would be a mere mote in comparison to the sun, and it would be impossible for one half of it to be darkness at any time; the rays from one side of the sun would overlap or go beyond the north pole on the one hand, and the rays from the other side of the sun would overlap or go beyond the so-called south pole! Try reader for yourself. Make your diagram of the globe on a sheet of paper, and take the whole size of one wall of your chamber for the sun; then draw your lines accordingly, that is if you can.

This tremendous exaggeration of the sun's size is a mere theory of the astronomers, and is bound up with the whole hypothetical system. To make its reputed size at all harmonize with the theory, the astronomers have to push the sun away from the earth 92 millions of miles, or more, to make it look small enough! This tremendous distance and size is the basis of their theory about sun spots. "Spots," indeed!

It is a gross misnomer, too, for the astronomers to call them sunspots, when they teach that they are thousands of miles wide. Holes so large, that as one of these scientists declares "the earth could easily drop in." This same astronomer—Mr. Garrett P. Serviss—who has been writing to *The American* (New York), is reported to have said:

> "If people had telescope eyes, so that they could see at a glance things hidden from all but the astronomers, they would leave the most exciting occupation of life, and stand gazing with *awe*—if *not with fear*—at the strange sights in the sun."

Yes, they want us to look with "telescope eyes" at these things, and not with the eyes which God has given us. If we were to look at a tiny insect with a telescope eye, or rather with a microscopic eye, we could IMAGINE it bigger than an elephant; but the little thing would not alter its actual size, would it?

It is this "telescope eye" which makes astronomers see in the sun "an immense globe of blazing gas," swaying the earth and the distant planets "as resistlessly as the ocean sways a floating chip"! The spots break out "on the distorted face of the solar giant like black soot." "Their centres are yawning holes, many thousand miles in depth"! That is to the "telescope eye," which magnifies depth as well as length and breadth.

Is it not wonderful? If we only had been created with "telescope eyes." But I think that the Creator of the world has done better for us, and given us natural eyes, wherewith we may see things in their natural proportions.

And yet a weekly paper, of Jan. 14th, 1904, which professes to honour the Creator, and advocates the Seventh Day Sabbath as the memorial of Creation, publishes the above absurd sentence as "a sign of the times," and publishes it with signs of prophecy.

Doubtless such teachings are a sign of the times in which we live; when men, and even professed Christians, are departing from the old paths which were found d up n *faith in the Divine inspiration of the Bible.* If the Bible be inspired, —and we believe it is—how can Christians consistently believe such extravagant astronomical theories, in the face of the first chapter of Genesis, the second and fourth Commandments, and the many references to the order of Creation which are interspersed in the Word of God. But I must draw this article to a close.

What I have tried to show is, that the globular theory is not consistent with known facts. And I have shown this especially in the last diagram by placing the sun where objectors have thus put it. And even there we have shown that this agrees neither with astronomical theories nor with Zetetic facts. In short it is impossible to represent the globular system of the universe on paper at all, for its assumptions are so extravagant and outrageous that even the astronomers themselves cannot represent them in their own books. And what is more, it seems evident that they dare not make the attempt, lest their diagrams strike their readers as suspicious and preposterous.

THE BOOK OF JOB
IN CONNECTION WITH "SCIENCE" TRULY SO-CALLED:
OR,
DIVINE ASTRONOMY.

" Hast thou perceived the breadth of the earth."

When speaking upon the subject of the Divine Cosmogony, revealed to Moses by the Creator and recorded by Moses, very frequently have I referred to the above passage of Scripture. I have reminded my hearers that there is great depth and much food for study in this portion of Holy Writ, *i.e.*, in the whole Book of Job. But even a casual reader may clearly perceive, that the lesson which the Lord God Jehovah intended to impart to mankind in the one eight-worded question, which appears at the head of this article, is, that true knowledge and wisdom can alone emanate from the Creator Himself.

Therefore I am rejoiced to announce to my readers that I have been favoured with an early copy of a new work by Dr. Bullinger, on the Book of Job.* Part I., giving its scope, as containing the "oldest lesson in the world;"† and Part II., giving a new translation of the whole book.

It is with the latter that we are principally concerned now; because, for the first time, that wonderful and beautiful book is made intelligible to English readers.

Both the A.V. and R.V. are, as is well-known, in many places quite incoherent—conveying and containing inadequate reasonable meaning. Of course, I need hardly add that there could be no improvement made to the original.

*NOTE.—Published by Eyre & Spottiswoode, 33, Paternoster Row, E.C.
Price, 5/- cloth; 7/6 leather (suitable for presents).
To be obtained also of the Ed. of *The Earth*.
†NOTE.—For the justification of the translation, where it differs from the A.V., our readers must consult Dr. Bullinger's work itself.

Dr. Bullinger's new translation is based on six great principles, which make it quite unique. (1) It is *Rhythmical*; (2) It gives the *Structure* of the whole and every portion. (3) It notes and translates all the *Figures of Speech*. (4) It is *Idiomatic*. (5) It is *Critical*, giving the notes from Dr. Ginsburg's Hebrew Text. (6) It distinguishes and preserves the various *Divine Names and Titles*.

The work will specially interest readers of *The Earth*, for they are aware how much the Book of Job has to say about Astronomy, and about the Earth.

Dr. Bullinger says however, that while the book contains a divinely inspired account of what the various speakers said, it does not follow that what they said was inspired. He passes over, therefore, the references to Astronomy made by Job and his friends, for though we may regard them as representing the current beliefs of themselves and their times, yet they cannot be cited as Divine utterances.

It is quite a different matter, however, when we come to the words of Jehovah Himself, in chapters xxxviii., xxxix., and xl. There we have the Creator of Heaven and Earth speaking of His own works. He who created all things is the only One who knows all about them and is able to reveal them.

It is wonderful to read His own words about His own works, and Dr. Bullinger has succeeded in making the words live before our eyes. We sometimes hear of young ladies' circles for reading Tennyson or Browning, &c. We shall be surprised, if among Christians there are not circles formed for reading this wonderful book, if only for the beautiful English in which it is presented.

When we come to Jehovah's own address to Job, we see that it is divided into two parts. Each is followed by an utterance of Job: the first time to say he could not answer; and the second time to answer to some purpose.

The two parts of Jehovah's great address are about two great departments of His creative work. The first about the *Inanimate* Creation; and the second about the *Animate* Creation.

It is with the former of these that the readers of *The Earth* will be specially interested.

As we read it in the A.V., it seems a jumble—and the various subjects seemed to be mixed up in hopeless confusion:

but Dr. Bullinger shows us the Structure, and we see at once, as it is presented to the eye, how perfect and yet how simple it all is. We give

THE STRUCTURE OF JOB XXXVIII. 4—35.

THE INANIMATE CREATION:

(JEHOVAH'S WISDOM EXHIBITED IN OUTWARD ACTIVITIES).

A | 4-7. The Earth.

    B | a^1 | 8-11. The Sea.   ⎫ Things
           b | 12-15. The Dawn.   ⎬ pertaining
                                                                         to the
        | a^2 | 16-18. The Springs of the Sea. ⎭ Earth.

        C | 19-. Light.   ⎫ Things pertain-
                                             ⎬ ing to both
                                                        the heavens and
        C | 19-21. Darkness.   ⎭ the earth.

    B | a_1 | 22, 23. Snow and Hail.   ⎫ Things
           | b | 24-27. Lightning.   ⎬ pertaining
                                                          to the
        | a^2 | 28-30. Rain, Dew, and Frost. ⎭ Heavens.

A | 31-35. The Heavens.

It will be seen that first we have the Earth (A. *vv.* 4-7.) This is balanced by the Heavens (*A. vv.* 31-35). Then, following the Earth, we have (B *vv.* 8-18) things pertaining to the Earth. Then again alternate, Liquids and Light.

Then in the centre we have Light and Darkness (C & *C*, *vv.* 19-21).

Then, balancing the things of the Earth, we have (*B, vv.* 22-30) the things pertaining to the heavens, leading up to the Heavens themselves, (*A. vv.* 31-35).

We will now take the liberty of giving our readers the whole of the first part of Jehovah's address, about

THE INANIMATE CREATION.

They will not fail to notice two great points :—

(1) How absolutely incompatible is the *theory* of Evolution in the face of this Scripture. One *must* go ; and the theory of Evolution must vanish, before the fact of Creation.

(2) How plainly the gropings of scientists stand exposed. Time was when they thought they knew so much, that, like school-boys, they thought "the Book was wrong." But to-day—every fresh discovery of fact shows the falsity of their old exploded theories, and the wondrous perfection of the Divine Word. Once, they thought they knew all about "Light," and they laughed at the statements in this chapter : Now they find they know so little they are beginning to discover something of the meaning of His words, when Jehovah speaks of "the parting of the light."

But we must now give the translation itself.

A, (*vv.* 4–7).—THE EARTH.

4. Where wast thou when I Earth's foundations laid?
   Say ; if thou know, and understandest it.
5. Who fixed its measurements (for thou wilt know)?
   Or, Who upon it stretchèd out the line?
6. On what were its foundations made to rest?
   Or, Who its corner-stone [so truly] laid?
7. When all the morning stars in chorus sang,
   And all the sons of God did shout for joy.

a¹, (*vv.* 8–11).—THE SEA.

8. Or, Who fenced in with doors the [roaring] sea,
   When bursting forth from [Nature's] womb it came?
9. What time I made the clouds its covering robe,
   And darkness deep the swaddling band thereof?
10. When I decreed for it My boundary,
    And set its bars and doors, and to it said
11.   "Thus far—no farther; Ocean, shalt thou come;
    "Here shalt thou stay the swelling of thy waves?"

b, (*vv.* 12–15).—THE MORN AND DAWN.

12. Hast thou called morning forth since thou wast born,
    Or, taught the early Dawn to know its place?
13. [Bid Morn] lay hold on outskirts of the Earth,
    [Taught Dawn] to rout the lawless from their place?

14. [Bid Morn] change earth as clay beneath the seal,
    [Bid Dawn] enrobe the beauteous world with light?
15. Thus Morning robs the wicked of their prey;
    And stays, arrested, the uplifted arm.

a^2, (vv. 16-18).—The Springs of the Sea.

16. The fountains of the Sea; hast thou explored?
    Or, Hast thou searched the secrets of the Deep?
17. The gates of Death? Have they been shown to thee?
    Or, Hast thou seen the portals of its shade?
18. The utmost breadths of earth: Hast thou surveyed?
    Reply, if thou hast knowledge of it all.

C and C (vv. 19-21).—Light and Darkness.

19. Where lies the way that leads to Light's abode?
    And as for Darkness: Where's the place thereof;
20. That thou shouldst bring each to its proper bound
    And know the paths that lead unto its house?
21. Thou know'st [of course]: thou must have been then born;
    And great must be the number of thy days.

a^1, (vv. 22, 23).—Snow and Hail.

22. The Treasuries of Snow: Hast thou approached?
    Or, Hast thou seen the store-house of the Hail?
23. Which 'gainst a time of trouble I have kept,
    Against a day of battle and of war.

b (vv. 24-27).—Lightning and Thunder.

24. By what way part themselves the rays of Light?
    How drives the East-wind o'er the earth its course?
25. Who cleft a channel for the floods of rain?
    Or passage for the sudden Thunder-flash?
26. So that it rains on lands where no one dwells,
    On wilderness where no man hath his house;
27. To saturate the wild and thirsty waste,
    And cause the meadows' tender herb to shoot.

a^2, (vv. 28-30).—Rain, Dew, and Frost.

28. The Rain: Hath it a father [besides Me]?
    The drops of Dew: Who hath begotten them?

29. Whose is the womb whence cometh forth the Ice?
And heav'n's Hoar-frost, Who gave to it its birth?
30. As, turned to stone, the waters hide themselves;
The surface of the deep, congeal'd, coheres.

A (vv. 31–35).—THE HEAVENS.

31. Canst thou bind fast the cluster Pleiades?
Or, canst thou loosen great Orion's bands?
32. Canst thou lead forth the Zodiacs monthly Signs?
Or, canst thou guide Arcturus and his sons?
33. The statutes of the heavens: Know'st thou these?
Didst thou set its dominion o'er the Earth?
34. The clouds: to them canst thou lift up thy voice
That plenteousness of rain may cover thee?
35. Canst thou send lightnings forth, that they may go
And say to thee ' Behold us! Here we are?

I wish I had space to give more, but the above will serve as a specimen of Dr. Bullinger's work; while it will be of special interest to all Zetetics, as bringing out the grandeur of Jehovah's words as He speaks of His own works.

THE ANTARCTIC EXPEDITION.

The narrative of the latest Antarctic Expedition, under Captain Scott of the ship *Discoery*, is on the table before me. I read this report with the greatest avidity. My first, and, indeed, my chief, anxiety was to find some genuine evidence of the midnight sun down south, in keeping with stories which were circulated by the men of the ship *Morning*, on their landing in England.

At the time that these curious reports were circulated, I judged them to be thoroughly irresponsible, and now that I have read Captain Cook's highly responsible narrative I remain in just the same state of mind as before, in that, from first to last there is not the faintest mention of any Midnight Sun, or of any *continued* daylight,—but the whole phraseology of the report is distinctly in opposition to any such conclusion, and I find the 24 hours invariably divided into the usual intervals of morning and evening, noon and night.

The sun also is spoken of as going and coming, and as rising, and as being *below the horizon*, and consequently as *setting also*. This appears to be about September, and is of the utmost importance, in that the point of sunrise is clearly indicated in one instance, showing that the sun actually rose about East North East—northerly—and lit up the sides of the mountains facing *north*. This point of sunrise is of the utmost importance, because, if the earth were a globe, the sun must have risen about due East, and the eastern sides of the mountains would have been lit up " by the glow of the sun" when " still below the horizon ; " whereas the passage goes on to say that " the other sides were *dark and shadowy.*"

The word "*shadowy*" has also great weight in it, because the shadow would of course be to the South, and just opposite to the point of sunrise. Here then—and in the words of the authorities themselves—we have clear proof that the sun not only rose above the horizon, but also,—and this is the crucial point—rose in addition to the *Northward*, and not East nor South as it must have done had the earth been a revolving and spinning globe and the sun itself a *stationary object*.

The sun is spoken of, on this occasion, as " *returning*." What does returning mean ? Clearly, that the sun was not stationary, but on the *move*, and that during the Antarctic winter the sun had moved into northern latitudes, and was, in September, "*returning*" south again, and thus made the Antarctic summer, or the resemblance of a summer, which is about as much as they really get in those extreme latitudes.

Further, I have now very much pleasure in pointing out that these extreme latitudes are possible only on one condition, and that condition is that the sun's path, during our winter months, is in fact a sharp ellipse. Of this I am certainly positive, and without *it* the latitudes down south could not, and would not, exist. It may encourage readers to believe me if I say how I obtained the knowledge of the sun's *daily* path being an ellipse during our winter months.

It occurred this way. When fifteen years of age, I went a voyage to Australia in my cousin's ship, *The Albemarle*. We had a first officer, named Mr. Merritt, and he was the smartest of the smart; the best trimmer of canvas I ever met with, and a gentleman and philosopher. When taking the sun, at 12 (midday), he would occasionally talk to himself aloud, and this more frequently as we approached Australia. As I was always looking on I used to hear his audible but no doubt (as he thought) private chat. His method of "taking the sun" was purely his own, and very clever. He usually commenced about ten minutes before noon, and kept constantly turning to the compass, so as to note the bearings (as I naturally suppose) of the ship's head and the sun.

On some of these occasions he would say: "The sun's path must have a great ellipse on it." This he would repeat several times nearly every day, until the sun came almost plumb overhead and the observation almost impossible.

Now reader, you will understand how I know that the sun's daily path must have a great ellipse on it; and many years of plan making confirms me in the belief that it must be so. And it is this ellipse alone which admits of southern latitudes in the direction of New Zealand; without it those latitudes are quite impossible on a Plane Earth, as the *distances* would be *preposterous* and convey the idea that the Plane Earth is nothing more than the globe flattened out.

This, of course, is not the case, and I am exceedingly obliged to Lady Blount, whose highly superior and clear discernment enabled her, without human aid, to use her God-given judgment and to abandon a former map of the earth which was a flattening of the globe, pure and simple. It looked very nice, I admit, but was hopelessly astray for all that, and only served as an amusing bogey for the scientific members of the Royal Geographical Society to poke fun at; and planists have to thank her ladyship for upholding my skeleton, but possible, plans in lieu of that costly affair. By means of my plans we shall, no doubt, arrive at something conclusive—finally genuine and acceptable—but perhaps, not in my life-time, as I well know the veritable difficulties of the case, and also know that hurry is not *speed* in this matter, which is *too big* for one life-time.

I do not wish to fall into the same error of *immensity* as the astronomers. Their whole idea of a God-glorious Creation is *immensity*—impossible and *impracticable* immensity —whereby they overlook the kernel to eat the shell. Speaking of our earth (the kernel), they say it has no significance whatever, but is dwarfed by millions of stars, suns, and other worlds, which apparently create themselves. In this they are very much mistaken, as our earth is by far the *largest*, which in itself is not a bad bid for significance. The sun is undoubtedly only 32 miles in diameter, and this can be so easily proved, that I wonder at their persistence in a hopeless and untenable position.

To recur more directly to the account of the Antarctic Expedition. There is positively no mention of, nor even suggestion of, the continuous daylight so often claimed for the Antarctic regions; but morning and evening, noon and night, is the repeated language of the narrative throughout.

Fine nights may have occurred now and then, but such must have been caused by the moon, which appears to leave these latitudes every month for a few days, and probably goes away south in the interval.

The "*flat horizon*" is also spoken of, and at the same time geographical miles. These expressions are inconsistent, one with the other. The horizon was of course *flat*, and the miles were also *statute miles* (as are all miles on maps). The map makers know this quite well.

I have now said sufficient to show that the Antarctic Expedition furnishes further proof that the earth is a plane surface; but sunset still admits of a few explanatory remarks. This feature is very important. The sun, without doubt, sets away to the northward, and not southerly nor due West, as it would do on a globe; thus, to the ship *Discovery*, it might make a Midnight Sun, much in the same direction as the Midnight Sun when seen from the North Cape—the main difference being that the sun is further to the southward, and therefore invisible to a spectator at North Cape, Norway. But for this difference of distance the two Midnight Suns —North and South—might be described as visible almost from the same place.

This is very fairly shown on my diagram. The midnight sun at North Cape is lost to view, owing to increasing distance from the observer, about August, and it gradually increases its distance still further South—till Christmas Day —but, owing to the ellipse of the sun, its return path Northwards from the *Discovery* crosses the meridian of Greenwich again, and eastward from the locality of the Midnight Sun as seen from North Cape.

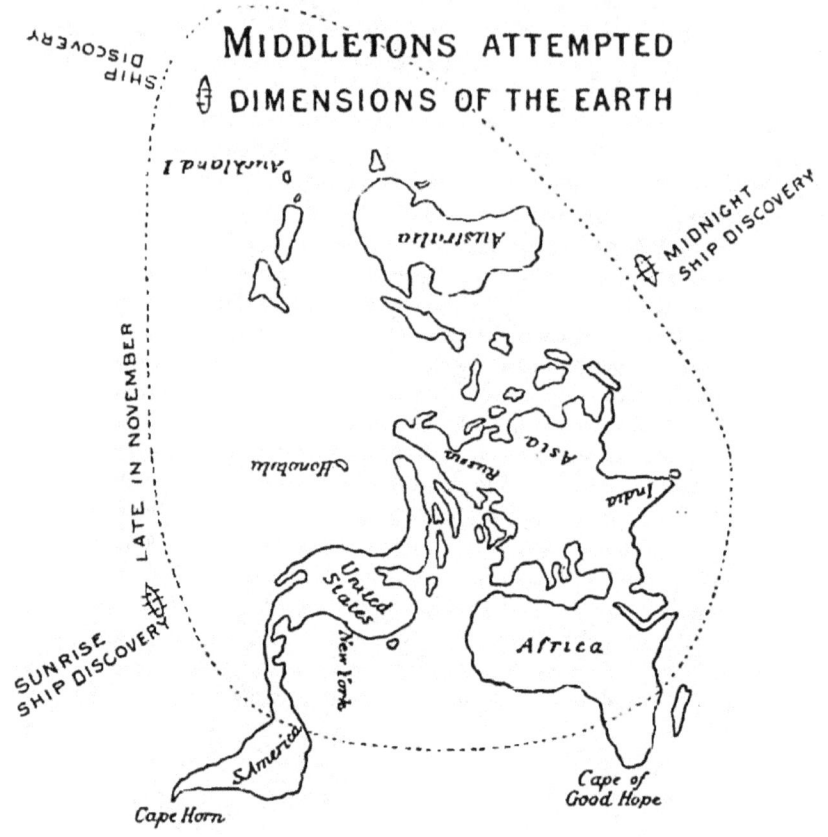

The elliptical path of the sun most certainly invites the closest observation, such as might perhaps be acheived from the observatory of Ben Nevis. This observatory I am told cost £5,000 to establish, but is now to be dismantled—may I ask why? Is it found to be too inconvenient, as telling tales? Or what other reason is to be put forward for the dismantling of an observatory, just calculated by its position to be of the highest service? Have they actually seen the Southern Midnight Sun from it? and is this the reason for its abandonment? We of the Planist School have no terror of any hidden secrets which such an observatory would disclose. As a set off to this abandonment, will some rich planist try a captive balloon from North Cape—to go up like the French Balloon—say a mile high?

Depend upon it, the aeronauts in such a balloon will see the Midnight Sun, such as may possibly be seen by South Poleites, in common with the crew of the *Discovery*.

E. E. MIDDLETON.

EXTRACT FROM DR. HASTINGS' BIBLE DICTIONARY.

"The accompanying diagram will enable the reader to comprehend the ordinary conceptions of an ancient Semite (whether Babylonian or Hebrew) respecting the universe in which he lived. The writer of this article sketched this outline from a study of numerous Old Testament passages,

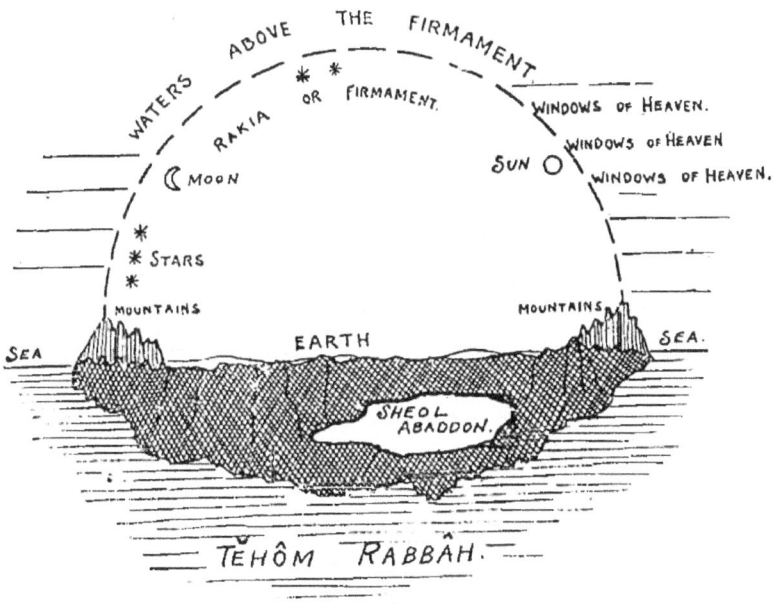

about twelve years ago, and found in Jensen's "Cosmologie der Bab," published in 1890, a diagram almost identical in character, descriptive of the universe according to Babylonian conceptions, and based purely upon the data of cunieform inscriptions. In both we have a heavenly upper ocean, and in both the earth was conceived as resting upon a vast water-depth or *Tehom* (called also in Babylonian, *apsu*). The Hebrews thought of the world as a disc [circle, of *Isa.* xl. 22); and to this earthly disc corresponded the heavenly disc (called also circuit, of *Job* xxii. 14, *Pr.* viii. 27).

Beneath the world rested the unknown and mysterious *Tehom Rabah* (of the language of *Ps.* xxiv. 3). The Flood not only descended through the windows of heaven but also ascended from the deep nether springs, called "springs of the great *Tehom*" (*Gen.* vii. 11), which were cleft open. These deep springs were accordingly called *Tehomoth* (*Pr.* iii. 20), and were believed to communicate through the depths of the earth, by means of passages, with the great *Tehom* which lay below. In a striking passage in *Amos* vii. 14, the prophet portrays a judgment in which the fire of Jehovah will devour this great water-depth. Within the earth itself lay the realm of the departed, Sheol or Hades."

The word "globe" does not occur in the Bible at all.

Himmel und Erde nach der Vorstellung der Babylonier.

J. R.

A PRACTICAL PROOF THAT THE EARTH IS NOT A GLOBE.

The annexed drawing was made especially for *The Earth* by an artist who, in 1892, drew in pencil the view from nature. The mountain represented is one of the Andes, in Venezuela (S. America). The distance from the end of the mountains to Ucuare is about 20 miles. The mountain referred to is 400 metres above the level of the sea; the highest point of the range is 500 metres. The distance from Maiketia (which is less than 10 feet above the level of the sea) to Ucuare is 100 miles). The mountains—which can be clearly seen from Maiketia mountain—are 80 miles distant at the Ucuare end. If the earth were a globe the amount of curvature would prevent the mountain being seen at a distance of 50 miles. When the height of an observer is 10 feet his line of sight would be a tangent at a distance of about 4 miles. This gives about 1,411 feet for the dip. Taking the height of the lower peak at 400 metres, or 1312 English feet, the peak of the mountain would be below the horizon about 99 feet. And taking the higher peak, 500 metres, or 1640 feet, and subtracting the dip for 46 miles, namely 1,411, this mountain point would be above the line of sight about 229 feet. So that the lower peak would be about 2539 feet below the line of sight if the earth were a globe; and similarly the higher peak would be about 2,211 feet below the line of sight. But as both peaks have been seen and drawn by my friend under the conditions named, this affords another clear proof that the earth is *not* a globe such as the astronomers hypothecate.

WHY THE EARTH IS NOT A GLOBE.

If the earth be a globe there can be no doubt that the air, or atmosphere, must rotate with it.

When travelling by the train, at the rate of 40 or 50 miles an hour, we have doubtless, on putting our head or hand out of the window of the carriage, felt and been surprised at the force and resistance of the air, and wondered what the force and resistance of the air would be if "the earth's swift and numerous motions," were a fact.

Could any conceivable thing resist the tremendous force of these "orbital and axial motions" ascribed to the earth?

Could anything stay on the earth, light or heavy; the mountains or hills; the seas or oceans; even the air, and things floating in the air? Nay. Nothing could possibly abide nor find a resting place upon it, if the earth and air were in opposition.

But notwithstanding all this, how can we account for the wind blowing in opposition to the earth's motion, and in every direction under heaven, carrying smoke, and dust, even the air itself, and very light and heavy things, and substances that can float or move with the air?

Of course, this is not all that could be said upon the subject; many proofs can be found, and various arguments can be produced. But there is only one thing to account for such an anomaly, and that is that the earth is *not* a globe, and has no motion at all.

And the truth of God's Word comes out *clearly* and *unshaken*, that God has fixed the pillars of the earth, and established the earth that it cannot be moved— and that it has foundations, and that its Builder and Maker is God.

Feb. 2nd, 1904. "TRUTH."

Extracts from an article, entitled
ASTRONOMY OF THE BIBLE : by Prof. Lewis Swift, F.R.A.S.

"The Bible is not a work on astronomy." [Bible astronomy is the only true astronomy.—Ed.] The sun is one great reservoir of heat and light to the earth, and yet, strictly speaking, neither comes from there; nothing in fact, but cold dark waves of the all pervading ether. How can heat reach us from the sun, passing as it must through 93,000,000 miles of space, probably a hundred degrees below zero?

"These waves pass through space without heating or lighting it, and plunge into our atmosphere without heating or lighting that except slightly; but when they strike the earth and are reflected quicker, if possible, than a flash of lightning, they are transformed into both heat and light. Light is the most rapid moving principle in nature, equal to 186,300 miles a second, or, while a person would say the words 'Christian Herald,' it would revolve seven-and-a-half times around the earth.

"I advise the reader to stop a moment, and reflect on what is involved in the mighty idea of a circuit of seven times and-a-half in one second. Light reaches us from the moon in one-and-three-fourth seconds; from the sun in eight minutes; from the nearest star in four years; from the Pole Star in forty-eight years; and from the most distant stars, that our great telescopes can see like atoms of diamond dust floating in the sunbeams, the light must have been many thousand years on its journey.

There are other waves which are a blessing to the human race; waves of sensation. which move very slowly, only about 100 feet in a second, producing pain, and taste, and smell, and pleasure, and hatred, and love, &c , but no taste or pain is felt till the waves reach the brain. If a babe, in its cradle, had an arm 93,000,000 miles long, and should insert its finger in the sun it would not know that its finger was burned until after the lapse of 140 years."

[This "babe" illustration—after Sir R. Ball's style—is too babyish! My space is too precious for me to insert any more of this fallacious rubbish. Nevertheless, I am grateful to Mr. H. Murray Bing for his kindness in copying out the article and sending it from America.—Ed.]

SOUTH AND NORTH.

T. H. A. queries the statement made in *The Earth* for December and January, p. 307, respecting the shining of the sun at places south of the equator; and he informs me that a friend of his—the captain of a steamer loading at Rosario, in the Argentine Republic—had been observing the different times of the day; and he writes under date, January 26th, 1904:

"I can't say I have seen the sun shine on the south side of the houses; in fact, I can't see how it is possible in this latitude, for the sun rises E S E, due North at noon, and sets about W S W. On Midsummer day one could almost say it shines all round; for the sun is nearly right above you in this latitude, so that the houses throw no shadow whatever."

Though this is apparently neutral evidence in respect to the sun shining on the south side of buildings, yet T.H.A. says the writer seems to confirm what Mr. Cook, of Perth Observatory, W. Australia, writes, viz: "that in the early morning, and late afternoon, the sun shines upon the south side of buildings between September 23rd and March 21st, at places more than $23\frac{1}{2}$ degrees south of the equator; the sun never shines on the south side at mid-day. At places between the equator and latitude $23\frac{1}{2}$ degrees S, the sun shines on the south wall throughout the day at midsummer (December)."

T.H.A. says his friend's statement appears to be at variance with the words printed on p. 301 in *The Earth*, viz : " The sun, without doubt, sets away to the northward, and not southerly, nor due west, as it would do on a globe," etc. This is an excerpt taken from an article on the "Antarctic Expedition," and is in keeping with the context, and with what actually occurs. The evidence of our senses tells us that the motion of the sun is a visible reality—for if it be observed from any latitude a few degrees north of the tropic of Cancer, and for any period before or after the time of southing, i.e., passing the meridian, it will be seen to describe an arc of a circle. By way of illustration: if I watch the sun's progress on any day during the summer months, say at the head of the new pier at Brighton, the sun's first appearance above the horizon will be observed to be at a point considerably to the north of East, or a line drawn at right angles to north and south, and it will be seen to ascend in a curve southwards until it reaches the meridian, thence descending in a westerly curve until it arrives at the horizon, setting considerably to the north of West, not southerly or due west, as it would do on a globe.

T. H. A. is exercised in his mind with reference to the remark on p. 19 of my pamphlet, *Celestial Phenomena*. He does not see how the stars characteristic of the southern and northern parts of the earth can revolve round their respective centres, and yet that the Southern Cross should be visible from every known and habitable point of the southern hemisphere.

Mr. Cook, writing from the Government Observatory, Perth, W.A., says: "there is a point in our sky round which all stars appear to revolve. There is not any star in this exact spot; but there is a small star (Sigma Octantis) situated very close to this spot, closer, in fact, than your Polaris is to your North Celestial Pole. The sun and moon always appear to revolve round this point the whole year through. Of course some allowance must be made for their gradual change in declination: i.e., their motion is more in the form of a spiral."

T. H. A. asks: "If the constellation called the Southern Cross revolves round its own centre, and *that* not the same as the northern centre, how can the southern Cross be seen, say at the opposite side of the plane earth?" Mr. Cook says that three stars of the Southern Cross never set; the fourth just goes below the southern horizon for a short time each day. The altitude of the South Pole is exactly the same as the latitude of the observer's locality, and if the distance of a star from the Pole exceeds this, the star will be below the horizon at its lowest transit. Thus in Cossack, lat. 20 deg. 40 min. S., the whole of the Cross will disappear as it swings round below the pole. The circum-polar constellations (meaning those which never set) depend upon the latitude of the observer. Octans, the constellation in which the Pole is situated, is truly circumpolar to us in Perth: i.e., all its stars are constantly above the horizon, or in sight at night time; each star describing a circle daily round the pole." But Mr. Cook makes this admission: "I do not know where the south magnetic pole is situated. We hope to find this out upon the return of the Antarctic Expedition."

Mr. Cook writes from a globularist standpoint; at the same time I believe in his honesty of purpose; and I am much indebted to him for photos of instruments used in making plain the astronomical instructions given in the supplement to the Education Circular; when he tells us how to find the sun's path in the sky for a particular day, and how to find a point in the sky which is the centre of the circle, it appears to me that he is describing a moving sun—*not* an earth moving round the sun—for to speak of the sun's path implies that the sun moves in that path; in fact he heads paragraph 16 with these words:

SUN'S ANNUAL MOVEMENT.

In this paragraph we are informed that when the sun's position in respect to the stars is measured by special instruments, it is found that the sun is steadily moving eastward among the stars, taking exactly a year to complete one revolution. I accept this statement as the statement of a matter of fact. If the sun appears to move, as astronomers confess it does so appear, why should we not believe that it does actually move? Some reasons ought to be given.

A greatly esteemed friend much desires that the kind account of myself (The Ed.) and my work in connection with *The Earth*, which appeared, with my portrait, in *Home Chat* and other papers, shall be reprinted in *The Earth*. And a great many others have made the same request; but I regret that I must disappoint my kind friends, as lack of space alone would preclude the possibility of doing so.

A mining engineer, just home from Columbia, S. America, amongst other things told me, that the cutting through Panama for the canal revealed the fact that the Atlantic and Pacific Oceans were on precisely the same level, and that no locks were required for the canal. He also told me that he had travelled 800 miles down a river in Colombia in an open boat, and that there was no danger from cataracts—but only from Alligators. He says that he has travelled all over the world, and that the countries are smaller than represented on the globe.
E. E. M.

THE EARTH: IS IT A GLOBE?

"The planetary system,' said Humboldt, in its relation of absolute magnitude, relative position of the axis, density, time of rotation, and different degrees of eccentricity of the orbits, "has, to our apprehension, nothing more of natural necessity than the relative distribution of land and water on the surface of our *globe*, the configuration of continents, or the elevation of mountain chains." No general law in these respects, is discoverable either in the regions of space or in the irregularities of the crust of the earth.

The foregoing describes, in Humboldt's language, the condition of the orthodox planetary system of his day. These remarks apply with equal force to the teachings of present-day astronomers of the globular school who call themselves scientists (from *scio :* " I know,") science (*sciens :* "knowledge "). But orthodox astronomy is admittedly a *theory*, founded upon speculation ; and the moment speculation becomes knowledge then speculation ceases. Therefore, it is a misnomer to designate such theoretical astronomers " scientists." On the other hand " plane-earthists" denominate themselves " Zetetics," from the Greek *zetéo :* " I seek, search for, investigate, inquire into " ; *zetetes :* "searcher, inquirer." Zetetics are, consequently, those who do n)t take for granted the theories which may be offered to them, but make investigations to see whether these things be true or not, and, if not, to endeavour to arrive at the truth. They therefore investigate the common statement that "the earth is round or spherical, like a ball or an orange, because ships have actually and repeatedly made the circuit of the *globe*." They naturally ask : " Are these deductions in accordance with facts ? " Vessels and steamers continually go round the Isle of Wight and the Isle of Man ; therefore, *if* the earth be a " globe " " *because* vessels and steamers go round it," then (by the same line of reasoning) the Isle of

Wight and the Isle of Man are globes *because* vessels continually go round them. Now we have positive evidence, founded upon personal knowledge, that the Isle of Wight and the Isle of Man are *not* globes; hence it follows that the earth is not necessarily a globe because ships have gone round it.

"If the earth is flat, when we go to the edge we may tumble over."

We have been met by this surmise so many times that we feel it is what comes naturally to the mind of the ordinary man, or woman, who has been brought up in the belief that the earth is suspended in space and, therefore, if it were a plane there would be the danger of our walking over the edge, and falling into space. But no one has been able to get to the edge. Beyond a certain point in the Antarctic regions God has not yet permitted men to go; therefore, no one has been placed in such a position where it would be possible to "tumble over." Commander R. F. Scott, R.N., and his party of explorers—who sailed from New Zealand in December, 1901, have undoubtedly penetrated further into the interior of extreme southern regions than any previous explorers, and after three years of courageous exploration their description of the furthest limits of the unknown continent is, that it was "found to be a bleak plateau, rising 9,000 feet above the sea, and stretching *interminably* to the south."

In travelling east or west we go round a centre. In going due south we may make for a certain point; but the voyager, when he has pushed his way as far as possible beyond all known land is stopped by mountains of ice; and he finds himself beyond the regular influence of the sun's rays, or "beyond the limits of light and darkness." And it is clearly evident that no man could possibly continue to travel due south much further than Commander R. F. Scott, reached, because he could not exist far beyond the limits of the sun's rays."

Zetetics when *instituting* inquiries as to the form of the earth, and the phenomena which pertain to the world or all that they can see, are met by a number of questionable theories, but on examining the *rationale* of these theories, which need examination, I shall as far as possible, use plain language.

On the threshold of the argument as to the earth *not* being a globe, we are told that wherever we look—either skyward or seaward—there is an appearance of an arc or the segment of a circle. Ships disappear and come in sight as if they were moving over a sphere—and it is asked: "Can we trust our senses?" Truth replies: "Yes;" but it is necessary that we should not allow our God-given senses to be deceived either by men or demons.

There are men living, people calling themselves "scientists," who would have us believe that they can fathom and define the ways of Eternal Providence better than God Himself—and have them "set in a note, learned and conned by rote," to cast in the teeth of all who would dare to think otherwise. They love to get hold of "a principle," to use their own language, and push it home to its ultimate conclusion. That is being thorough according to their form of judgment. It is being consistent, and they worship consistency. They echo the words of one of Shakespeare's characters:

"Consistency, thou art a jewel!"

It does not occur to these people that it is more important that their deductions shall not be false to facts, and that consistency in pursuing a fixed course which has diverged from good common sense and sound reasoning, is wrong-headed and guilty error. They will not admit that, as a matter of fact, there are even very few wise general principles that are universally true in the sense that they can be made into rigid rules; hence the necessity of bringing all theories to the tests of experience and experiments, and the danger of leaning on general rules for the understanding of the great problems of the world.

The theory that the earth is a globe, whizzing through space at the rate of 63,000 miles every hour, with many additional motions, is based upon an assumption in direct opposition to our God-given senses. If the earth moves at this extraordinary speed, how is it that we do not perceive that it moves? The pseudo-scientists being wedded to the globular theory, instead of reconsidering their deductions, provide us with figures the magnitude of which no intellect can grasp, and the great majority accept their *dicta* because it is considered "the proper thing to do."

If we travel by land or sea from any part of the earth in the direction of any meridian line, and towards the northern central star Polaris, we come to a region of ice, where the star, which has acted as our guide, is vertical to our position, *i.e.*: directly above us. This is not necessarily the centre of the *earth*. We may describe it as a vast lake sea, about 1,000 miles in diameter, surrounded by an immense barrier of ice close upon a hundred miles in breadth. From this region we can trace outlines of lands which project from it, and the surfaces of which are above the water—and we see it demonstrated that the present form of the earth ("dry land" as distinguished from the waters of the "great deep") partakes of an irregular mass of islands, capes, and bays, terminating in huge bluffs (or headlands) projecting in a direction away from the north, and principally towards the south; and this is in accordance with the teaching of the Holy Scriptures. By sailing with our backs continually to the central star Polaris we arrive at another region of ice; in fact, upon whatever meridian we sail (keeping north behind us) we shall be ultimately checked in our progress by vast cliffs of ice acting as barriers. There is evidently a boundary of ice encircling the southern seas, with irregular masses of land stretching out towards the south, engirdled by packs of ice and frozen barriers, the depth and breadth, and whole extent of which have not been ascertained. How far the ice extends has not, even by the recent *Discovery* expedition, been approximately discovered. The earth and sea's extremity has not been penetrated by the most powerful telescope yet invented, and the "beyond" is still hidden in gloom and darkness from the human eye.

What the superficial extent or magnitude of the earth is, from northern regions to the south, can only be indefinitely stated so far as actual measurement is concerned. Of course we know that, in 1866, in laying the Atlantic cable from the Great Eastern steamship, the distance from Valentia, on the S.W. coast of Ireland, to Trinity Bay in Newfoundland, was ascertained to be 1665 miles, and we before knew that the longitude of Valentia is 10° 30' W., and Trinity Bay 53° 30' W, 43° representing the difference of longitude between the two places, and the whole distance round the earth being divided into 360°, if 43° are 1665 nautical miles (equal to 1916 statute miles) *then* 360° will be equal to 13,834 nautical miles (15957 statute miles).

THE MUTUAL RELATIONS OF THE SUN AND EARTH.

(A Review.)

The above is the title of a book which has lately been published. Although the writer of this book is a globularist, he differs from the teaching of orthodox Newtonianism in such a marked manner, that Zetetics can but welcome it because of its reasonableness. It deals with that which has been a most difficult problem to many, viz: "What is the actual size of the sun?"

The book is written with clearness and common-sense, the matter being put in a reasonable form. The object of the writer is to prove that the actual size of the sun corresponds with the geographical area of vertical solar rays, and we are reminded that the troops engaged in the South African War, leaving the south coast of England in north latitude about 50°, were carried to Cape town in south latitude 33°, a distance of over 5,000 geographical miles.

Sailing over the northern tropic they crossed the equator, sailed over the southern tropic and into the south temperate zone; when the officers of the transport took their daily observations of the sun's altitude it was noticed that the sun bore south and that the spectator's shadow was projected north. After a few days, the sun still bore south, but at a higher degree of altitude, and the shadow projected north was shorter. The sphere of the heavens seemed slowly to rotate from east to west, the northern constellations seeming to sink lower and lower in the heavens, new constellations appearing above the southern horizon. In about a week after leaving England the transport arrived near the tropics, a zone (or belt) circling at $23\frac{1}{2}°$ of latitude on either side of the equator. On some parallel of latitude in this zone the sun is vertical—*i.e.*, the sun at noon is in the zenith. Whenever that particular parallel is passed the sun's position is passed.

A quotation is made from a newspaper correspondent, who wrote: "On that Sunday we passed to the south of the sun. At noon on that day the ship was in Latitude $14°\ 30'$ north, and the sun in latitude 14° 28'. Henceforth we were to look at him......with our back to the south instead of north." Here at noon the ship was in the area of vertical rays; the sun was in the zenith, and the spectator's shadow was projected downwards, and hence invisible; but by the next noon the ship passed south of the area, the bearing of the sun and the projection of the shadow being reversed—*i.e.*, instead of viewing the sun with the back to the north, it was viewed with the back to the south.

If (as has been stated) the volume of the sun is 1,407,000 times that of the earth—how would it be possible for an observer on the earth's surface to pass the position of the sun as recorded by Dr. Robertson? The utter impossibility is self-evident, and it is palpable that compared with the earth's magnitude, the sun is a small body. The doctor demonstrates that the whole area of the sun is contained within 32 miles. The appearances observed in passing the sun's position are somewhat similar to those seen when we pass beneath an electric lamp in the street. If the direction in which one passes under one of these lamps be from north to south it will be observed that when north of the lamp's position, the light will be seen to bear south and our shadow to be projected northward; when immediately beneath it the light will be in the zenith, and the shadow projected downwards—being thus invisible; but, having passed the position of the light, we must turn round in order to see it, *i.e*,, "face it"—for it bears north and our shadow is projected southward. Now whatever be the physical constitution of a self-luminous body, the rays of light emitted by that body are propogated in straight lines in every direction from every part of the luminous surface so long as the rays of light traverse the same medium. Whenever a luminous body illuminates an opaque surface there must always be an area of central (*i.e.*, perpendicular) or vertical rays.

Proceeding on the lines of Euclid, which no mathematician can deny to be based on true mathematical principles, Dr. Robertson proceeds to fully demonstrate by diagrams that the diameter of the area of vertical rays must always be equal to the diameter of the luminous body. The author thus has grasped the very essence of what we have thought for years, and that is: " The real size of the sun may be found in the area of vertical solar rays." My meaning might be made a little more explicit by representing a globular body thus :

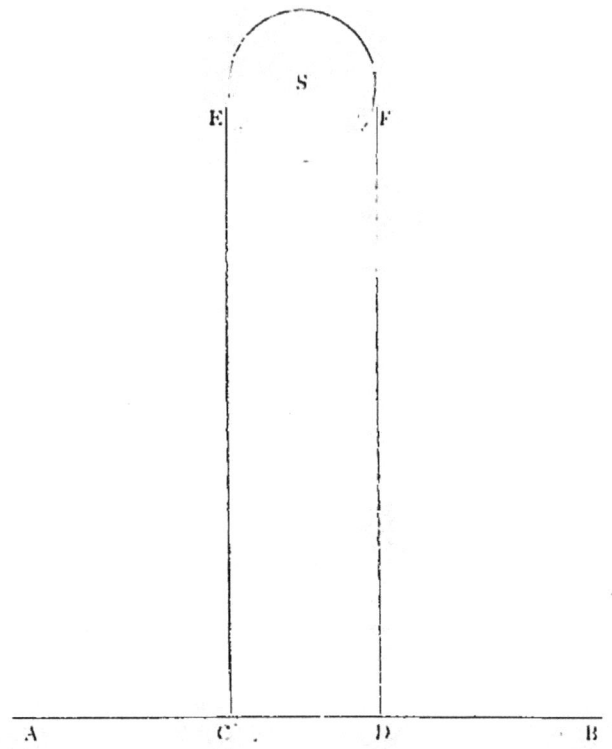

Let S represent the sun, and E C and F D the vertical rays of light falling upon the plane earth, A B. The vertical rays of the globular illuminated body of the sun would be represented by the straight lines continued down to C D. Now it is *self-evident* that C D is the diameter of the globular body S. The base A B on the outer sides of C D will receive the oblique rays, which rays are perceptibly oblique by their casting of shadows.

There is a deflecting influence to which the other solar rays are liable in their passage to the earth, and that is atmospheric refraction ; but near and at the zenith it is so slight as to be practically non-existent. In endeavouring to find out the area of vertical solar rays, we are told that "the ancient geographers found themselves considerably embarrassed in their attempt to fix the northern tropic, for though they took a very proper method, namely, to observe the most northerly place where objects had no shadow on a certain day—yet they found that on the same day no shadow was cast for a space of no less than 300 stadia."*

Dr. Robertson says that the sun has a real annual orbit, and that it actually moves in the plane of the ecliptic; consequently its size must be very small when compared with the earth. There is no evidence to show that the sun is of stupendous magnitude. The principle that "size is as the distance" is applicable only to opaque bodies seen under equal illumination, and is totally inapplicable in the case of the sun, which is "the great source of light."

In respect to eclipses, the author of this unique little book shows conclusively that they are not determined by the above supposed *motions* of the various orbs, though we very much question whether they are calculated on the basis of an orbital motion of the sun; not that we doubt the sun's orbital motion, but if they are so calculated, how did Regiomontanus calculate the exact time of an eclipse, at least twenty-eight years before Columbus saw it in the West Indian Islands in 1504? We contend that there is no calculation needed—simply keeping records of past eclipses, and watching for the recurrence of eclipse cycles.

We must do the Doctor justice by pointing out that he does not admit there is an orbital motion of the earth; yet he says that "the apparent rotation of the celestial sphere" is a term used to denote that appearance, seen on a clear

---

*The stadium was 630·93 feet, and 7½ stadia were equivalent to a Roman mile which is equal to 4,732 English feet; and 300 stadia would be equal to 31·15 geographical miles. The apparent diameter of the sun, as measured on the sphere of the heavens, is given in the Nautical Almanac for March 22nd, or September 23rd, as 32' 2", the average of the two observations. In the practice of nautical astronomy 32' of arc on the sphere of the heavens is equal to 32 geographical miles on the surface of the earth.

night, when all the stars appear to move slowly from east to west, across the concave sphere containing all the heavenly bodies, slowly rotating round two fixed pivots or poles. Now this appearance can only have one of two possible interpretations—*i.e.*, the celestial sphere is a real sphere containing all the heavenly bodies fixed in some mysterious manner, and actually rotates on the celestial poles, and an imaginary axis (termed the celestial axis) the earth remaining absolutely motionless in the centre—or : that the earth in a stationary position, has a motion of rotation on its own axis in an opposite direction, from west to east. But why does Dr. Robertson consider it *impossible* that the sun, moon, and stars could actually revolve round the earth every twenty-four hours ? Why should he interpret the apparent rotation of the celestial sphere to mean a real rotation of the earth in the opposite direction ? Of course we see the difficulty of his position by his retention of the view that the earth is a globe, partly because a ship disappears hull first (owing to a cause which we have explained many times) having nothing to do with the assumed rotundity of the earth's surface.

However, personally, I am grateful to the author for his lucid explanation regarding his conclusions as to the actual size of the sun, and showing so clearly the utter fallaciousness of the popularly accepted theory.

With the following interesting paragraphs, taken from one of the learned Dr.'s letters, I must conclude this short review :

> " Modern Astronomy is divisible into two distinct parts or systems, which are not only different but directly antagonistic to each other. There is first, Practical Astronomy, with that most useful and important sub-division, usually called Nautical or Geographical Astronomy. This is the astronomy handed down to us from remote ages, and has been gradually and constantly improved during the lapse of centuries, until now it has reached a state of great perfection. Regarding the accuracy, indeed the extreme accuracy, of this system there can be no reasonable doubt, its practical problems are daily verified by thousands of independent observers."

> " The other system, which is variously termed by different authors, The Copernican, Newtonian, or Speculative Astronomy, is the astronomy of hypotheses. It was originated by Copernicus. It was his idea of how the Universe should have been made ; he thought that the lamp of the world, the sun, should be placed ' In the midst of the beautiful temple of nature, ruling the whole family of circling stars that revolve round him." No doubt this idea makes a very pretty little diagram on paper, which is about all that can be said in its favour. It assumes the sun to

be stationary, whereas the first system assumes the sun to travel every hour in the ecliptic. The two systems accordingly, are in direct antagonism. This system of hypotheses quite independent of facts, has gradually expanded since the time of Copernicus until now it has reached a pitch of extravagance which is truly wonderful."

DEGREES.

A Reply.

A "degree," according to the *Encyclopædia Britannica*, is the 360th part of the circumference of a circle, which part is taken as the principal unit of measure for arcs and angles. The degree (°) is divided into 4 minutes of 60 seconds. Astronomers assume that the earth is a globe, and each 360th part of a circle all round it is called a "degree;" it is also assumed that the earth moves, though it is the sun which appears to do so.

The position of the sun in regard to the earth's surface is changed one degree in four minutes; in other words: 15 degrees per hour, and 180 degrees (or half the circumference of the earth) in 12 hours; hence the hour on a time-piece is divided into 60 minutes. The sun's time varies, but clock time does not vary.

In regard, however, to the earth's surface, we are told that the distance between parallels of latitude in different latitudes is not uniform, the length of the degree being greater at the equator than at the poles.

The length of a degree perpendicular to the meridian has been computed and compared with the length of a meridional degree in the same latitude, giving the proportion of the poles to the equatorial axis. The result differed considerably from that obtained by meridional degrees.

Degrees of longitude radiating from the North have been stated to gradually increase in extent as they approach the equator, beyond which they are again said to converge and gradually diminish in extent towards the south. This is the popularly accepted theory.

The matter might be decided by measuring some distance to the south of the equator at right angles to a given meridian (with non-expanding rods), and between two points where the sun is vertical at an interval of 4 minutes of solar time—*i.e.*, as one degree is a 360th part of the sun's whole path over the earth so is the period of 4 minutes a 360th part of the whole 24 hours, which the sun repuires to complete his course ; therefore, whatever space on the earth is contained between any two points (where the sun is on the meridian at 12 o'clock and 4 minutes past 12) will be one degree of longitude. If we know the approximate distance between any two places in the South, on or about the same latitude, we can calculate the length of a degree of longitude.

No shadow of doubt rests in my mind that the degrees South converge the same as they do in the North, so that the length of a degree South grows less as we go further from the equator. " Parallax " taught otherwise. I believe I have seen the quotation from his book, but have not read the work through.

If meridians converge south of the equator (as I believe they do) then a degree would measure less at 30° south than at the equator. But taking the ratio of the supposed globe degrees at that distance, both north and south would be about 47 miles long.

I have heard such scientists as decide these things admit that they are far from their measurements of degrees ; and I am convinced that no man has yet "perceived the breadth of the earth," nor measured it practically.

The globular idea must be stamped out from a man's mind before he can see things from a true position, and think of the sun's rays, as he proceeds above his pathway in the heavens, falling upon and directly touching the earth's surface so as to form a circle, and the extreme limits of his rays of light forming a larger circle within a larger circle, where sunlight comes to an end, and beyond the limits of day and night.

NOTICE.—In the next issue of *The Earth*, we shall (D.V.) give an illustration showing the photograph taken on the Old Bedford Canal, under circumstances recounted on another page of this issue.

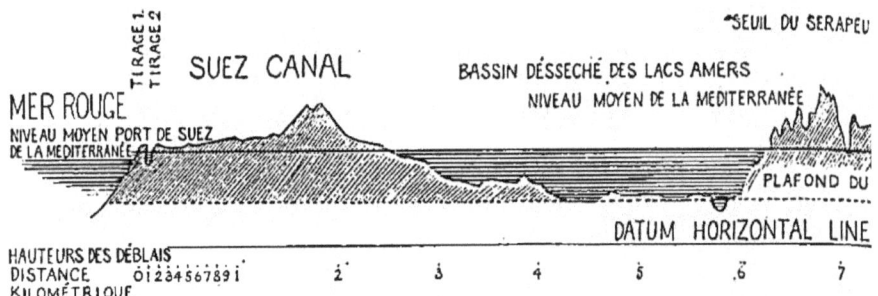

The diagram now presented to my readers of *The Earth*, bears evidence that the earth is a plane. It has reference to the great ship canal which connects the Mediterranean Sea with the Gulf of Suez on the Red Sea; Port Said (Egypt) being the entrance to the canal in that direction. The canal is 100 English statute miles in length, and is entirely without locks. The water within it is really a continuation of the Mediterranean Sea to the Red Sea.

The average level of the Mediterranean is 6 inches above the Red Sea; but the flood tides in the Red Sea rise 4 feet above the highest (and its ebb fall nearly 3 feet below the lowest) in the Mediterranean. The datum line is 26 feet below the level of the Mediterranean, and is continued horizontally from one sea to the other, and, throughout the whole

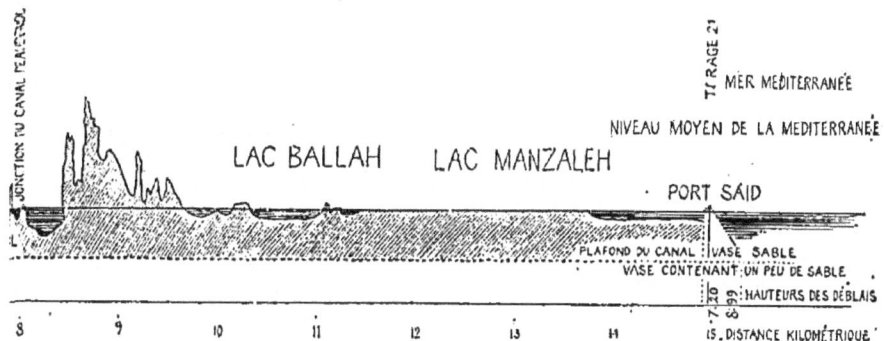

length of the canal the surface of the water runs parallel with this datum. The 100 miles of water in the canal and the surface of the Mediterranean Sea are a continuation of the same horizontal line.

If the earth were globular the water at one end of a canal one hundred miles in length would be 1 mile 1380 feet below the other end. But who has seen such a fall in the Suez Canal, or any other long stretch of water? No one. And who, but a globularist could believe in such a thing against all the evidence of known facts, and against the evidence of our God-given senses?

This canal affords another illustration that the surface of the great waters of the earth are horizontal, and therefore, disproves the theory of the earth's rotundity.

SOUTH POLE.

The Daily Mail for April 2nd, 1904, contained an article (under the above heading) which had a special interest for those who believe in a plane and motionless earth, as the return of the *Discovery* in search of the "South Pole" was professed to be described in a full narrative by Capt. Scott, R.N., Commander of the Antarctic Exploring Expedition, which sailed from New Zealand in December, 1901.

From the account it appears that the *Discovery*, and the relief ships—*Morning* and *Terra Nova*—arrived at Lyttleton, N.Z., on April 1st, after voyaging $2\frac{1}{2}$ years in the Antarctic regions. As the Expedition was assisted out of the National Exchequer, whatever facts may have been elicited ought to be made public.

It may be stated for our readers' information that it was in January, 1902 (six months after sailing from Cowes) that the *Discovery* entered the ice-field; a month later, when 2,000 miles south of New Zealand, she became gripped in the ice. This occurred in a region near the volcano Erebus, an active crater, named after the leading ship of the expedition commanded by Sir James Clarke Ross, about 60 years ago.

In that voyage the evidence pointed to the fact of the earth being a plane, the extremities of which are bounded by vast regions of ice and water, and irregular masses of land.

I believe when a true plan of the earth is known, it will be found to have four "corners." Three of them are known, and the fourth exists, possibly under the water. It has not yet been discovered. Mr. E. E. Middleton, I am convinced, is on the right track.

That the sun's path has been moving southwards in a concentric course may reasonably account for the changes in temperature that must have taken place on portions of the earth's surface, where remains of verdure that could only have existed in a different climate are found. We may note the discovery that certain specimens of flora found in the North polar regions exist in the southern ice fields. The fossil remains of plants discovered by certain explorers, are thought to point to the fact that, at some period in the past, the now icy south was once warmer. At present, at the point to which the *Discovery* expedition penetrated, the mean temperature for the year is below zero ; they once experienced 100 degrees of frost. In such a locality scanty moss, with a few lichens, form the only plant life.

When the explorers were sleighing through a blizzard, we are told that " if their gloves were not securely fastened on, they would instantly be blown away" (!) This corroborates our personal conclusions, viz. : that the furthest south being beyond the vision of light and darkness—or day and night —is piercingly cold, and subject to boisterous winds, which sweep with intense force across the clashing icebergs.

The narrative given in the above named paper, is, as far as it goes, favourable to the deduction that the earth rests upon and within the waters of the great deep, and that it is a floating island, or series of islands, buoyed up by the waters, and probably supported by submarine land connected with other land beneath the ice in the extreme south.

Commander Scott, in describing his winter sojourn in the wild Antarctic regions of solitude, was most persevering in his attempt "to look on the frozen page of God, and see what the letters meant." By his sledge journeys into the interior of the unknown continent, he says, he has succeeded in finding it to be a bleak plateau (elevated plain) rising 9,000 feet above the sea, "*and stretching interminably to the south.*" This goes far to put the stamp of proof upon what we have expressed as our belief in respect to what exists far south.

Captain Scott, with Mr. Skelton and party, found a new route to the West, and established a depôt 2,000 feet up the glacier, 60 miles from the ship. On October 6th, 1903, one section of the explorers started for the strait in lat. 80 S, and they found it contained a large glacier formed from the inland ice ; and they obtained information as to the point of junction between the barrier-ice and the land. A depôt, established the previous year, was found to have moved a quarter of a mile to the north. Six of the party reached a point 160 miles S E of the ship, travelling continuously over a *level plain.* No trace of land, and no obstacles in the ice were encountered, "and evidence was obtained showing this *vast plain to be afloat.*"

When the party crossed the 80th parallel (for the first time in the world's history) the compass pointed the wrong way.

It is something to know that the expedition was within 500 miles of the so-called " South Pole," and that all this way off the compass was reversed.

During the return journey the Possession Islands were found to be more numerous than shown on the charts ; but Wilke's Land, Ring Gold Knoll, and other lands marked on the chart, were apparently *not* in existence ; and the *Discovery* sailed right over the spot where they were supposed to be located.

When steaming along the great Ice Barrier, discovered by Sir James Clarke Ross in 1842, at the furthest easterly point Captain Scott discovered new land, which His Majesty has been pleased to have called after himself, viz. : " King Edward VII. Land."

We may note that the *Discovery*, in settling down into winter quarters in February, 1902, was frozen in, "and endured *a long dark winter, with a night of 122 days*, when the temperature fell to 62° below zero, and it was unsafe to venture from the ship, for even a mile, because of the blinding blizzard that raged almost continuously." This quotation is an excerpt from the statement of Lieut. Shackleton, of the *Discovery*. " Does the phrase, 'a night of 122 days' mean that the sun was not seen for that long period ? " was a question put to me ; and I replied, " Certainly." And as such is undoubtedly the case, I ask, how would it be possible to experience "a night of 122 days," if the earth be a globe careering round the sun, as they say it does ?

It is with decided satisfaction that I read of a western route having been located, and a depôt being established up the glacier, 2,000 feet above sea level ; but it appears to me a little premature for the Royal Society Executive to determine that the form of the presentation to Capt. Scott shall be a silver *globe*, with the route marked to scale, seeing that the *Discovery* expedition has in no way strengthened the hypothesis that we are living on a globe. If a globe is a suitable present on which to mark the route of the ship, why was the ship, in the first instance, not navigated by the aid of a globular chart ? We do not think that the gallant captain would have ventured out to sea with a globular chart ; had he attempted to do so, I feel assured that the voyage would never have been as successful as it has.

I would suggest, therefore, as a more suitable form of

present, a large silver block in the form of the great "level plateau," which he discovered, "stretching interminably to the south."

THE SUN AND MOON MIRACLE.

Under the above heading an article appeared in *The Messenger* for May, 1904. *The Messenger* is a monthly paper which professes to teach advanced religious truth. It also advertises, on its covers, criticisms of popular and sectarian doctrines; also a reply to the higher critics, and "Joshua's address to the sun and moon."

It might be expected that a paper of such pretentions would adhere to strict Bible teaching at all hazards, as against so-called scientific theories respecting the universe of God's Creation. But the article under consideration betrays more concern to reconcile Bible statements with the perverted "science" of the day, especially of course the so-called "science of astronomy, than to find out what is the truth of God on such matters. Hence the editor is more ready to quote from men who uphold modern theories of astronomy, than from Christian Zetetic writers who can give the only explanation of the miracle which is at all consistent with Bible teaching. But the editor of *The Messenger* seems to ignore these, and tries with others to make the divine records bend into something like harmony with the science of the day. This seems highly inconsistent of one who professes a superior standard of Bible exposition.

The editor of *The Messenger* quotes from *The Glasgow Herald*, which was reviewing an article on the subject in the *Church Quarterly*. From his article I will give the following extracts :

> "One of the most frequently discussed of these difficulties has been, to all appearance, solved with admirable scholarship, and in the most conclusive manner, by a writer in the current number of the *Church Quarterly*. Probably no miraculous intervention of providence has presented a more formidable problem to human reason than that at Gibeon, when the sun stood still in the midst of the heavens in order that Joshua and his people might avenge themselves on their enemies. That the sun was actually and literally arrested in his mid-day course

> is verified by the writer of the book of Ecclesiasticus, the Rabbins, and the Greek and Latin Fathers.
>
> "When the stupendous import of the miracle was understood by divines acquainted with the mechanism of the solar system (?) this literal interpretation was regarded as untenable, and while the German theologians, in explaining the occurence as an *impression* produced on the mind of Joshua, the sun *seeming* to stand still because the day's work had been achieved between noon and sunset, our English exegetists *conjecture* that the refracted light of the sun was sustained in the heavens after the disc had gone down."

This extract shows how the miracle was literally accepted by ancient worthies, who had no need to "reconcile" it with modern atheistic theories of the universe; and that it was only when modern "divines" accepted astronomical theories about "the mechanism of the universe" that they felt constrained to find some other "explanation" in harmony therewith. It is a curious fact that all such "divines," including the editors of professedly religious papers, seek for an explanation in harmony with what they believe to be the facts of "science," rather than one in harmony with the inspired account and the general teaching of the Holy Scriptures on Creative truth. Really they, for the most part, throw discredit on the Bible account, as we have seen, by assuming it was "an impression" merely on Joshua's mind (what about the Amorites—had they the same impression?), or, that the phenomenon was caused by "refraction," or that the account is a myth, or a mere poetic license due to the exuberance of the Eastern imagination!

But let us notice what is this last, though probably not final, speculation as to what the miracle really is supposed to be, if miracle at all.

The plain and unvarnished account of the Bible is to be discredited, and so the writer asks: "Why should the moon pause in the valley of Aijalon when at mid-day her light would have added no increasement to that of the sun?"

Now the fact that the moon's motion was also arrested, instead of being a difficulty with Bible students ought rather to show them that Joshua spoke by the inspiration of God. Had he only spoken as a man the sun's light was all that he needed to enable him as an Israelitish general to pursue, and complete the conquest of, his enemies. Then why did he command the moon also to stand still? Because if the moon had not been stayed, as well as the sun, the Israelitish calendar would have been put out of order; the months would have been disarranged, and the reckoning of time and the cycles of time would have been interfered with. Hence we may see that Joshua spoke under the direct inspiration of God, and that the sun did stand still, otherwise the moon need not have been interfered with at all.

But what is this supposed

NEW EXPLANATION.

I will again quote from the article in *The Messenger*.

> "To the writer in the *Church Quarterly* the miracle appears to have been one of 'protracted darkness rather than of protracted light.' The sun and moon did indeed 'stand still'" (though they are trying all the time to prove they did not!) "But the English words fail to convey the significance of the expression in the text." (Oh!) "In the language of the ancient poem, the Book of Jasher, Joshua addressed the sun, and said: 'Be thou silent! Be dumb!' And the sun stood still; silent; and the moon stayed; that is *stopped or ceased shining.*" (Italics mine).

So that according to this the sun did *not* "stand still" at all, neither was the moon actually stayed over any valley; they both simply "stopped shining," and went their ways! So that Joshua was wrong, the ancient Israelitish worthies were all deceived, and eminent men of God since those days, until in modern times some "German theologians" arose who were acquainted with the mechanism of the solar system"!

Well may the Bible cry out: "Save me from my friends"! if this is the way they handle the Holy Writings.

We might certainly have expected something better than this from a prominent Christian editor. He quotes from *Rig Vida*, to show how a solar myth sprang up, but in his testimony fails! because he does *not* quote from the Word of God. He goes into Sanscrit, and says that the root *ark* means "to make bright, to cheer, to gladden, etc. From this root "one of the names of the sun, *Arkab*," was derived meaning "a hymn, a song of praise, etc. Hence the myth that hymns of praise proceeded from the sun. Therefore, Joshua's command amounted to no more than this, the sun must stop singing, that is, it must put out, or hide its light; it must "cease shining." That was all, and so astronomy, or rather the Bible is saved from the reproach of science! The darkness was "protracted rather than the light"! Although it was light to begin with, the "darkness was protracted"! And this protracted darkness enabled Joshua to pursue his enemies!

Well, those who can accept this explanation are easily led, but it seems apparent that it requires more credulity, or scientific gullibility, shall I say, than the ordinary Bible account demands of simple faith in Him who is able to do things past our human comprehension—even the Creator of heaven and earth.

BE SILENT.

The great point insisted upon by these would-be expounders, is, that the original Hebrew for "stand still" means also "be thou silent." But this is no new discovery. The translators of the Old Testament must have understood the Hebrew language. And in both the "Authorized Version," and in the Revised Version, the old words are written: "Stand thou still;" while in the Revised Version the words: "be silent" are placed in the margin. There is no objection to either reading if properly understood. It is a wellknown fact that all motion produces sound, and as the sun is in rapid motion, there must be some sound attending on that motion. So that here again we have another proof of the scientific accuracy of the Bible.

The Bible clearly teaches that the sun's motion is accompanied by a sound; therefore if this sound must cease, the motion also, which causes it, must cease. Hence the record of this miracle is quite in keeping with Bible teaching, and true science.

To command the sun to "be silent" is equivalent to the command "stand thou still;" as there was no other way of stopping the sound but by arresting the motion. This at once harmonizes the difficulties. If we believe as the Bible teaches, and observation proves, that the sun is in motion over a motionless earth, all is clear; but if we have more faith in so-called "science" than in the Bible, we shall have to accept all sorts of tricks of interpretation, resorted to in order to get rid of the difficulty of reconciling the statements of the Bible with modern ideas of the mechanism of the universe. Besides, those who try to explain away this miracle because of the idea involved of a moving sun, to be consistent, ought also to explain away every other passage in the Bible (and they are many) which speaks of the motion of the sun.

Of course those who object to the account because of the miraculous are practically unbelievers in Bible inspiration. But professed Christians, and editors who pose as teachers of a high Christianity, ought, in all consistency, to accept the Bible and the Word of God before the fanciful speculations of modern theoretical science.

The book of Nature, also, itself reveals the true order of Creation to those who have eyes to read it aright. Only lately, in a daily paper, was another decisive proof of our contention. But this I hope to give in another article, under the title of *The Land Proof.*

THE PLANET EARTH.

Another paper which professes to speak in a religious manner, gives an article with the heading of the *Planet Earth.* The writer, under "Medical Talk," says, "that the earth, in common with all the planets, revolves around the sun is a fact which rests upon the clearest demonstrations of philosophy. That it revolves like them upon its own axis, is a truth which every rising and setting sun illustrates." "Either the earth moves around its axis every day, or the whole universe moves around it in the same time."

This is the style of their reasoning. The "fact" of the earth's revolution rests, we are told, upon philosophy! We were taught to believe that "facts are stubborn things," but this gentleman has evidently found one of another kind. It "rests upon philosophy." This is a poor basis for a fact to rest upon; and strange to say this fact does not seem to rest long at a time. We are continually shaking it off its philosophic pedestal. Yet the writer adds, "it pleased the all wise Creator to assign the earth its position amongst the heavenly bodies." Oh! we should like to know where he learned this. Has the Creator given him a special revelation? We cannot find it in the Book which He has given us containing the revelation of His Will and His Works. We freely admit that the earth either moves round its axis of rotation "or" the whole of the heavenly bodies move around the earth. But which is it? Our opponents affirm that it is earth which rotates; hence they are so anxious to explain away Joshua's miracle, from which it is evident that he thought it was the *sun* which moved! If they could prove that the sun, moon, and stars, were gigantic bodies at immense distances, it would seem incongruous to suppose that they all revolve about the earth. As this writer says: "To *suppose* the latter case to be a fact would be to cast a reflection on the wisdom of the Supreme Architect, whose laws are universally harmonious." But if an architect built a house, and then made an "electric globe to light it a million times bigger than the house, would not this reflect upon his skill? So their suppositions are unsound, and their philosophy is at fault.

If we profess to believe the Bible we should stand by its teachings against all the "vain philosophy" in the world. Bible writers, and Bible readers, for over five thousand years believed that the earth was stationary, and that the sun moved around it, and so caused the alternation of night and day. This agrees with the evident meaning of the account of Joshua's miracle. If any one could give good proof that the earth moves, and that the sun is relatively stationary to it, then it would be time enough to attempt to "reconcile" the account with the "facts" of astronomy. But no man in the world has ever been able to do this.

That the account was believed by the world of old, as a miraculous intervention on the part of the Creator, Who surely has all power over His Own Works, is testified by Josephus, whose *statements* I hope to give in a future issue.

CELESTIAL MOTIONS.

All that is really known of the motions of the heavenly bodies (sun, moon, stars, planets, and comets,) is from observation. No other source of information is open to man.

Whatever calculations may be made, they are all based upon observed phenomena.

Whatever laws are laid down they should be the result of observation.

Eclipses can be known and foretold only by carefully observing and noting all their past occurrences. The tables for future eclipses of the sun and moon could never have been made by mere calculation apart from the tabulation of the motions of these bodies in past ages.

Hence it is that buildings for the purpose of discovering these celestial motions are called "Observatories," and the science is called "Astronomy," which means the *Star Laws*, or the laws of the stars; *i.e.*, the laws which govern the motions of the heavenly bodies.

Our English word "science" is the Latin word *scientia*, and means *knowledge:* and our English word "knowledge" is from the Greek word *gnosis*.

We must distinguish, therefore, between *science* and *hypothesis;* between facts and theories; between what we know and what we think. If we do this there will be quite a number of so-called sciences which are no sciences at all. Geology is in no sense a science. Chemistry is. Astronomy is a mongrel science, being partly knowledge and partly hypothesis. Geometry is a science. What we *know* can never be altered. But hypotheses must be constantly changed in order to accommodate or modify them with newly acquired facts.

A fact is a *thing done*, and is unchangeable.

We approach the subject of the star motions, therefore, apart from all hypotheses. So-called astronomy, instead of consisting of the collection of observed facts, is a changing system of theories invented in order to explain the observed phenomena.

When theories are put forward where only a few facts are known, they have to be altered as the knowledge is accumulated. And no true theory can be formed unless and until we have all the data before us.

It is on these lines that the subject of astronomy is treated in the *Encyclopædia Britannica*. There we read:

"Whether the earth rotates within the star sphere, or the star sphere rotates round the earth, or both the earth and the star sphere rotate, it is known that *relatively to the earth*, the star sphere rotates from east to west in twenty-four sidereal hours. This rotation whether apparent or real, takes place without any appreciable change in the relative position of the fixed stars."

Here then we have a plain statement of facts. But here is the parting of the ways. Men agree to theorize as to these observed motions, and at the outset, before they acquire a single additional fact, they lay it down that these motions of the star sphere are only *apparent*, and not *real*.

But this is just the point which has to be proved.

Instead of waiting to prove it, they beg the whole question; and quietly assume a conclusion which is absolutely at variance with our senses, which alone are capable of judging the matter at all.

And astronomers reach this conclusion, in spite of the fact that they have to make, in the same article, the following confession of truth:

"Thus far there is nothing in the observed celestial motions which opposes itself to the belief that the earth is a FIXED CENTRE around which the celestial bodies are carried."

We are content with this presentation of the subject. We accept it at its full value; and this value cannot be overestimated.

If we follow the authors of this statement, we should have to ignore all our senses; and abandon fact for theory, what is clearly apparent for what is clearly unreal (though it is called "real.") And we shall have to accept the theory of a "fictitious sun" as a substitute for the "real sun."

Thus do astronomers juggle with words; and draw fictitious diagrams in order to support their theories.

No matter what new facts may be observed, they are useless to them; for they must be *forced* into harmony with their great hypothesis. The hypothesis must not be adapted to the newly discovered facts.

Astronomers are therefore just in the condition of an animal tethered for his food. No matter what good food may lie beyond the length of the tether, it is impossible for the animal to get at it. So these astronomers may go to their so-called "south pole," but they start, securely tethered by their dominating hypothesis, which will control the interpretation of whatever they may see. And if the newly discovered facts do not agree with the theory, so much the worse for the facts, which will be severely suppressed or contorted.

But to return to Celestial Motions. The laws which

govern them all can be ascertained only from *observation*. Any subsequent calculations of the motions of the heavenly bodies must be based on the previously known observations.

There can be no calculations independent of previous observations.

As to the moon: it is a matter of observation that she moves in the same direction as the sun, and that it takes her $27\frac{1}{3}$ days to make her circuit of the star sphere. She is observed to be in the same place among the fixed stars every $27\frac{1}{3}$ days. Relatively to the sun, she is rather more than 2 days later. So that the lunar month is about $29\frac{1}{2}$ days.

The planets are observed to have their own independent paths among the fixed stars; and they vary in their respective circuits. Observation shows that they appear to move in loops: *i.e.*, their circuits are made round moving centres. The purpose of astronomy is not to explain these motions but to accept the observations which explain the motions to us.

The Sun has his own observed motions relatively to the fixed stars and the earth. And in the absence of any proof to the contrary we must believe that these motions are his *real* motions. If any wish to persuade us that these motions are not real, but only "*apparent*," the burden of proof lies with them, and they must give us indisputable evidence that our senses are not to be trusted. But this is exactly what is wanting. Instead of proofs we are asked to accept hypotheses. But this we decline to do.

It is idle to demand a hypothesis from us. It is not for us to *explain* the phenomena, but to observe them, and record them, and believe them.

When what we see agrees with the Word of God, there is still stronger ground why we should not be moved from our position. In *Ps.* xix. 4-6, He who made the sun has told us that:

"For the sun He hath set a tent among them (*i.e.*, the stars),
And he, as a bridegroom, is going forth from his chamber.
He rejoiceth as a mighty one to run his course.
From [one] end of the heavens is his going forth,
And his revolution unto the ends of it.
And there is nothing hid from his heat."

The statement in the *Encyclopædia Britannica* (Art. Astronomy) accords exactly with this:

"The observations of the sun's motions and place among the fixed stars during so great a number of years, furnish complete evidence that the sun moves through a great circle, and its path is always the same, and among the same stars.

"In a period of about 365 days the sun traverses the whole of this path; and this period fixes for us the length of the year."

From the most ancient times this path of the sun was called the "ZODIAC." This word is popularly supposed to refer to animals or living creatures, from the Greek word *zao, to live.* But this is quite a mistake. The Greek word *Zodiac* comes from a primitive root, through the Hebrew, *Sodi,* which in Sanscrit means *a way* or *a step.* As the word *Zodiac* means the *way* or *path* of the sun as it moves among the stars in the course of the year, the name was obviously derived from this observed fact.

This path (or Zodiac) was divided into twelve parts. Why, we know not, unless it arose from the observation of twelve full moons in the successive parts of it in the course of the year. These twelve parts were called "mansions" or "houses" in which the sun was looked on as dwelling for one month in each.

The antiquity of these observations is shown by the "Fifth Creation Tablet," now in the British Museum. It reads as follows—our own remarks being put with brackets:

"Anu [*i.e., the Creator*] made excellent the mansions of the great gods [*twelve*] in number [*i.e., the twelve signs of the Zodiac, or mansions of the sun.*]

"The stars he placed in them. The lumasi [i.e., *the groups of stars* or *figures*] he fixed.

"He arranged the year according to the bounds [i.e., *the twelve signs*] which he defined.

"For each of the twelve months three rows of stars [i.e., *constellations*] he fixed.

"From the day when the year issues forth, unto the close, he marked the mansions [i.e., *the Signs of the Zodiac*] of the wandering stars [or *planets*] to know their courses that they might not err or deflect at all."

Thus the most ancient observations agree with the most modern; and both certify to the truth of God's Word.

The path of the sun through the Twelve Signs of the Zodiac is called the *Ecliptic*, because it is the line in which *eclipses** take place.

But the sun does not come back to quite the exact point in the sign, at the same moment when he commenced the year. This difference goes on every year; but it is so small that it takes about 71 or 72 years to make a difference of one three-hundred-and-sixtieth part of the whole Zodiac. In other words it would take no less than 25,579 years for the sun to complete this vast cycle.

This path of the sun, i.e., the *Ecliptic*, if it could be viewed from immediately beneath the " Polar " Star, would be seen as a complete and perfect circle, in which the Sun would be exactly the same distance from the horizon during the whole 24 hours: so that there would be no rising or setting. When, however, the observer is removed from this central position, the path of the sun is seen to be necessarily oblique, and the sun will be seen to rise and set obliquely.

When the observer is standing beneath the sun, instead of the " Polar " star, the sun would be seen to rise and set *perpendicularly*, and no longer *obliquely;* the *obliquity* depending on the position of the observer relative to the point of the sun's polarity. At this point, and on this line there would be *no shadow*, no matter what latitude he might be in.†

But this line, called the *Equator* appears to be constantly moving North or South, because there is another motion of the sun to be considered. Indeed, there are *four* motions to the sun. Two relative to the Fixed Stars, and two relative of the Earth. There is

 I.—The motion relative to the *Fixed Stars:*
  (1) The *annual* cycle.
  (2) The *great cycle* of 25,579 years.
 II.—The motion relative to the *Earth.*
  (1) The *daily* circuit of the hours.
  (2) The *annual* movement North and South, caused by the daily motion being spiral.

---

*Eclipse means a *failure*; by one body being left or blotted out for a time through another body being interposed.

†When we are in this position with God, we experience the truth of *James* i. 17. With Him as our only object there is "*no parallax.*" But it is the introduction of any other second object which causes *parallax*. If we keep *polarity* with Him then there is no *parallax*, and no shadow caused by His turning.

It is needful to add a few words with regard to the last of these four motions caused by the daily circuit being spiral. On or about June 21st the sun's daily path relative to the earth reaches its furthest point North ; and on about December 21st it reaches its furthest point South.

The line immediately beneath the sun when at its northern limit is called the "Tropic of Cancer." The word "tropic" is from the Greek *tropos, a turning*. On reaching that northern point it *turns* back, until in about six months it reaches its most southern limit, and again turns North. The southern point is called "The Tropic (or turning) of Capricorn." And the parts of the earth between the circles thus marked out are called "*The Tropics.*" The northern limit is known as the "Summer Solstice," and the southern limit as "The Winter Solstice." The days are longer in the former, and shorter in the latter. *Solstice* is derived from *Sol, the sun,* and *sisto, to make to stand*. When the sun is midway between the two *Solstices* the days and nights are equal. This occurs about March 21st, when going North, and about September 23rd, when declining South ; and the days and nights being then equal, those points are called the *Equinoxes* (from the Latin *nox, night*).

Further observation reveals the fact that the sun's path, called *the Ecliptic*, is not at these points immediately concentric with the *Equator*: that is to say, the *Equator* does not have the *Ecliptic* for its zenith ; but the two circles are what is called *Eccentric*.

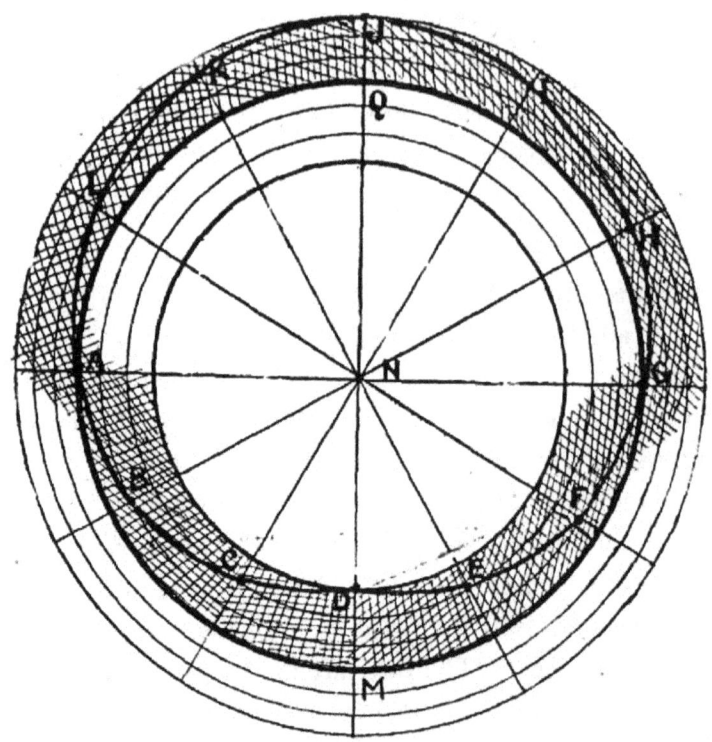

In this diagram A M G Q represents the earth's *Equator*, and A D G J represents the sun's *Ecliptic*; and the points A and G will be the Equinoctial points, while points M and J will be sun's furthest point North and South. The twelve parts into which both are divided, correspond with the twelve signs of the Zodiac, which are represented as follows:

A B : Aries (Equinox)	G H : Libra (Equinox)
B C : Taurus	H I : Scorpio
C D : Gemini	I J : Sagittarius
D E : Cancer	J K : Capricorn
(M the sun's furthest North)	(J the sun's furthest South)
E F : Leo	K L : Aquarius
F G : Libra	L A : Pisces.

As to the origin and meaning of these signs we must refer our readers to Dr. Bullinger's *Witness of the Stars*.* One

---

*Eyre & Spottiswoode, 33, Paternoster Row, London ; price 7/6 illustrated.

thing seems evident, that their origin was Patriarchal, and served for 2000 years to perpetuate the great primeval promise and prophecy of *Gen*. iii. 15; and to keep alive the hope of the Coming "Seed of the Woman." The first sign, "Virgo," shows Christ as the Virgin-born; while the last shows Him as the "Lion of the tribe of Judah" leaping forth to crush the head of the Dragon beneath His feet.

When the Scriptures of Truth were afterwards written, the need for the Heavenly witness would be no longer needed. Hence the Babylonian and Greek Mythology was not virtually some newly invented system of error, but rather the perversion and corruption of primitive truth, after its great lesson had been lost to the nations.

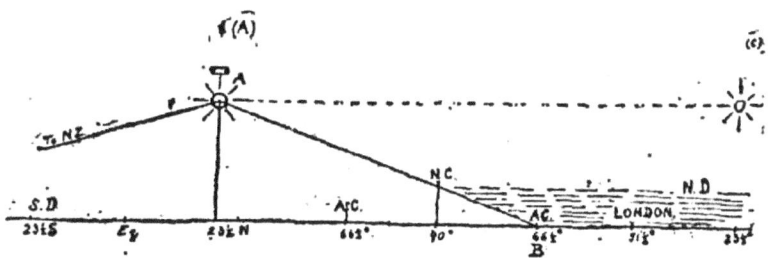

I have drawn a rough diagram showing how the sun can be seen when in Cancer. It is not true to scale, but it will illustrate my meaning. When the sun is in Cancer, at A, it could be seen at midnight over the North Centre (N.C.) as far as B, $66\frac{1}{2}°$ north latitude; that is 90° degrees away. It cannot be seen in London, nor far beyond the 66th parallel N, because the rays are refracted above the atmosphere; so that all that is beyond is in darkness at night. But at the same time the sun can be seen at N.Z. (New Zealand) because New Zealand is less than 90° away. Twelve hours later, when the sun arrives at C, it is of course daytime in London, $51\frac{1}{2}°$ N, and then the sun is too far away to be seen in New Zealand, and they have darkness there. But the sun can still be seen at what is called the North "Pole," because it is less than 90° degrees away, hence the sun can be seen over this area all the 24 hours, and *vice versa* in the extreme south during our winter.

The distance the sun can be seen, I believe, is about 90°.

CONVERGING MERIDIANS DOWN SOUTH.

Shortly, I do not believe in converging meridians down south, to a South Pole. Nor do I believe in any converging meridians to a North Pole either.

I do not believe in a North Pole, and still less in a Southern Pole of the globular hypothesis. I do believe in a North Polar Ocean, and also in southern ice; but to what extent both, or either, prevail is certainly a very fair question. For instance: in this plan which now appears in Lady Blount's magazine, the meridians do not converge to any North Pole, but to Greenwich, and this I find to be a very long step in a right direction.

This possibility of convergence to Greenwich was pointed out by myself many years ago, in a publication I then issued, and in which I then charged the astronomers with radiating their longitudes actually from Greenwich, and shifting them, and the point of radiation, to a suppositious north pole situated on the top of an imaginary globe. What I thought then, some thirty years ago, I now hold with greater force than before. I see that *it must be so*, and England may be proved to be the centre of the earth, and this accounts for its comparatively *equable* climate.

Other countries have both much colder winters, and also much hotter summers. Northern China, for instance, is intensely hot in summer, whilst its winter is of such severity that the rivers freeze for months at a time, and even the sea freezes. Japan, again, has a warm moist summer, and grows rice, and even tea, but in winter it suffers greatly from snow and the cold is intense.

Yet only a few hundred miles makes all this difference, owing to the sun being of very moderate dimensions, and travelling instead of standing stationary, and, in addition, being tolerably close to the earth. If the sun were to remain stationary we should very soon be burnt to cinders. As it is, matters are dangerous enough in some countries, where immense forests become like touchwood and ignite themselves, threatening destruction to cities and towns in their locality.

This is not at all in keeping with a spinning globe, which should be constantly cool, if not indeed cold, with a sun situated ninety-five millions of miles away.

In North America, at such a moderate place as Chicago, sunstroke prevails in summer, whilst people are sometimes frozen to death in winter. These extraordinary extremes can arise only from the motion of the sun—and a sun fairly close to the earth.

Then take St. Petersburg, in Russia. It is about fifteen hundred miles from London, and its climate is temperate when compared with Manchuria or Port Arthur, in the latter we see that the ocean actually freezes for miles outside the harbour.

Mosquitoes abound in some places which have very severe winters, and this must arise from the fact that the sun travels to them in summer.

Converging meridians are not to be depended upon entirely from any point; but a radiation from Greenwich certainly seems to be more helpful than from any other spot, and when dealing with southern longitudes, radiation requires to be tempered with distance and judgment.

The longitudes are, however, representative, and more so directly from Greenwich, and on the *English* side, than any other way.

As regards these especially southern and converging meridians to a supposititious south pole, some such appearance may occur in the months of November and December, but I feel certain that they can be traced to the long *loop* of the sun which runs down south of New Zealand and returns around and of course south of Australia. Something of this sort does occur, beyond a doubt, in those particular months, but how about the other months.

If there is any weight in this claim for a South Pole, and converging meridians to it, the appearance should last for about *five months* and *not two only*. Does it? that is the question.

Now-a-days when numbers of people have friends in Australia this point should not long remain in doubt. For my own part. I did not see anything of the sort, and I was there on Christmas day, in Adelaide, whilst the whole record of our voyage was opposed to any such appearances, even in November and December. Anything of the kind must be *confined* further east and south of New Zealand. And this narrowing of the subject is totally opposed to a South Pole, *such as is claimed for a globe*. A hole in the barrier of ice there may be, but no South Pole of the earth's supposed axis.

At the same time it does not matter to myself personally, as I have so many different ways of laying out the earth, that if this plan from Greenwich does not hit it another one is sure to do so ; but I must say for this one, now advanced, that it gives the very best account of the climates of the earth. Even the paddy fields of Japan can be accounted for by the ellipse of the sun, which simply runs up towards Japan, and makes that country almost tropical in summer, whilst it is terribly cold in winter, and much given to snow, as might be expected from its humid summer.

Then, again, the present expedition to Tibet shows that country to be miserably small and quite an arctic region, whereas according to the globe it should be very expansive and quite tropical in latitude, 30° North. Of course the elevation of Tibet would account for a little cold, but not the excessive frost which has been met with ; and nothing can get over the smallness of a country which should be very expansive, and also thickly populated. Instead of this being the case, we find a very small region and few inhabitants.

Tibet is simply in keeping with Asia generally—the more you look at it the more it disappears.

E. E. MIDDLETON.

THE EARTH: IS IT A GLOBE?

In whatever direction men have travelled from the north towards due south it has been found mechanically impossible to proceed beyond a certain distance. This is admitted by Professor Marienburg, of the University of Chicago, who is a startling theorist, declaring that the end of the world is near at hand- -having "jumped its orbit," so to speak, and is wobbling round in space like a drunken man, and this irregularity will cause our summers to become hotter and hotter, whilst our winters will be colder and colder, so that in the course of 20 years, at the outside, we shall be either roasted or frozen to death. Even this would be better than being comet-stricken after the fashion so graphically depicted by Mr. H. G. Wells. Professor Dewar goes even further and says that such a catastrophe may occur at any time.

If ever the Almighty permits man to reach further than where day and night now come to an end, he will then have more knowledge of what is at present unknown to us all. But I believe this will not be accomplished until "the day of the Lord," which will come "as a thief in the night," at the appointed time. "And the heavens shall pass away with a great noise."

The intensified power of the sun (as prophesied) may be the appointed means in the future (we know not how near) of melting the great ice barriers which we believe form an unsurpassable boundary to the sea. But if it be the will of God that these immense ice barriers shall be melted, and the waters thus permitted to pass hence, as it is written "there was no more sea," man may then pass unto "the Great Beyond." This we know not now.

The accumulation of ice round the South and North Poles (the "ice caps" geologists call them) are becoming thicker and thicker. More ice is stated to form by freezing during the long arctic and antarctic winters than is lost through melting in the short summers.

In *Science Siftings*, under date February 7th, 1903, it was stated that Sergeant Julius Fredericks, a survivor of the Greely Expedition, and one of the three who also made the sensational dash for the North Pole, reaching (to use the language of the journal under notice) the 83rd parallel—will leave the United States, in order to make another attempt to reach the coveted goal—intending to make the venture in an air-ship being constructed for the express purpose. A voyage will be made to the 70th parallel in ships. Here a provision camp will be established. Then the voyagers will start north in the airship—represented in the annexed illustration:

It is argued that the distance between the 70th parallel and the pole may be overcome in such a ship. Whether this proposed voyage over arctic ice and snow will be successful is, in our opinion, open to grave doubt. Mr. Fredericks' opinion is that the country at the 83rd parallel was once inhabited by a highly civilized people, and that as the pole is approached the climate moderates, and that " immediately around it there are all kinds of vegetation to sustain animal life." He (Mr. Fredericks) also believes that the people who once " lived at the 83rd parallel " have moved further north to escape the increasing cold—and that they now inhabit the country immediately round the " pole." He says that during his last expedition he passed the remains of villages, and that " in every case the woodwork was petrified. What remained of the houses showed superior handiwork in design and architecture." The night before they started to return, the aurora appeared in dazzling brilliancy in the northern sky, and they saw a great city in the distance composed of buildings of various heights, whilst around it were hills and valleys covered with verdure. They all looked upon it and wondered at it; but there it was, as clearly outlined as though it were but a mile away.

Mr. Fredericks says: " I am sure this was not a mirage, and I *believe* I have seen the city, which is now inhabited by the people who migrated from the south to the warmer climate around the pole."

In this last quotation Mr. Fredericks says he *believes* he has " seen a city, with verdure clad, around the north pole ;" but, if he has seen the city, why does he want to say that he believes he has seen it ?

Referring to the statements and deductions of Mr. Fredericks—the editor of the journal under notice says that, if they prove to be correct, " generally accepted beliefs will be considerably shaken."

This brings to my mind a striking paragraph in *The Leader*, June 24th, 1904, the wording of which was as follows :

> "THE SOUTH POLE."
>
> "An ingenious theory is advanced by Carsten Borchgravink, the Antarctic explorer, in support of his belief that the South Pole is surrounded by a continent of land, about twice the size of Europe. The northern hemisphere of the earth measures far more than the southern hemisphere, as far as is known. Therefore, since we are agreed that the earth is a sphere, if there was not a great continent south of the Pole, how, he asks, could the earth maintain its balance ?
>
> "And on this belief the explorer bases another in the existence of strange animals, formations, new people, new civilizations, new religions, new developments—all awaiting beyond the unknwon sea. For which reason naturally South Pole exploration is infinitely more important than the finding of the North Pole."

The above speculations may be very interesting, but they are merely speculations, and unlikely to prove true. But if any people should be found there, it would be safe to predict that they will not be " new peoples," nor " new civilizations ;" for, as the Bible shows, all the inhabitants of the earth have come from one source, and whenever, in the past, any new inhabited island, or continent, has been discovered, the people were always found sinners and mortals like the rest of mankind, and generally less civilized, and more degraded.

"STRETCHED OUT UPON THE WATERS."
By E. H. RICHES, LL.D., F.R.A.S.,
Member of the " London Mathematical Society," late Cantab, etc.

It may be stated here, that, experiments tending to show that the earth is fixed and free from all motion, have been brought under my notice, which are of a somewhat interesting character. It is taught in modern schools that "the motion of the earth with its accompanying atmosphere, is not perceptible to us; but the sun *appears* to us to move. Therefore we will now suppose this apparent motion of the sun to exist in reality, and in doing so we will regard the locus of its motion as a circle, at a certain distance from the *plane* of the earth's surface, concentric with the North Pole. It is at once acknowledged, that if the apparent motion of the sun be noticed from any northern latitude, and for any period before and after the time of its passing the meridian (or southing) it will appear that, in its motion to describe the arc of a circle.

Now any object, moving in an arc, cannot possibly return to the centre of that arc without describing a circle. It would seem then, that the sun does this daily, and that visibly.

To support this, we might call to mind the observations of the arctic navigator, Captain Parry, who, with several others with him, upon ascending high land at the North Pole, saw the sun *describing a circle upon the northern horizon*, and that more than once. Regarding the earth's surface as a vast plane this can be readily conceivable, and also that the circular path of the sun's daily motion be over some countries of this plane. In performing its journey, the sun may travel at just such a rate as to afford light to those countries within its reach, for the period of time called a day. And we believe the *extent* of land and water thus receiving light to be such as to admit of this idea. It is well known that those parts of the earth's surface, in the vicinity of the north pole, have no light from the sun in some months of the year. This is by no means a difficulty to be accounted for, in the theory which we are supposing, the diameter of the sun's path is constantly changing, —diminishing as it does from Dec. 21st to June 15th, and enlarging from June to December. There is no doubt of this fact, for it is proved by the northern and southern declination; in other words, that the sun's path, is nearest the north pole in summer, and in the winter it is furthest away from it.

In the following table by Mr. Glaisher, the difference of altitude caused by the difference in position, as noted at different times of the year, may be seen.

Sun's altitude, at the time of southing, or being on the meridian :—

Date.		Sun's altitude.		Time of Southing.
June 15	...	62 deg.	...	0 m. 4 s. before noon
,, 30	...	61¾ ,,	...	3 m. 18 s. afternoon
July 15	...	59⅔ ,,	...	5 m. 38 s. ,,
,, 31	...	56½ ,,	...	6 m. 4 s. ,,
Aug. 15	...	52½ ,,	...	0 m. 11 s. ,,
,, 31	...	47 ,,	...	0 m. 5 s. ,,
Sept. 15	...	38¾ ,,	...	4 m. 58 s. before noon
,, 30	...	35½ ,,	...	10 m. 6 s. ,,
Oct. 31	...	24 ,,	...	16 m. 14 s. ,,
Nov. 30	...	17 ,,	...	10 m. 58 s. ,,
Dec. 21	...	12 ,,	...	0 m. 27 s. ,,
,, 31	...	15 ,,	...	3 m. 29 s. after noon
Jan. 1	...	15½ ,,	...	3 m. 36 s. ,,
,, 15	...	17 ,,	...	9 m. 33 s. ,,
,, 31	...	21 ,,	...	13 m. 41 s. ,,
Feb. 15	...	25 ,,	...	14 m. 28 s. ,,
,, 29	...	30½ ,,	...	12 m. 43 s. ,,
Mar. 15 { On the equator at 6 a.m. }		36 ,, / 38½ ,,	... / ...	9 m. 2 s. ,, / 9 m. 0 s. ,,
,, 21	...	42½ ,,	...	4 m. 10 s. before noon
April 15	...	48 ,,	...	0 m. 8 s. ,,
,, 30	...	53 ,,	...	2 m. 58 s. ,,
May 15	...	57 ,,	...	3 m. 54 s. ,,
,, 31	...	60 ,,	...	2 m. 37 s. ,,

Briefly then, it may be observed, that the 6 months' darkness at the North Pole is at once accounted for, by noting the change in the length of the diameter of the circular line of motion, of the sun's course. The sun travelling over the plane of the earth, at once too, decides the question of *why* some countries should be warmer than others. Those immediately under the influence of the sun's rays, must naturally be warmer than those more remote. We have supposed then, the sun to travel in a circular course parallel to the earth's surface, and perform the whole circle of its journey once in 24 hours. Thus in 24 hours, every part of the earth experiences day and night, sunrise and sunset. At whatever place on the earth's surface an observer may be, it will appear to him that the sun seems to him to *rise* in the east (with respect to his position) and *set* in the west.

According, though, to one supposed theory, however, the sun is supposed to be *always* at the same distance from the earth's surface, and the *apparent arc* which it makes from our sunrise and sunset is only natural if the earth *be* a plane. Optics prove this. Let us compare the sun to a balloon sailing away from us. As the distance between us and the balloon increases, although its altitude may not increase, it will appear to us gradually to approach the horizon. So it is with our view of the sun; when at sunrise it first appears to our view, it would *seem* to be rising from the horizon. By the same rule in optics, at the close of our day, whenced the sun is travelling away in the distance, *sunset* will come to us, as the sun appears again to dip beyond the horizon; so, as sunset is coming on with us, sunrise is coming on to others. This is plain and consistent, and worthy of consideration.

THE BEDFORD LEVEL.

The Ed. of *The Earth* has received the following characteristic letter from a correspondent, dealing with her recent visit to the Old Bedford Canal, and experiments thereon.

"Dear Lady Blount,—Many thanks for your kindly sending me *The Earth*—and especially wherein is the visit to your Bedford Level; and I find you are greatly in error in your calculations, for according to the curvature of the first mile the *maximum* is in the centre, or *half-mile* only, and so the centre of the 6 miles is at the third mile, which *is* the greatest curvature. Hence $3 \times 3 = 9$, 9×8 in. $= 6$ ft. only, and not 24 feet as you fancy it to be; and if I recollect right this was to a few inches what was proven in court in John Hampden's case, when the £500 wager fell through, because it was well known as a matter of fact, and so no wager or cavilling could stand in law, therefore the Newtonian wagerer had to return the money.

"When a young man I served $3\frac{1}{2}$ years in the office of a ship-owner's firm in Ann Street, Belfast, *quite near to the notable 'Long Bridge'* which commanded a sea view over towards the Isle of Man, and we used to look out for the lumber ships' arrival from Canada, and with a powerful telescope we could see the *masts* of our ship flying her private signal while the hull and lower parts of the masts were invisible, and from my knowledge of astronomy I recollect calculating the distance to be 28 miles at the time of the morning, which I noted, and asked the captain if I was right and he corroborated my calculation from his log book.

"Of course I knew Parallax and John Hampden both, and after having one hour's talk with the former he gave up all his ideas of the flat earth, and disposed of his astronomy and literature a few days after, and ceased, ever after, to promulgate the idea. I am sorry that you should be so *hypnotized* with the idea, for there is a factor which astronomers all are aware of, viz. : 'refraction' of the air, which so to speak bends the horizon, or rising sun, upwards before its time, and makes the curvature of the sea to a *certain extent level*, and hence we *see the sun seemingly before her real rising*. There can be no question of the *sun being beneath our feet at midnight*, and the moon's light is a reflection of the sun's indisputably, which it could not be if your teaching of a flat and motionless earth were right. Parallax acknowledged to me he conceived his ideas when seven years of age on hearing a lecture on astronomy.

"Dear Lady Blount, I know you to be a genuine *honest*-minded woman and most desirous of promulgating the truth in Nature, and of the most advanced views possible, and thus I feel it my duty to write you, to keep my *conscience clear*, for I feel for you. Although I had no special interest in astronomy beyond seeing the infinite *goodness, justice, and mercy in God*—who has placed the Kingdom of Heaven in each and all of us who choose to accept and develope it, and thus see his eternal life in every living thing as a spiritual living entity.

With best wishes, I am, very truly yours, W.

Where the writer of the above letter gets his notion in respect to rate of curvature, when applied to an assumed globular earth 25,000 miles in circumference, we need to surmise.

There is no need for us to follow our correspondent in his arguments, as he stands alone in supporting a globular earth while at the same time he ignores the fundamental phenomena required and necessitated by a globe.

If W. is a globist he must accept the principles of globists, and reason on the data of his own authorities. He attempts to ignore at one stroke both *their theories* and *our facts*.

It is useless to argue with anyone who takes up such an untenable position as this.

All that is necessary for us to do is to explain to him fully what his own theory is, what it rests on, and what a globular earth requires and necessitates.

We have nothing to do with his greatest rise being in the first three miles. He must settle that point with Sir Isaac Newton and Greenwich Observatory. It has nothing to do with us.

We stand by our experiments of *facts*. He must stand

by globular *theories*. He cannot have it both ways. He cannot throw them over in order to knock us over.

Now the globularists demand two positions; and we willingly concede them. For if the earth be a globe, they are absolutely necessary to that theory. They all stand or fall together. .

I. The first is, that on a globe the observer, wherever he stands, is, and must be, *at the top*.
II. From this "top" the surface of the globe must *fall away in every direction*, and this falling away from the tangent is called *the depression*.

In our various papers we have spoken of the rate and amount of this depression as being ascertained by multiplying the *square of the distance* by 8 and treating the product as inches. This though accurate enough for moderate distances is not a scientific way of expressing the calculation. It is only a rough "rule of thumb," very useful and handy for quick calculation.

But as our correspondent seems to ignore it or at any rate not to understand it, we feel it necessary to explain more fully the phenomenon *required* by his position.

We will explain it by this diagram.

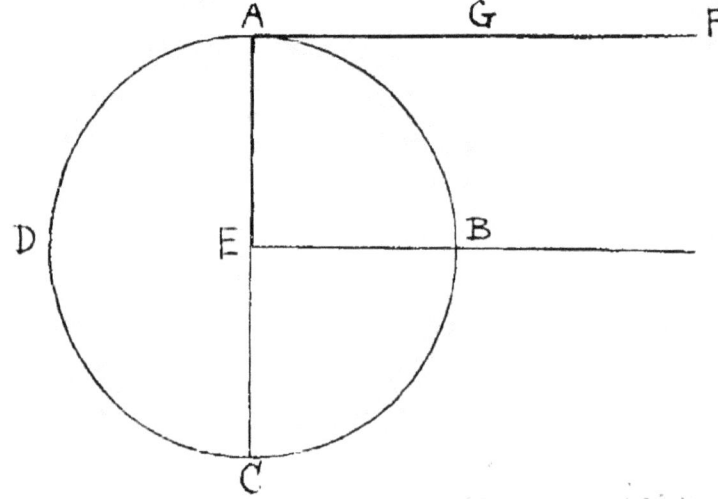

Let A B C D be the outline of the globe. The diameter is stated to be about 8,000 miles, and the circumference roughly about 24,000 miles. In this statement the further supposed phenomenon of the so-called "flattening of the poles" has to be taken into account.

Let A G F be the tangent, or the line of sight of the observer, A being his position "at the top." From this line the surface of the globe must fall away in the direction of A B.

But A B is about 6,000 miles and A E is 4,000 miles. Therefore in travelling from A to B the observer will be, on reaching B, 4,000 miles below the point G on the tangent line. The distance, 4,000 miles, will be the amount of the *Depression*.

Now the law which governs the calculation of the amount of this *Depression* is given by Sir Isaac Newton as fundamental to globular phenomena. He shows and proves the law, that,

The square of the *Distance* is equal to the product of the *Diameter* multiplied by the *Depression*.

He expresses it thus:—

$$(Distance)^2 = Diameter \times Depression.$$

This is given as a corollary of Lemma XI. in Newton's *Principia*. It has nothing to do with us: but is what is necessary if the earth be a globe.

Now, that statement of the law being correct it follows that the square of the Distance, divided by the Diameter equals the Depression. Or:—

$$\frac{(Distance)^2}{Diameter} = Depression.$$

If therefore we wish to find the *Depression* for any given distance we must first square the distance, and divide it by 8,000, (that being the number of miles in the earth's diameter). This will give us the amount of the *Depressson*.

The fact that 8 times this particular distance is 64,000, and that the number of inches in one mile is very nearly the same, viz.: 63,360, accounts for the "rule of thumb" which simply multiplies the square of the distance by 8, and reckons the answer as inches.

It will be seen by working out the Distance experimented

on in the Bedford Level, how nearly this rough and ready rule corresponds with the exact mathematical calculation.

The distance in question is 6 miles. Now

$$\frac{6^2}{8,000} = \text{the } \textit{Depression}.$$

But we cannot divide 36 miles (6 × 6, or 6^2) by 8,000 without reducing the 36 to some smaller dimension. Let us reduce 36 miles to inches, and then we have

$$\frac{2,280,960}{8,000} = 285 \text{ inches } (\textit{Depression}).$$

By the rule of thumb method we have

$$\frac{\begin{array}{r}6 \times 6 = 36\\8\end{array}}{288}$$

So that by the two methods of calculation, the "rule of thumb" method (288 inches) is only 3 inches in excess of the Newtonian mathematical method, which is 285 inches.

In the Bedford Level experiments the signal, which was visible at 6 miles distance, ought to have been 23-ft. 9-in. below the tangent line of the eye on the horizon.

This is a complete answer to W.'s objections; although I doubt not that they were proffered with good intentions.

When it comes to calculations of the globular theories we prefer Sir Isaac Newton's own *Principia* to the calculations of his would-be followers.

The phenomenon of the surface of the horizon rising to the tangent level of the eye, is apparent only, and is due to perspective. Astronomers make a great distinction between "*apparent*" motion and "real" motion. We claim the same liberty here and ignore the *apparent* horizon in discussing the *real* horizon.

THE SUN'S MOTIONS NORTH AND SOUTH.

Zetetics, who derive their name from *Zeteo*, to search out or to investigate, may fairly claim that they have frequently and practically proved that the surface shape of the earth and sea are, generally speaking, horizontal. Every copy of *The Earth* gives proofs of this fact. Then, when tangible proofs are given, objectors, instead of considering the evidences brought forward, go off into celestial phenomena. Even some whose education would lead us to suppose that they had, to some extent at least, cultivated the logical faculty, act in this manner. Thus we are to some extent driven to consider celestial phenomena with a view to meet objections, or answering enquirers. Education, as conducted on modern lines, does not always conduce to the bringing out of the logical faculty. So by way of introduction we must emphasize the fact that if we can give only one proof that the earth is a motionless plane, no other fact in Nature can controvert or overthrow that primary fact ; but the fresh fact must be explained, if explained at all, in harmony therewith.

Now Lady Blount's late photographic experiment on the Bedford Canal, with a Dallmeyer photographic lens, conducted by an expert photographer under the direction of her ladyship, has undoubtedly given Zetetics printed proof of their basal fact, namely that water is level, and the earth therefore a plane. This was a great service rendered to the truth, for which due credit should be given to her, both by Zetetics and Globularists. The experiments were conducted openly by an expert photographer at considerable expense of time and money to her ladyship, for no personal gain ; but that of the praiseworthy object of illustrating the truth. This should show our opponents that we are sincere, whether they are so or not. It is hard to believe that some of our critics are sincere, for they make no effort and are at no expense nor trouble to find out the truth in this matter. But sitting perchance in an editorial chair, or may be simply writing as private and irresponsible critics, they urge their weak and sometimes fallacious objections. For instance one editor of a photographic journal speculates as to what the account of the experiments may have arisen from, as though to suggest that he was not sure that the experiments were made! He should acquaint himself with the subject before he writes upon it. Then, on the supposition that the experiments were performed, he proceeds to explain away the results saying: "On the other hand, unusual or special atmospheric conditions of refraction often step in, and render objects visible which are considerably below the horizon." This is the old trick of mere partisans, who are foresworn, as it were, to their own views whatever evidence is produced. Another, a private correspondent, who professes to be critical, though he is not always logical, ignoring the zeal, trouble, and expense of conducting the experiment, writes, coldly harping upon the same monotonous strain, "refraction." He confesses that as a globularist he was somewhat staggered by the conclusive evidence there obtained, until he was reminded by a letter in *The Earth* of some "mathematical tables," giving tables of "correction for refraction"! Though as the Ed. then very properly added in a foot-note, "proof should be first given that any correction was needed over a level surface, where the rays of light would travel through a medium of almost unvarying density." But though they have no proof that there was any correction needed, they seem to think that the possibility of such is enough without any evidence and so they sit still and cry out "Refraction"! It is amusing. But when the ship disappears at sea, *that* is not caused by refraction but "curvature"! But when the ship is shown through a good glass, or a signal close to the water's edge six miles away, they then again shout "Refraction"! Thus, like the man in Æsop's Fables, they can blow both hot and cold. But we must leave the dishonest critics to their delusions, and try as best we can to help true enquirers.

The questions most commonly asked of late, are such as the following. Has a midnight-sun been seen in the south? Is it reconcilable with the plane-earth teaching? Do degrees converge or diverge south of the equator? And what then must be the motion, or motions, of celestial bodies, and especially of the sun in southern latitudes? In the following articles we shall try to answer these questions according to the best light we have received up to the present, and of course in harmony with the ascertained fact that the earth is a plane. We must start with facts, and endeavour to make logical deductions from them; and we must remember that we are dealing with celestial phenomena rather than with terrestrial.

THE SUN'S MOTIONS NORTH.

Let us start with the motions of the sun North, for it is with these that we are most familiar. On June 22nd, this year, 1904 A.D., the sun entered the tropical sign of Cancer. It then attained its furthest North declination, or distance from the celestial equator, 23° 27'. It also then attains its highest noon altitude in countries situated like England, and those still further north. Hence the northern summer then begins. But the sun does not remain at this declination but for a short time. It begins to enlarge its daily circuit round the northern portion of the earth. We will illustrate its motion by a diagram.

DIAGRAM I

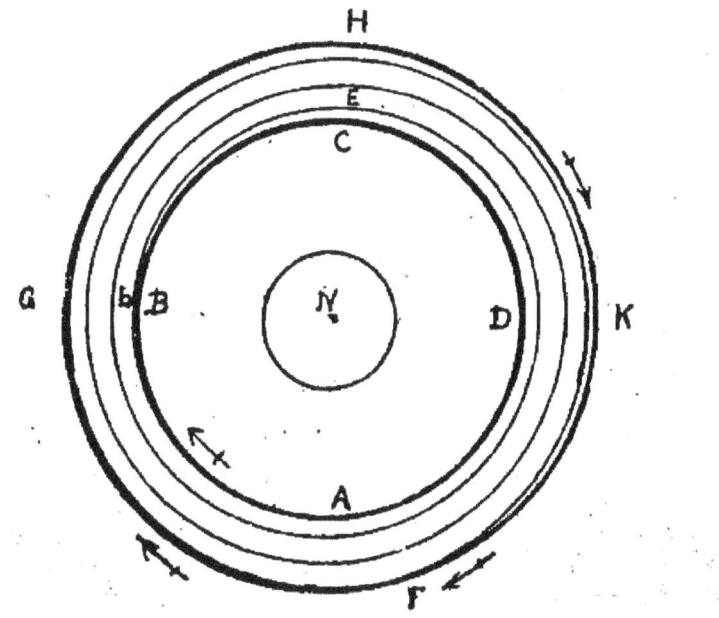

In the above diagram A B C D represents what is usually called the Tropic of Cancer.

It would be more correct to say it represents the path of the sun for that one day when the sun enters the first degree of the celestial sign Cancer. The sun moves round in its northern circuit in the direction of the arrows, that is, supposing it to start at A, it goes on to B in six hours, to C in twelve, to D in eighteen, and back again to A in twenty-four. But when the sun gets back to A it begins to leave the circle A B C D, and gradually recedes further from the centre N, which Zetetics call the North Centre.

In fact the circle A B C D is the only circle which the sun makes for six months, until it makes a similar circle in the South. The circle the sun makes in Cancer then begins to enlarge, and leaving the circle A B C D, the sun next courses from A towards (b) and on to E, &c., in a spiral movement which is almost circular but not quite so. Its declination varies one or two minutes per day to the end of June, and more rapidly afterwards, until the sun gets back to the equator F G H K, when of course it has little or no declination. Thus in three months the sun arrives at the equator, making in this time about eighty-nine daily revolutions round the northern parts of the earth. We have only shown three spiral lines in the diagram between the tropic of Cancer and the equator, because it would manifestly overcrowd the diagram to make eighty-eight or eighty-nine circles. But if we remember that the sun makes about thirty different revolutions per month, we shall see that it is a very fine spiral line which would be required to exhibit the sun's path for this period.

That the sun moves daily round us anyone can see from his own observation; and though many tests have been applied by Zetetics, the earth has never been found to have any motion, that is, such as astronomers call its "diurnal motion." That the sun moves in a *spiral* motion needs closer observation, with daily comparisons of its position when rising, culminating, and setting.

But even impartial globularists have confessed to this spiral-like movement of the sun, when, forgetting their globular theories, they honestly describe Nature as they really see her. For instance, in an interesting book by Paul B. du Chaillu, entitled *The Land of the Midnight Sun*, he says: "The sun at midnight is always NORTH of the observer (fact) on account of the position of the earth" (theory).

> "It seems to travel in a 'circle' (fact)...... At the pole the observer seems to be in the centre of a GRAND SPIRAL MOVEMENT OF THE SUN, which further south takes place north of him."

This agrees well with the plane truth, but it is out of harmony with the globular theory, as was shown many years ago by "Zetetes" in his pamphlet on *The Midnight Sun* (north).

THE SUN'S MOTION SOUTH.

We next proceed to give some evidence of the sun's motions in southern regions. Here we shall have to depend upon the evidence we have gathered for some time past, both from Zetetics in southern latitudes, and also from others who are globularists.

One correspondent in E. Australia, an intelligent Zetetic, and formerly a teacher says:

> "When I stand with my face to the North, the sun rises in the south-east, and travels from my right hand to my left almost straight overhead but a little in front of my face, and then sets in the south-west. This is in the height of our summer—Christmas-time. The south side of buildings gets the sun in the mornings and evenings in summer, but not in winter, as the sun rises more north-east and sets more north-west; and it does not rise nearly so high overhead."—R.A.

This is good general testimony, and it agrees with other reliable, and perhaps more "scientific" testimony we received from the Perth Government Astronomer in West Australia, some of which lately appeared in *The Earth*. It also agrees with evidence from a Zetetic, printed in *The Earth (not a globe) Review* so far back as 1893. That Zetetic, Mr. George Revell, further said:

> "The Southern Cross and all other Constellations do most certainly appear to revolve around a southern point or centre. I have proved this beyond doubt by close observation......the circle seems to narrow in winter, and expand in summer."

This is important testimony, and we quote it from Zetetics in the South because we believe it will appeal more forcibly to Zetetics in the North than would the testimony of those opposed to the plane truth.

These southern Zetetics know that the earth and sea are horizontal and stationary, yet they are candid enough to testify to celestial motions which some illogically think are opposed to this great fact. But one fact can never contradict another fact: both must be true. Zetetics therefore in the North must be candid enough to accept the facts on celestial motions in the South, just as we wish globularists to be candid, and reasonable enough to accept the well-known fact that water is level, and the earth therefore a plane. Only those who are candid and sincere will arrive at all the truth. And they may not obtain it "all"; but they certainly will obtain much more than those who are not candid.

We shall (D.V.) give some further evidence respecting southern celestial phenomena in our next article, and attempt to illustrate the same by further diagrams. As we write chiefly for Zetetics we shall close this chapter with a quotation from one who, according to our Lord was inspired by the Spirit of God when he wrote:

> "The heavens declare the glory of God, and the firmament showeth His handiwork. Day unto day uttereth speech; and night unto night showeth knowledge. There is no speech nor language where their voice is not heard. Their line (margin: *rule*) is gone out through all the earth; and their words to the end of the world. In them hath he set a tabernacle for the sun, which is as a bridegroom coming out of his chamber; and rejoiceth as a strong man to RUN A RACE. His going forth is from the end of the heaven, and his circuit unto the ends of it: and there is nothing hid from the heat thereof."—*Ps.* xix. 1-6.

These wonderful words contain some valuable hints which we may further explain as we proceed with later chapters.

By LADY BLOUNT and ALBERT SMITH.

DARWIN'S THEORIES DENIED AND REFUTED BY HIS OWN FOLLOWERS.

St. Louis, September 23rd, 1904.

"Professor Hugo De Vrees, Professor of Botany at the University of Amsterdam, declared yesterday evening, before the Congress of Arts and Sciences here, that he having followed carefully similar investigations with Darwin, disproved Darwin's theory of the Origin of Species. The statement caused a great sensation."—(Vide the *Daily News*, and other papers).

[As time advances the labours of Truth-seekers *must* bear fruit; the idols of men's vain imagination will topple off their pedestals, leaving their deluded worshippers amazed in a thick foggy intellectual dust. What with the recent telescopic-photographic smashing-demonstration of the Globular Theory on the Bedford Canal, and the public demolishment of the idol of Evolution, which ramified from that lunatic Theory, the much belauded fabric of Modern Theoretical Elementary (so-called) Science is reeling to its destruction, and the sooner this is complete, and the fragments swept into limbo, the better it will be for everybody.]

WHAT DID "THE DISCOVERY" MEN DISCOVER?

According to the reports anent the reception of Captain Scott and his colleagues, on their return from the South Polar Expedition, it would seem that: they discovered that which is being kept a great "secret," which must not be revealed during the present year; but, Shakespeare truly says: "By indirections find directions out." Acting on the bard of Avon's principle it is fairly evident that the "secret" which threatens to revolutionize present orthodox theories, lies in the fact that *where* the fossils were found *there* could not be any pole. Fossils always indicate that the sea must have existed where they were found. Well may the men be sworn to secrecy for twelvemonths and a day!

> What a voyage! and what a long way
> To go only to find a stray—

Homley Talks
The Real Reason Why

In the New Testament there are many warnings given us against false teachers and false doctrines.

The apostle Peter says that as there were false prophets amongst the Israelites of old, so "*shall* there be amongst you (Christians), who privily shall bring in damnable heresies." "And many," he adds, "shall follow their pernicious ways; by reason of whom the way of truth shall be evil spoken of."—2 *Pet.* ii. 1, 2.

We see the fulfilment of this prophecy in our own day. Ungodly teachers, and even some who profess to be Christians, are bringing in speculations, and doctrines in the name of religion and "science" which are rapidly undermining belief in the inspiration of the Bible, and the doctrines of the Christian religion.

When the enemy of mankind tempted the Man Christ Jesus he first tried to corrupt Him morally, by offering the world's wealth if He would only bow down and worship him. Failing to so ensnare Him, he afterwards tried to take away His life, and he succeeded through human instruments in his awful and murderous aim. But God raised Jesus from the dead, and afterwards took Him bodily to heaven.

With the followers of Christ he reversed the plan. He at first caused many of them to be put to death; but still the truth of God flourished.

Then through the Emperor Constantine he offered them wealth and worldly advancement. This corrupted many, and true faith seemed to decline. The "dark ages" closed in over the world; and the true faith was hidden in the secret and inaccessible places of the Earth.

Still the truth existed; and some "chosen vessels" in *all* ages kept the Commandments of God, and the faith of Jesus, and held fast also to Bible Cosmogony.

Then the Reformation blazed forth "and light was sown for the righteous." Truth came to the front again. Printing was discovered. The Bible was printed in the vulgar tongue. The enemy prompted his human instruments and agents to seek out, and even to buy up, all the Bibles they could find, so that the Word of God might be destroyed, and the Bibles be burnt that the light of Truth might be extinguished. But better Bibles were printed and more of them. Thus the truth triumphed and so the plan failed.

Profitting by past experience, the enemy grew more subtle and a deeper plan was laid. The multiplication of Bibles could not be stopped, therefore let them be corrupted. So "Catholic" translations of the Bible were printed, corrupting the original texts in favour of idolatry and Mariolatry. Still faithful translations multiplied and they did good work, strengthening the minds of men and purifying their hearts by faith.

And now a deeper scheme came into operation. Men sprang up in our seats of learning, and in famous universities on the continent, who threw doubt upon the Bible as a mere human book or compilation of books, uninspired and unscientific. This scheme prospered and has attained vast proportions, and it seems now likely to succeed where other plans have failed.

So that an earnest word of warning is needed against these false doctrines which are everywhere corrupting the minds of men from the simplicity of the truth and faith in God. Take for instance the writings of the so-called

HIGHER CRITICS.

They first attack the authority of the New Testament. But this was happily defended by men equally learned, who proved the epistles of Paul were genuine and the four gospels faithful narratives of historical events.

The downgrade critics now attack the Old Testament, and their chief attacks are directed against *Genesis* in particular and the Pentateuch in general. They see in *Genesis* the foundation of the Hebrew and Christian religions, so its authority must be overthrown.

The account of Creation is supposed therefore to be a "myth." Men raised up by God, such as Abraham, Isaac, Israel, Moses, and others, are supposed to be merely poetical "heroes." No proof is given, but their theory requires these hypotheses. The Bible is said to be only a human production, and the miraculous history of Israel in the past is nothing more than poetical fiction and exaggeration.

These so-called "higher critics" are trying to lower everything that is good in connection with our most holy faith. They would level everything downwards which is connected with a Personal God. They ought therefore to be called the Downgrade Critics, or the *Lower Critics*.

But the fault with these destructive critics is that they are not critical enough; they attempt to pull down but do not try to build up. Their objections against the Bible and its divine inspiration are for the most part subjective, and elaborated out of their own inner consciousness and not from historical evidence, or even from the facts of Nature.

But these Lower Critics are being answered by educated men who have formed themselves into what is called "The Bible League." We wish them God speed in their good work. Their efforts in support of the Inspiration of the Bible are in harmony with the work of true Zetetics. We only wish that they could see the truth of Zeteticism. We think it would help them much in their good work.

In a book entitled *Criticism Criticised*, containing a number of addresses on Old Testament Criticism, one writer, F. E. Spencer, M.A., well says: "It is well for those who steer to see rocks ahead. In some subtle and diffusive influences there lurks manifest danger. It is a STRONG DELUSION which boasts to see no more the glory and the handiwork of Jehovah in the universe, but only the glory of Copernicus, and Newton, and Darwin, and others......The fear of God is gone when there remains no traces of His action."

THE REASON OF IT ALL.

There are some who cannot see this, and who, therefore, give these destructive critics credit not only for honesty, but for godliness! Yet these men attack the Word of God while the astronomers are secretly undermining our faith in the Creation of God. In fact God is being spirited away as it were both from His Word and His Works.

If a fortress was being attacked on all sides, time after time, persistently and with all the force of modern inventions and science, we should be poor logicians if we could not see that there must be some strong reason for all this expenditure of time, talent, and force. The fortress must be something very valuable, or it must contain something very valuable. Or, on the other hand, it must be peculiarly obnoxious to the beseiging party and obstructive to their final aims, to call forth such persistent and powerful attacks.

If you enjoyed this book, take a look at our other titles on our website at theonomos.com.

Fifty Reasons: Copernicus or the Bible – F. E. Pasche

Philosophy and vain deceit or true science? Which is right? The Bible and Practical Astronomy or the Babel of theoretical, poetical, Newtonian fiction? Contains the original pages first published by Rev. Pasche in 1915. All pages have been digitally restored and published in this beautiful little 60 page booklet.

Is the Bible from Heaven? Is the Earth a Globe? – Alex Gleason

Does Modern Science and the Bible Agree? Includes an Accurate Chronology of All Past Time Containing a Classification of All the Eclipses from Creation. Alexander Gleason, creator of the Gleason New Standard Map of the World, makes the case for a flat earth. Includes an accurate chronology of all past time containing a classification of all the eclipses from creation. – Written in the late 1800's. Over 400 pages of Gleason's original text and illustrations, diagrams and charts.

Worlds Beyond the Poles – Physical Continuity of the Universe – F. Amadeo Giannini

The enclosed pages contain the first and only description of the realistic Universe of land, water, oxygen, and vegetation, where human and other forms of animal life abound. This is not a work of fiction nor is it a technical analysis of anything. It is a simple recital of fact which transcends the most elaborate fiction ever conceived. It projects man's first understanding of the factual and endless universe which contains human life throughout its vast length and width – regardless of all abstract theory to the contrary. - It is the view of the author that the earth is "flat" (a plane) and stationary as the Bible also depicts it to be.– Includes Original Images.

The Principled Legal Standard for the First Genuine Doctrinal Reformation of the Church – Alleman/Tellez

Does God bear false witness? Is it ever good to punish the innocent? Is the Righteous Judge ever unjust? The answers to these questions should be academic. Yet our unchanging perfectly righteous God is accused of such crimes all the time, in all denominations and religions across the board. The Principled Legal Standard deals with these issues head on, providing irrefutable arguments against all lies and false allegations against God. Everything is put under the legal lens from the angelic rebellion and Adam's trespass to the incarnation and Christ's death on the cross. Satan is the originator of all lies, deceptions and false doctrines, which places the supporters and rejecters of the traditional penal substitute doctrine into two camps: one is telling the truth about God and His absolute righteousness, whereas the other is bearing false witness about God and the ordeal on the cross. All those teaching false doctrines accuse God of acting unlawfully by putting on a pretense, altering court records and committing an injustice against an innocent party, thereby asserting that the accusations made by Satan are true, while arguing that he won his case against God. Discover how all who teach false doctrines attempt to witness against God as they sit at Satan's defense table. Get ready for the shock of your life as the case for the truth is irrefutably made!

www.ingramcontent.com/pod-product-compliance
Lightning Source LLC
Chambersburg PA
CBHW031425160426
43195CB00010BB/613